PHYSICO-CHEMICAL PROPERTIES OF SELECTED ANIONIC, CATIONIC AND NONIONIC SURFACTANTS

PHYSICO-CHEMICAL PROPERTIES OF SELECTED ANIONIC, CATIONIC AND NONIONIC SURFACTANTS

N.M. van Os

Koninklijke/Shell Laboratorium Amsterdam
Amsterdam, The Netherlands

J.R. Haak

Enschede, The Netherlands

L.A.M. Rupert

Thornton Research Centre
Chester, UK

ELSEVIER
Amsterdam – London – New York – Tokyo   1993

ELSEVIER SCIENCE PUBLISHERS B.V.
Sara Burgerhartstraat 25
P.O. Box 211, 1000 AE Amsterdam, The Netherlands

ISBN: 0-444-89691-0

CONTENTS

FOREWORD

The number of physico-chemical investigations of surfactants in solution, whether aqueous or non-aqueous, has dramatically increased in recent years. Four reasons can be advanced for this upsurge. First, surfactants are full-fledged industrial products which are now manufactured by millions of tons, have invaded almost every sector of industry, and affect many aspects of our daily life, so that any study aimed at improving an industrial process or a formulation involving surfactants may be expected to have an economic impact. Second, attempts are being made to use surfactant-containing systems for solving energy-related problems, such as those encountered in enhanced oil recovery and the storage of solar energy via photodecomposition of water. Third, several new techniques which have become available in many laboratories happen to be particularly suited to the physico-chemical investigation of surfactants in solution; among these, quasi-elastic light scattering and time-resolved fluorescence quenching are now the most widely used. Last but not least, the ability of surfactants to adsorb at interfaces and to combine into differently organised assemblies and lyotropic phases in solution has always fascinated scientists and stimulated research in this field. It is remarkable that new organised assemblies are still being discovered some seventy years after the first report of lyotropic phases appeared in the literature.

Literature reports on surfactants in solutions are scattered over a plethora of scientific journals and books which differ widely in scope and readership. They contain a staggering number of physico-chemical data on these systems. Most scientists involved in surfactant research have at one time or another felt the need for numerical values for properties such as critical micelle concentration (CMC), micelle aggregation number or radius, Krafft temperature or cloud point, and micelle ionisation degree, to cite but a few. Such data, however, are often difficult to retrieve because there have been no systematic compilations, with the exception of those for CMCs by Mukerjee and Mysels in 1971 and for micelle aggregation numbers by Grieser and Drummond in 1988.

The present compilation would thus seem timely. Rather than tabulating data for a single property and a large number of surfactants, as has been done in the above two compilations, the authors have tried to cover as completely as possible the physico-chemical properties of selected series of homologous surfactants. These surfactants are in most cases isomerically pure, are well known, and have been used in numerous academic and industrial studies. The properties include aggregation number, cloud point, CMC, ^{13}C-NMR, correlation length, counterion binding, density, enthalpy of micelle formation, entropy of micelle formation, Gibbs' free energy of micelle formation, head group area, ^{1}H-NMR, hydration number, Krafft temperature, micelle radius, microscopic viscosity, melting point, miscibility curve, partial molar volume, phase inversion temperature, phase diagram, refractive index, self-diffusion coefficient, surface tension, and upper critical temperature.

The solvent is water in most cases. Some data refer to properties in D_2O, electrolyte solutions, and non-aqueous solvents. The major variables recorded are temperature and concentration. Data on the purity of the compounds and the accuracy of the measurement methods are not included, as these can easily be found in the original sources, which mostly date from the period 1970-1991 and are given at the end of each chapter.

I am convinced that this precious collection of data will be of great use to anyone involved in Colloid and Surface Science, academics as well as industrial workers, and that it will stimulate further work.

Raoul Zana
C.N.R.S., Strasbourg, France

January 1993

INTRODUCTION

THE RISE OF COLLOID AND INTERFACE SCIENCE

Although chemistry had established itself as one of the natural sciences during the second half of the 18th century, it was not until 1850-1870 that chemists started to realise that the physical properties of a substance were related to its chemical composition. Among them was the Scottish chemist Thomas Graham, who was the first to recognise the properties nowadays regarded as characteristic of the colloidal state, and who in 1861 coined the term colloid (from the Greek kolla, meaning glue) and invented many other terms still used in colloid science.

Today, colloid and interface science is a flourishing branch of learning: it penetrates many fields of study such as biology, physics, physical organic chemistry, and inorganic chemistry; it bears upon numerous modern technical developments such as enhanced oil recovery, mineral flotation, and paper de-inking, and it can be called truly interdisciplinary. Over the past 20 years, the pace of change in colloid and interface science has been very fast and wide ranging, and one can even speak of a Renaissance of the discipline. Yearly, some 8000-9000 publications dealing with colloid and interface science appear in the literature, an estimated one tenth of which describe physico-chemical properties of surface active substances. Several international journals are largely devoted to colloid and interface science such as the Journal of Colloid and Interface Science, Colloids and Surfaces, Advances in Colloid and Interface Science, Langmuir, Tenside-Surfactants-Detergents, the Journal of Dispersion Science and Technology, and the Journal of Surface Science and Technology. Useful reviews (1-12), conference proceedings (13-17), and (text)books (18-25) have also appeared in the literature.

Several factors can explain the rise of colloid and interface science in recent times. The increased use of substances of a colloidal nature in households and industries has stimulated fundamental and applied research in this branch of science. New experimental methods (26-29) have come to the fore which allow the researcher to probe deeper and more accurately into the structure of particles and surfaces of colloidal size. On the theoretical side, self-consistent field and lattice methods (30-34), and numerical simulations (35-41) enable the scientist not only to make new generalisations but also to use them in the search for molecules with predictable properties.

Colloidal particles can exist in many forms but an important class is that of the association colloids, also known as surface active agents or surfactants. They are being manufactured on a very large scale and are being applied in many thousands of formulations for everyday household and industrial uses. The physical chemistry of such compounds is a fascinating subject.Scientists working in this field often prefer to work with well-defined model substances. Indeed, it has been the availability of isomerically pure surfactants that has enabled the study of controlled aggregation in great detail, and it is therefore only fitting to acknowledge the contribution made by synthetic organic chemistry (42-44). A field which is becoming increasingly important is that of functionalised surfactants useful for micellar catalysis (45) and that of polymeric surfactants for applications such as photochemistry and oil recovery.

HOW THIS COMPILATION CAME ABOUT

The action of a surfactant in most interfacial phenomena depends on its ability to self-aggregate; this in turn depends on the chemical structure of the surfactant. We have for a number of years been interested in such structure-performance relations and at one point wanted to calculate the hydrophobic free energy and enthalpy of micelle formation and the ion binding parameter for the homologous alkyl sulphates by means of the nonlinear Poisson-Boltzmann equation (46-48). For this we needed the experimental critical micelle concentrations and aggregation numbers of sodium octyl sulphate through sodium tetradecylsulphate; we also needed experimental values of the enthalpies of micelle formation for comparison with the calculated values. Finding such information in the literature took a few hours, and left us with no doubt that it would be very useful to have a compilation of such data for future reference. This led to other questions such as: "Should not such a compilation also include Krafft temperatures, diffusion coefficients, micelle radii?" And, "Would it be of interest to have cationic and nonionic surfactants included as well?" From that time on, we started to gather such data ourselves from the literature; our selection criteria were that the properties should reflect our own research interest in the physical chemistry of surfactants in solution, and that the surfactants should be well-defined and well-known products. Very soon, our collection started to grow and its increasing size made us wonder whether other workers in this field might not benefit from it too... the answer to that question is this book: a single source containing a range of physico-chemical properties of widely used model surfactants to be consulted by industrial and academic workers in colloid and interface science, detergency, oil field chemistry, emulsion technology, the food industry, the mineral recovery industry, and related areas.

THE SCOPE OF THIS COMPILATION

The compilation consists of five parts. The first part deals with physico-chemical properties of anionic surfactants with chapters on alkyl sulphates, alkanesulphonates, and alkylarenesulphonates. The main counterion is sodium; occasionally, reference is made to counterions other than sodium.

The second part deals with physico-chemical properties of cationic surfactants with chapters on alkyltrimethylammonium salts and alkylpyridinium salts. The latter chapter is supplemented with a figure section showing mainly variations in relaxation times, diffusion coefficients, and microscopic viscosities with surfactant structure and solution composition. The major counterions are bromide, chloride, and iodide, but references to others such as hydroxide, nitrate, and tosylate can be found as well.

The third part deals with physico-chemical properties of nonionic surfactants with chapters on alcohol ethoxylates, and alkylphenol ethoxylates. Each chapter includes phase diagrams showing the rich polymorphism of binary mixtures of nonionic surfactants and water, binary mixtures of nonionic surfactants and oil, and ternary mixtures of nonionic surfactants, water and oil.

The fourth part is the index. It starts with explanatory notes, telling what can be found in each of the five indexes that follow: a compound index, a property index, a general formula index, a chemical formula index, and a cross index. The complete compilation is contained in these five indexes by means of keywords, and any user is advised to consult this part first as it provides easy access to the data.

The fifth part is a glossary of abbreviations used in the tables of the compilation.

The physico-chemical properties include critical micelle concentrations, Krafft temperatures, cloud points, thermodynamic parameters of micelle formation, aggregation numbers, surface tensions, micelle dissociation degrees, micelle radii, diffusion coefficients, heat capacities, NMR data, and a few more. The compounds are in most cases well-defined substances; where unavoidable, technical mixtures were included as in the case of the alkylphenol ethoxylate compounds. All references pertaining to the properties presented in a particular chapter, are given at the end of that chapter. Almost all of the properties were taken from tables in the references, some were taken from the text, and only a few were read from graphs.

Other useful sources which the reader might wish to consult include the compilation of Critical Micelle Concentrations by Mukerjee and Mysels (49) which contains values published prior to 1966. Surfactants and interfacial phenomena are treated in a book by Rosen (50), which also contains a review of a limited number of surfactant properties obtained before 1977. A useful source on aggregation numbers is that published in 1988 by Grieser and Drummond (51). A review of thermodynamic quantities of micelle formation prior to 1979 was given by Stenius et al. (52), while Van Os et al. have given a guide to the more current literature on this subject (53). Reviews on Krafft temperatures and correlations with surfactant chemical structure have been given by Gu and Sjöblom (54). Finally, we would like to mention the mini-encyclopedia of Colloid and Surface Science (55) which provides information on many aspects in this field, including definitions of various physico-chemical properties.

HOW MOST OF THE TABLES ARE ORGANISED

Many of the tables in the compilation are arranged in similar fashion: they give the numerical value of a certain property of a compound in aqueous solution at various temperatures, as reported by one or more authors. Many tables display homologous series of compounds, and sometimes the solvent is not water, but an aqueous electrolyte solution, D_2O, an electrolyte solution in D_2O, or an organic solvent. The column labelled "Method" contains an abbreviation of the method by which the property was determined; all abbreviations are explained in part five of the book. Sometimes a method is common to all entries in a table; in that case the method is given in the table heading or at the bottom of the table. The column labelled "Ref" contains a figure which refers to the original source quoted at the back of each chapter; whenever one particular reference applies to all entries in a table, that reference is given in the table heading or at the bottom of the table.

The unit in which a property is expressed is shown in the table heading or in the table itself. Whenever authors have chosen to use units different from those in the table, we have indicated this in a footnote.

We have not indicated the purity of the compounds, or the accuracy of the experimental method. As all original literature sources are given, this information can be readily retrieved by the user.

CONCLUDING REMARKS

Although we have been careful in quoting the original values, errors are perhaps unavoidable in an undertaking of this size, and we will be grateful to accept revisions, comments, and additions to the material presented here. We can accept no responsibility for mishaps arising from the use of values in this compilation.

We are also aware of the fact that several publications must have been overlooked altogether, and that some of these may include valuable data on the surfactants tabulated in this compilation. Any user who notes such omissions is kindly invited to report them to the authors for possible inclusion in a second edition.

We hope that the present compilation will promote discussion and stimulate further research in the fascinating area of association colloids.

N.M. van Os   **J.R. Haak**   **L.A.M. Rupert**
Haarlem,   *Enschede,*   *Upton by Chester,*
The Netherlands   *The Netherlands*   *United Kingdom*

January 1993

REFERENCES

Reviews

(1)   "Surfactant Science Series" (M. J. Schick, and F. M. Fowkes, Consulting Eds.), Marcel Dekker Inc., New York, from 1966 onwards.

(2)   Israelachvili, J. N., Mitchell, D. J., and Ninham, B. W., J. Chem. Soc., Faraday Trans. 2, 72, 1525 (1976).

(3)   Vrij, A., Pure & Appl. Chem., 48, 471 (1976).

(4)   Wennerström, H., and Lindman, Phys. Rep., 52, 1 (1979).

(5)   Lindmann, B., and Wennerström, H., Top. Curr. Chem., 87, 1 (1980).

(6)   "Aggregation Processes in Solution", (E. Wyn-Jones and J. Gormally, Eds.), Elsevier, Amsterdam, 1983.

(7)   Pratt, L. R., Owenson, B., and Sun, Z., Adv. Colloid Interface Sci., 26, 69 (1986).

(8)   Kahlweit, M., Strey, R., Haase, D., Kunieda, H., Schmeling, T., Faulhaber, B., Borkovec, M., Eicke, H.-F., Busse, G., Eggers, F., Funck, Th., Richmann, H., Magid, L., Söderman, O., Stilbs, P., Winkler, J., Dittrich, A., Jahn, W., J. Colloid Interface Sci., 118, 436 (1987).

(9)   Schambil, F., and Schwuger, M. J., Surfactants Consumer Prod., 133 (1987).

(10)   Hoffmann, H., and Ebert, G., Angew. Chem., Int. Ed. Engl., 27, 902 (1988).

(11)   Lindman, B., and Ninham, B., Ettore Majorana Int. Sci. Ser.: Phys. Sci., 41 (Progr. Microem.), 85, (1989).

(12)   "Micellar Solutions and Microemulsions", (S.-H. Chen and R. Rajagopalan, Eds.), Springer Verlag, New York, 1990.

Conference & Symposium Proceedings

(13)   International Congress on Surface Active Substances: 1st Proceedings, Paris, 1954; 2nd Proceedings, London, 1957, 3rd Proceedings, Cologne, 1960; 4th Proceedings, Brussels, 1964; 5th Proceedings, Barcelona, 1968; 6th Proceedings, Zurich, 1972; 6th Proceedings, Moscow, 1976.

(14)   International Symposium on Surfactants in Solution: 1. Micellization, Solubilization, and Microemulsions (K. L. Mittal, Ed.), Plenum, New York, 1977; 2. Solution Chemistry of Surfactants (K. L. Mittal, Ed.), Plenum, New York, 1979; 3. Solution Behavior of Surfactants: Theoretical and Applied Aspects (K. L. Mittal and E. J. Fendler, Eds.), Plenum, New York, 1982; 4. Surfactants in Solution (K. L. Mittal and B. Lindman, Eds.), Plenum, New York, 1984; 5. Surfactants in Solution (K. L. Mittal and P. Bothorel, Eds.), Plenum, New York, 1986; 5. Surfactants in Solution (K. L. Mittal), Plenum, New York, 1989; 6. Surfactants in Solution (K. L. Mittal and D. O. Shah, Eds.), Plenum, New York, 1991.

(15)   "Modern Trends of Colloid Science in Chemistry and Biology" (H.-F. Eicke, Ed.), Birkhauser Verlag, Basel, 1985.

(16)   Proceedings of the International School of Physics "Enrico Fermi", Course XC, "Physics of Amphiphiles: Micelles, Vesicles, and Microemulsions" (V. Degiorgio, and M. Corti, Eds.), North-Holland, Amsterdam, 1985.

(17)   Progress in Colloid and Polymer Science Series, ECIS Proceedings, (H.-G. Kilian and G. Lagaly, Eds.), Steinkopff Verlag, Darmstadt from 1987 onwards.

(Text)books

(18) Gaines, G. L., "Insoluble Monolayers at Liquid-Gas Interfaces", Wiley, New York, 1966.

(19) Tanford, C., "The Hydrophobic Effect: Formation of Micelles and Biological Membranes", Wiley, New York, 1973.

(20) Adamson, A., "Physical Chemistry of Surfaces", Wiley, New York, 1976.

(21) Israelachvili, J. N., "Intermolecular and Surface Forces", Academic Press, London, 1985.

(22) Hiemenz, P., "Principles of Colloid and Surface Chemistry", Marcel Dekker, New York, 1986.

(23) Hunter, R. J., "Foundations of Colloid Science", Volumes I & II, Oxford Science Publications, 1987.

(24) Moroi, Y., "Micelles, Theoretical and Applied Aspects", Plenum, New York, 1992.

(25) Lyklema, J., "Foundations of Interface and Colloid Science", Marcel Dekker, New York, NY, 1992.

Experimental methods

(26) A useful review on NMR and force methods is given in reference (11).

(27) Thermodynamic methods, scattering techniques, rheological methods, and luminescence methods are reviewed in the Surfactant Science Series, Vol. 22, "Surfactant Solutions, New Methods of Investigation" (R. Zana, Ed.), Marcel Dekker, New York, 1987.

(28) Neutron small angle scattering, NMR methods, laser light scattering, kinetic methods, and rheological methods are reviewed in (16).

(29) Thermochemical methods to study surfactant solutions are reviewed in: Desnoyers, J. E., DeLisi, R., and Perron, G., Pure and Appl. Chem., 52, 433 (1980); Archer, D. G., Albert, H. J., White, D. E., and Wood, R. H., J. Colloid Interface Sci., 100, 68 (1984); Denoyel, R., Rouquerol, F., and Rouquerol, J., in "Adsorption from Solution" (R. Ottewill, Ed.), Academic Press, New York, 1983.

Self-consistent field and lattice methods

(30) Scheutjens, J. M. H. M., and Fleer, G. J., J. Phys. Chem., 83, 1619 (1979).

(31) Dill, K. A., and Flory, P. J., Proc. Nat. Acad. Sci. USA, 78, 676 (1981).

(32) Widom, B., Langmuir, 3, 12 (1987).

(33) Szleifer, I., Ben-Shaul, A., and Gelbart, W. M., J. Phys. Chem., 94, 5081 (1990).

(34) Mouritsen, O. G., in "Molecular Description of Biological Membrane Components by Computer Aided Conformational Analysis", Vol. I (R. Brasseur, Ed.), CRC Press, Boca Raton, 1990.

Numerical Simulations

(35) Gruen, D. W. R., Progr. Colloid Polymer Sci., 70, 6 (1985).

(36) Egberts, E., and Berendsen, H. J. C., J. Chem. Phys., 89, 3718 (1988).

(37) Smit, B., Hilbers, P. A. J., Esselink, K., Rupert, L. A. M., Van Os, N. M., and Schlijper, A. G., Nature, 348, 624 (1990).

(38) Karaborni, S., and Toxvaerd, S., J. Chem. Phys., 97, 5876 (1992); Karaborni, S., and O'Connell, J. P., "Surfactants in Solution" (K. L. Mittal, Ed.), Volume 11, 83 (1991).

(39) Klein, M. L., J. Chem. Soc., Faraday Trans., 88, 1701 (1992).

(40) Smit, B., in "New Perspectives on Computer Simulations in Chemical Physics" (M. P. Allen and D. J. Tildesley, Eds.), Proc. NATO ASI, Kluwer, 1993.

(41) Siepmann, J. I., and McDonald, I. R., Phys. Rev. Lett., 70, 453 (1993).

Synthetic Chemistry

(42) Engberts, J. B. F. N., and Nusselder, J. J. H., Pure and Appl. Chemistry, 62, 47 (1990).

(43) Menger, F. M., Angew. Chem., Int. Ed. Engl., 30, 1086 (1991).

(44) Kunitake, T., Angew. Chem. Int. Ed. Engl., 31, 709 (1992).

(45) Pelizetti, E., in "Physics of Amphiphiles: Micelles, Vesicles and Microemulsions" (V. Degiorgio and M. Corti, Eds.), North-Holland, Amsterdam, 1985.

On the Poisson-Boltzmann Equation

(46) Parsegian, V. A., Trans. Faraday Soc., 62, 848 (1966).

(47) Gunnason, G., Jonson, B., and Wennerström, H., J. Phys. Chem., 84, 3114 (1980).

(48) Evans, D. F., Mitchell, D. J., and Ninham, B. W., J. Phys. Chem., 88, 6344 (1984).

Physico-chemical Properties

(49)   Mukerjee, P., and Mysels, K. J., "Critical Micelle Concentrations of Aqueous Surfactant Systems", Nat. Stand. Ref. Data Ser., Nat. Bur. Stand. (U.S.), publication No. 36, (1971).

(50)   Rosen, M. J., "Surfactants and Interfacial Phenomena", Wiley, New York, 1989.

(51)   Grieser, F., and Drummond, C. J., Supplementary material belonging to J. Phys. Chem., 92, 5580 (1988).

(52)   Stenius, P., Backlund, S., and Ekwall, P., in "Thermodynamic and Transport Properties of Organic Salts" (P. Franzosini and M. Sanesi, Eds.), Pergamon, Oxford, 1980.

(53)   Van Os, N. M., Daane, G. J., and Haandrikman, G. J., J. Colloid Interface Sci., 141, 199 (1991).

(54)   Gu, T., and Sjöblom, J., Colloids Surf., 64, 39 (1992).

(55)   Becher, P., "Dictionary of Colloid and Surface Science", Marcel Dekker, New York, 1990.

ACKNOWLEDGEMENTS

Our sincere thanks are due to Raoul Zana for the favour of writing the foreword to this compilation; to Jerry Degeling-Cooke, Katherine Kartoredjo and Margot Spaargaren-Minnesma for typing most of the tables and for having done therefore, most of the work; to Magda van Nispen-Kluitenberg and Ruud Molthoff for typing the index; to Marleen de Mie for editorial assistance; to Saskia Haagsma for improvements in language and style; to Arie van Zon who helped clarify some points on organic nomenclature; to Roderick Groesbeek for putting together most of the illustrations; to Herman Holterman and Cor Kind for their patient cooperation in cross-checking the indexes; to various publishing companies and authors for their permission to reproduce figures. Perhaps most importantly our thanks are due to the experimentalists who determined the values assembled herein; their names can be found in the reference sections of this compilation.

We acknowledge with gratitude the permission of Shell Research BV to make the results of this study available to the scientific community.

PART I: ANIONIC SURFACTANTS

I.1 Alkylsulphates

Table I.1.1. Krafft temperature of sodium alkylsulphates in water

Alkylsulphate	Krafft temperature °C	Method	Reference
$C_{11}H_{23}OSO_3$	7	Sol	2
$C_{12}H_{25}OSO_3$	8	Sol	2
	9	Sol/C_m	1
	16	Sol	5
	16	Sol	7
	9	Co	8
$2-MeC_{11}H_{23}OSO_3$	< 0	Sol	6
$C_{13}H_{27}OSO_3$	20.8	Sol	2
$C_{14}H_{29}OSO_3$	21	Co	8
$C_{14}H_{29}OSO_3$	20.5	Sol	2
$C_{14}H_{29}OSO_3$	30	Sol	5
$2-MeC_{13}H_{27}OSO_3$	11	Sol	6
$C_{15}H_{31}OSO_3$	31.5	Sol	2
$C_{16}H_{33}OSO_3$	31	Sol	2
$C_{16}H_{33}OSO_3$	45	Sol	5
$2-MeC_{15}H_{31}OSO_3$	25	Sol	6
$C_{17}H_{35}OSO_3$	38.2	Sol	2
$C_{18}H_{37}OSO_3$	40.5	Sol	2
$C_{18}H_{37}OSO_3$	56	Sol	5
$2-MeC_{17}H_{35}OSO_3$	30	Sol	6

Table I.1.2. Krafft temperature of metal alkylsulphates
in water

Me alkylsulphate	Krafft temperature °C	Method	Reference
$C_{12}H_{25}OSO_3K$	34	not spec.	10
$C_{14}H_{29}OSO_3K$	41	Sol	4
$(C_{12}H_{25}OSO_3)_2Mg.6H_2O$	25	Sol	3
$(C_{14}H_{29}OSO_3)_2Mg$	38.5	Sol	4
$(C_{12}H_{25}OSO_3)_2Ca$	50	Sol	3
$(C_{12}H_{25}OSO_{32}Ca$	50	Co	8
$(C_{14}H_{29}OSO_3)_2Ca$	67	Sol	4
$(C_{14}H_{29}OSO_3)_2Ca$	70	Co	8
$(C_{12}H_{25}OSO_3)_2Sr$	64	Sol	3
$(C_{14}H_{29}OSO_3)_2Ba$	>100	Sol	4
$(C_{12}H_{25}OSO_3)_2Ba$	**	Sol	3
$(C_{14}H_{29}OSO_3)_2Mn$	29	Co	8
$(C_{12}H_{25}OSO_3)_2Mn$	16	Co	8
$(C_{12}H_{25}OSO_3)_2Mn$	16	Sol	3
$(C_{14}H_{29}OSO_3)_2Co$	29	Co	8
$(C_{12}H_{25}OSO_3)_2Co.6H_2O$	23	Sol	3
$(C_{14}H_{29}OSO_3)_2Cu$	40	Co	8
$(C_{12}H_{25}OSO_3)_2Cu.4H_2O$	24	Sol	3
$(C_{14}H_{29}OSO_3)_2Zn$	32	Co	8
$(C_{12}H_{25}OSO_3)_2Cu.4H_2O$	19	Co	8
$(C_{14}H_{29}OSO_3)_3Al$	>100	Sol	4
$(C_{12}H_{25}OSO_3)_2Zn$	11	Co	8
$(C_{12}H_{25}OSO_3)_2Pb$	53	Sol	3
$(C_{12}H_{25}OSO_3)_2MV^*$	25	Sol/Cm	9

* $(C_{12}H_{25}OSO_3)_2CH_3-N^+$⬡—⬡$N^+-CH_3$

** not observed below 100°C

Table I.1.3. Krafft temperature of sodium alkylsulphates in solvents other than water.

Alkylsulphate	Solvent	Krafft temperature °C	Method	Reference
$C_{12}H_{25}OSO_3$	Formamide	55	Sol	7
$C_8H_{17}OSO_3$	Hydrazine	13.2	Sol	20
$C_{10}H_{21}OSO_3$	Hydrazine	17.8	Sol	20
$C_{12}H_{25}OSO_3$	Hydrazine	31.4	Sol	20

Table I.1.4. Critical micelle concentration (CMC, in mmol/L) and the enthalpy of micellization (ΔH_m, in kJ/mol) of sodium alkylsulphates at various temperatures.

Alkylsulphate	Property	Temperature (°C)											Method	Ref.
		5	10	15	20	25	30	35	40	45	50	55		
$C_8H_{17}OSO_3$	CMC				134[a]								LS	11
									135				Co	12
							142						I	13
						72							Ca	14
		147.8	141.6	136.8	133.3	130.2	131.8	133.7	135.9	138.6	142.1	146.6	Co	15
			137[c]	135[b]	133	136							Co	16
						130	131	134	137	141			UR	17
											98		ST	18
	ΔH_m		4.7	4.0	3.0	0.8	-1.2	-2.6	-3.3	-3.9	-4.3	-4.9	calc	19
					5.9[a]	4.8	3.8	1.5					Ca	14
						3.3							Ca	26
			9.2	7.1	5.8	1.4	-3.3	-3.9	-4.8	-6.3	-8.2	-9.6	calc	15
$C_9H_{19}OSO_3$	CMC				64.6[a]								LS	11

Table I.1.4. (continued)

Alkylsulphate	Property	Temperature (°C) 5	10	15	20	25	30	35	40	45	50	55	60	Method	Ref.
$C_{10}H_{21}OSO_3$	CMC					24.5								Ca	14
							28							I	13
					30[a]									LS	11
								33.5		44.1				Fl	20
		36.5	35.0	33.9	33.3	33.0	32.9	33.3	33.7	34.5	35.5	36.9		Co	15
						31.1[d]								Co	21
						33.52								Co	22
						33.2								Co	23
						33.6								Co	16
						33								UR	24
										33				ST	18
	ΔH_m				4.9[a]	3.1	1.3	0.4						Ca	14
			2.9		2.5		0		-2.5		-5.0		-7.5	calc	25
			5.0	2.5	2.1		-0.6	-2.3	-3.7	-5.2	-6.3	-7.5		calc	19
						2.1								Ca	26
			8.7	5.6	3.2	1.5	-0.8	-3.1	-5.1	-7.3	-9.9	-12.2		calc	15
$CH_3(CHOSO_3)C_8H_{17}$ or 2 C_{10}	CMC					45.6								Co	25
	ΔH_m		6.7		4.2		1.7		-0.4		-2.9		-5.4	calc	25
$C_{11}H_{23}OSO_3$	CMC				16[a]	16								LS / Co	11 / 24

Table I.1.4. (continued)

Alkylsulphate	Property	5	10	15	20	25	30	35	40	45	50	55	Method	Ref.
$C_{12}H_{25}OSO_3$	CMC					8.27							Co	23
						8.16							Co	25
						3.8							Ca	14
						8.3							Ca	27
						8.0							Ca	28
						8.1							Co	29
						8.09							Co	22
						7.08							ST	30
						8.20							Co	16
						8.2							Co	24
						8.0							Co	21
						8.2								31
						8.07							Co	32
						8.2							Co	33
						8.0e							Ca	34
						7.24							ST	35
						8.12								36
						8.4	8.0						I	13
						8.14	8.25	8.2	8.53	9.0			Ca	37
						8.16	8.24	8.38	8.56	8.85	9.18	9.61	Co	10
					8.0								LS	11
		8.98	8.66	8.43	8.25								Co	15
									8.65	8.7			Ac	38
					8.1e	8.3e	8.4e	8.3e					Co	12
					7.6	7.4	7.4		7.9		8.4		Ca	39
						7.9e		8.0e	8.2e				Co	40
						7.4							Fl	41
												9.2	ST	42
					7.83a								calc	43
										8.1			ST	18

Table I.1.4. (continued)

Alkylsulphate	Property	Temperature (°C)													Method	Ref.
		10	15	20	25	30	35	40	45	50	55	60	65	70		
$C_{12}H_{25}OSO_3$	ΔH_m				0.68										Ca	27
					-0.20				-9.2f						Ca	37
					2.2										Ca	28
					1.7										Ca	28
					0.36	-2.55									Ca	44
					-1.26		-5.1								Ca	45
				1.09a	-2.13	-4.02	-7.07	-7.12	-9.30	-11.41					Ca	14
			5.12	2.81	0.47	-2.28	-4.78								Ca	46
							-7.45								Ca	39
						-2.6		-7.4							Ca	47
					0.1				-12.9						Ca	48
		5.0		1.29	-0.75	-3.30	-5.79								calc	49
		4.6	3.0	1.7	0.0	-1.3	-4.2	-4.6	-6.1	-7.5	-7.7	-10.9			calc	25
		7.5	5.7	3.8	0.1	-3.3	-4.9	-7.3	-9.8	-12.1	-13.9				calc	19
						-2.3		-4.9							calc	15
						-4.2	-25.5	-7.5		-7.2					calc	20
						-4.2		-7.5						-20.9	calc	50
$C_{14}H_{29}OSO_3$	CMC			2.1a	1.85										ST	4
															Sol	4
					2.05										Co	16
								2.21							Co	24
										2.1					ST	18
							2.21								Co	51
						2.08	2.14	2.22	2.31	2.43	2.59				Co	12
					2.05			2.4							Co	15
					2.05			2.21							Co	25
						2.1									I	13
								2.06							ST	30

Table I.1.4. (continued)

Alkylsulphate	Property	Temperature (°C)										Method	Ref.
		25	30	35	40	45	50	55	60	65	70		
$C_{14}H_{29}OSO_3$	ΔH_m					-6.3					-23.0	calc	50
			-2.9		-6.3		-9.6		-13.0		-16.3	calc	25
			-5.4	-7.9	-10.3	-12.7	-16.8	-20.7				calc	15
$CH_3(CHOSO_3)C_{12}H_{25}$	CMC	3.27			3.30				4.04			Co / Co	25 / 12
or $2C_{14}$	ΔH_m		-0.8		-4.6		-7.9		-11.3		-14.6	calc	25
$CH_3CH_2(CHOSO_3)C_{11}H_{23}$ or $3C_{14}$	CMC				4.30							Co	12
$CH_3C_2H_4(CHOSO_3)C_{10}H_{21}$	CMC	5.12			5.15				5.85			Co / Co	12 / 25
or $4C_{14}$	ΔH_m		0.8		-2.9		-6.3		-9.6		-13.4	calc	25
$CH_3C_3H_6(CHOSO_3)C_9H_{19}$ or $5C_{14}$	CMC				6.75							Co	12
$CH_3C_5H_{10}(CHOSO_3)C_7H_{15}$ or $7C_{14}$	CMC				9.70							Co	12

Table I.1.4. (continued)

Alkylsulphate	Property	Temperature			Method	Ref.
		30	40	50		
$CH_3(CHOSO_3)C_{13}H_{27}$ or $2C_{15}$	CMC		1.71		Co	12
$CH_3CH_2(CHOSO_3)C_{12}H_{25}$ or $3C_{15}$	CMC		2.20		Co	12
$CH_3C_3H_6(CHOSO_3)C_{10}H_{21}$ or $5C_{15}$	CMC		3.40		Co	12
$CH_3C_6H_{12}(CHOSO_3)C_7H_{15}$ or $8C_{15}$	CMC		6.65		Co	12
$C_{16}H_{33}OSO_3$	CMC	0.45	0.59 0.58	0.62	ST Co I Co	18 24 13 12
$CH_3C_2H_4(CHOSO_3)C_{12}H_{25}$ or $4C_{16}$	CMC		1.72		Co	12
$CH_3C_4H_8(CHOSO_3)C_{10}H_{21}$ or $6C_{16}$	CMC		2.35		Co	12
$CH_3C_6H_{12}(CHOSO_3)C_8H_{17}$ or $8C_{16}$	CMC		4.25		Co	12
$CH_3(CHOSO_3)C_{15}H_{31}$ or $2C_{17}$	CMC		0.49		Co	12
$CH_3C_7H_{14}(CHOSO_3)C_8H_{17}$ or $9C_{17}$	CMC		2.35		Co	12

Table I.1.4. (continued)

Alkylsulphate	Property	Temperature 40	50	60	Method	Ref.
$C_{18}H_{37}OSO_3$	CMC	0.16	0.20		Co ST	12 18
$CH_3(CHOSO_3)C_{16}H_{33}$ or $2C_{18}$	CMC	0.26			Co	12
$CH_3C_2H_4(CHOSO_3)C_{14}H_{29}$ or $4C_{18}$	CMC	0.45			Co	12
$CH_3C_4H_8(CHOSO_3)C_{12}H_{25}$ or $6C_{18}$	CMC	0.72			Co	12
$CH_3C_3H_6(CHOSO_3)C_{14}H_{25}$ or $5C_{19}$	CMC	0.33			Co	12
$CH_3C_8H_{16}(CHOSO_3)C_9H_{19}$ or $10C_{19}$	CMC	0.94			Co	12
$C_{20}H_{41}OSO_3$	CMC			0.052	EM	52
$C_{22}H_{45}OSO_3$	CMC			0.014	EM	52
$CH_3C_{13}H_{26}(CHOSO_3)C_{14}H_{29}$ or $15C_{29}$	CMC	0.08			Co	12

a) 21°C   b) 17°C   c) 12°C   d) 23°C   e) mmol/kg   f) 43°C   g) 37°C

Table I.1.5. Critical micelle concentration (CMC, in mmol/L) of sodium alkylsulphates in D_2O at various temperatures.

Alkylsulphate	Temperature (°C)					Method	Ref.
	20	25	30	35	40		
$C_8H_{17}OSO_3$		130				Co	16
$C_{10}H_{21}OSO_3$		32.5				Co	23
		31.4				Co	16
$C_{12}H_{25}OSO_3$		8.05				Co	23
		7.60				Co	16
$C_{14}H_{29}OSO_3$					1.97	Co	16
				1.94		Co	51

Table I.1.6. Critical micelle concentration (CMC, in mmol/L) of sodium alkylsulphates in hydrazine, aqueous hydrazine, and aqueous hydrazine/NaCl mixtures at various temperatures.

Alkylsulphate	X_{N2H4}	20	25	30	35	40	45	50	Method	Reference
							Temperature (°C)			
$C_8H_{17}OSO_3$	1				300				F1, NMR	20
$C_{10}H_{21}OSO_3$	1		48		80		110	114[a]	F1	20
									NMR	20
$C_{12}H_{25}OSO_3$	0.0		7.9		8.0		8.2			
	0.200		7.6		8.1		8.8			
	0.354		7.7		8.7		9.7			
	0.50		8.0		9.4		11.0		F1[b]	41[b]
	0.60		8.5		10.2		12.3			
	0.756		9.4		11.8		14.7			
	0.850		11.4		14.4		18.4			
	1.00				22.0		29.1			
	1				23				NMR	20
(0.05 M NaCl)	1				15.2		29.8		F1	20
	1				12.7		21.0		F1	20
(0.1 M NaCl)							17.5		F1	20

a) 47.5°C   b) CMC in mmol/kg

Table I.1.7. Critical micelle concentration (CMC, in mmol/L) of sodium dodecylsulphate in formamide.

Alkylsulphate	Solvent	Temperature (°C)						Method	Reference
		40	45	50	55	60	65		
$C_{12}H_{25}OSO_3$	Formamide					220		ST	7

Table I.1.8. Critical micelle concentration (CMC, in mmol/L) and enthalpy of micellization (ΔH_m, in kJ/mol) of sodium dodecylsulphate in aqueous NaCl at various temperatures.

[NaCl] mol/l	Property	10	20	25	30	40	50	60	Method	Reference
				Temperature (°C)						
0.001	CMC			8.2[a]					Ca	27
		7.8	7.3	7.2	7.3	7.6	8.3		Co	40
0.00316	CMC	6.9	6.4	6.2	6.4	6.6	7.4		Co	40
0.01	CMC	5.2	5.0	4.9	5.1	5.3	5.6		Co	40
				6.8[a]					Ca	27
			5.60[b]			5.37			LS	11
			5.13					6.17	ST	53
0.02	CMC			3.90					LS	29
0.03	CMC			3.13					Co, DS, Tu	29
				3.13					LS	11
0.1	CMC			1.9[a]					Ca	27
			1.48	1.46					Fl	33
			1.47[b]						DS, Co	29
			1.51			1.62			LS	11
								2.04	ST	53
0.2	CMC		7.59			8.71			LS	29
				0.92			1.45		ST	53
0.3	CMC		0.66[b]						ST	11
0.4	CMC			0.58					LS	29

Table I.1.8. (continued)

[NaCl] mol/l	Property	Temperature (°C)							Method	Reference
		10	20	25	30	40	50	60		
0.001	ΔH_m			0.59					Ca	27
0.01	ΔH_m			-0.80					Ca	27
0.023	ΔH_m			-0.64	-2.84				Ca	44
0.1	ΔH_m			-2.15 -2.97					Ca Ca	27 14
0.3	ΔH_m				-5.2				Ca	47
0.6	ΔH_m				-5.5	-10.2			Ca	47

a) mmol/kg   b) 21°C

Table I.1.9. Heat capacity of micellization (ΔC_{pm}, in J/mol K) of sodium alkylsulphates at various temperatures.

Alkylsulphate	Temperature °C			ΔCpm	Method	Reference
$C_8H_{17}OSO_3$	21	–	35	−310	calc	14
		25		−280	Ca	54
$C_{10}H_{21}OSO_3$	21	–	35	−402	calc	14
		25		−394	Ca	54
$C_{12}H_{25}OSO_3$	21	–	35	−561	calc	14
		25		−516	Ca	54
	25	–	43.4	−490	calc	37
	30	–	40	−478	calc	47
	20	–	35	−497	calc	49
	15	–	20	−462	calc	46
	20	–	25	−468	calc	46
	25	–	30	−550	calc	46
	30	–	35	−500	calc	46
	35	–	40	−468	calc	46
	40	–	45	−436	calc	46
	45	–	50	−422	calc	46

Table I.1.10. Aggregation number of sodium alkylsulphates at various temperatures

Alkylsulphate	Concentration mol/l	Temperature (°C) 20	25	30	35	40	45	50	Method	Reference
$C_6H_{13}OSO_3$	CMC		17						UR	24
$C_7H_{15}OSO_3$	CMC		22						UR	24
$C_8H_{17}OSO_3$	CMC		27						UR	24
	CMC	23.7[a]							LS	11
	CMC			12-17					I	13
$C_9H_{19}OSO_3$	CMC		33						UR	24
	CMC	30.5[a]							LS	11
$C_{10}H_{21}OSO_3$	CMC		50[b]						UR	24
	CMC	38.1[a]							LS	11
	CMC			27-32					I	13
	CMC					41			LS	21
$C_{11}H_{23}OSO_3$	CMC	45[a]							LS	11
$C_{12}H_{25}OSO_3$	CMC	57.3[a]							LS	11
	CMC			52-58					I	13
	CMC		62						LS	55
	CMC		64.4						calc	56
	CMC		64						LS	57
	CMC					54			LS	21
	CMC	66	66[f]	61[g]		62[h]		58[i]	Fl	64

Table I.1.10. (continued)

Alkylsulphate	Concentration mol/l	Temperature (°C)							Method	Reference
		20	25	30	35	40	45	50		
$C_{12}H_{25}OSO_3$	0.01	60							FI	33
	0.016					77			NS	72
	0.02	62							FI	33
	0.015–0.04			59–64					FI	58
	0.04		65						FI	59
	0.040					88			NS	72
	0.05	60							FI	33
	0.070					88			NS	72
	0.070		63						FI	70
	0.070		63						FI	73
	0.070		63						FI	64
	0.1	65							FI	33
	0.1	75		68[c]		59[d]		49[e]	FI	60
	0.10					98			NS	72
	0.2	64							FI	33
	0.280		70						FI	64
	0.3	64							FI	33
	0.4					120			NS	72
	0.490		79						FI	64
	0.5					122			NS	72
	0.730		91						FI	64

a) 21°C   b) 23°C   c) 31.7°C   d) 41.5°C   e) 51.4°C   f) 24°C   g) 32.7°C   h) 42.5°C   i) 51.5°C

Table I.1.11. Aggregation number of sodium dodecylsulphate in aqueous NaCl at various temperatures.

$C_{12}H_{25}OSO_3Na$ concentration mol/l	NaCl conc. mol/l	Temperature (°C)									Method	Reference
		20	25	30	35	40	45	50	55	60		
CMC	0.01		76.7								LS	61
CMC	0.03		75.6								LS	56
CMC	0.05		95[a]								LS	62
0.04	0.05					107					NS	72
0.10	0.05					109					NS	72
0.40	0.05					130					NS	72
0.60	0.05					128					NS	72
CMC	0.1		92[a]								LS	62
CMC	0.1		94			85					LS	63
CMC	0.1		88.3								LS	56
CMC	0.1		101								LS	61
CMC	0.1		93[a]								FI	64
CMC	0.1	91									FI	33
0.018	0.1	85									FI	33
0.05	0.1	90									FI	33
0.07	0.1	90									FI	33
0.1	0.1	90									FI	33
0.10	0.10					117					NS	72
0.18	0.1	92									FI	33
0.40	0.10					138					NS	72
0.60	0.10					135					NS	72

Table I.1.11. (continued)

$C_{12}H_{25}OSO_3Na$ concentration mol/l	NaCl conc. mol/l	Temperature (°C)									Method	Reference
		20	25	30	35	40	45	50	55	60		
CMC	0.1	93.4[b]									LS	11
CMC	0.03	70.8[b]									LS	11
CMC	0.01	64.2[b]									LS	11
CMC	0.3	123									LS	11
CMC	0.1		90.6								SD	65
CMC	0.1		93.0								SE	66
CMC	0.1		88.5								MO	67
CMC	0.1		92.7								MO	68
CMC	0.1					103				65	MO	69
0.07	0.15		82								F1	70
0.069	0.15		80								LS	71
CMC	0.20					112					MO	69
CMC	0.20		90[a]								LS	62
CMC	0.20		97			91					LS	63
0.04	0.20					120					NS	72
0.10	0.20					126					NS	72
0.40	0.20					146					NS	72

Table I.1.11. (continued)

C$_{12}$H$_{25}$OSO$_3$Na concentration (mol/l)	NaCl conc. (mol/l)	Temperature (°C)									Method	Reference
		20	25	30	35	40	45	50	55	60		
CMC	0.30		104			98					LS	63
CMC	0.30		110[a]								Fl	64
CMC	0.30					117					MO	69
0.069	0.30		120								LS	71
0.070	0.30		100								Fl	70
0.070	0.30		104							72	Fl	73
CMC	0.40		106			104					LS	63
CMC	0.40					130					MO	69
CMC	0.45		108			105					LS	63
CMC	0.45		130[a]								Fl	64
0.069	0.45		205								LS	71
CMC	0.50		110			107					LS	63
CMC	0.50					138					MO	69
CMC	0.55					109					LS	63
0.069	0.55		600								LS	71
CMC	0.60					112					LS	63
CMC	0.60					183					MO	69
CMC	0.60		182[a]								Fl	64
0.00867	0.60					66					LS	16
0.0173	0.60					92					LS	16
0.0347	0.60					127					LS	16
0.0693	0.60					185					LS	16
0.069	0.60		950								LS	71
0.070	0.60		260								Fl	73
0.070	0.60		154							95	Fl	70

Table I.1.11. (continued)

$C_{12}H_{25}OSO_3Na$ concentration	NaCl conc.	Temperature (°C)									Method	Reference
mol/l	mol/l	20	25	30	35	40	45	50	55	60		
CMC	0.70					346					MO	69
0.070	0.75		370								Fl	73
0.00867	0.80					216					LS	16
0.0173	0.80					398					LS	16
0.0260	0.80					502					LS	16
0.0347	0.80					588					LS	16

a) 24°C   b) 21°C

Table I.1.12. Aggregation number of sodium alkylsulphates (excluding sodium dodecylsulphate) in aqueous NaCl at various temperatures.

Alkylsulphate	NaCl conc. mol/l	Temperature (°C)	Agg.no.	Method	Reference
$C_8H_{17}OSO_3$ (CMC)	0.00	21	23.7	LS	11
	0.03	21	25.0	LS	11
	0.1	21	29.3	LS	11
	0.3	21	31.0	LS	11
	0.1	21	47.8	LS	11
$C_9H_{19}OSO_3$ (CMC)	0.00	21	30.5	LS	11
	0.03	21	30.9	LS	11
	0.1	21	36.6	LS	11
	0.3	21	41.9	LS	11
$C_{10}H_{21}OSO_3$ (CMC)	0.00	21	38.1	LS	11
	0.1	21	51.0	LS	11
	0.3	21	61.5	LS	11
$C_{11}H_{23}OSO_3$ (CMC)	0.00	21	45.0	LS	11
	0.03	21	50.4	LS	11
	0.1	21	59.4	LS	11
	0.3	21	72.8	LS	11
$C_{14}H_{29}OSO_3$ (CMC)	0.01	23	138	LS	21

Table I.1.13. Aggregation number of sodium dodecylsulphate in H_2O-D_2O mixtures and in D_2O-NaCl mixtures at various temperatures.

$C_{12}H_{25}OSO_3Na$ conc. mol/l	H_2O/D_2O ratio	NaCl conc. mol/l	Temperature °C		Method	Reference
			25	40		
0.058	0.01		70		NS	74
0.058	0.10		69		NS	74
0.058	0.20		69		NS	74
0.058	0.30		67		NS	74
0.058	0.41		66		NS	74
0.00867	0.00	0.6		70	LS	16
0.0173	0.00	0.6		122	LS	16
0.0347	0.00	0.6		211	LS	16
0.0693	0.00	0.6		320	LS	16
0.00434	0.00	0.8		194	LS	16
0.00867	0.00	0.8		416	LS	16
0.0173	0.00	0.8		666	LS	16
0.0260	0.00	0.8		785	LS	16
0.0347	0.00	0.8		928	LS	16

Table I.1.14. Diffusion coefficient (in 10^{-6} cm^2/s) of sodium dodecylsulphate in water and in aqueous NaCl at various temperatures.

$C_{12}H_{25}OSO_3Na$ concentration mol/l	NaCl conc. mol/l	Temperature °C															Method	Ref.
		10	15	20	25	30	35	40	45	50	55	60	65	70	75	85		
< CMC					8												Co	75
CMC					2												Co	75
0.2					5												Co	75
CMC	0.10	0.559[a]			0.96			1.37									LS	63
0.069	0.15		0.667	0.832	0.969		1.33										LS	71
CMC	0.20				0.95			1.35									LS	63
CMC	0.20				0.975			1.40		1.64							LS	16
0.069	0.30		0.544	0.678	0.823	0.992											LS	71
CMC	0.30				0.92			1.33									LS	63
CMC	0.40				0.88			1.33									LS	63
CMC	0.45				0.88			1.32									LS	63
0.069	0.45		0.284[b]	0.365	0.569	0.800	0.998	1.21		1.61		2.10		2.56			LS	71
CMC	0.50				0.84			1.30									LS	63
CMC	0.55							1.23									LS	63
0.069	0.55		0.156[b]	0.190	0.277		0.575	1.17	1.08		1.56						LS	71
CMC	0.60																LS	63
0.017	0.60		0.177[c]	0.236	0.342		0.713		1.17		1.55						LS	71
0.035	0.60		0.140[b]	0.178	0.263		0.577		1.03		1.49		1.92				LS	71
0.069	0.60			0.134	0.203		0.396	0.592	0.801	1.12	1.43	1.72	1.90		2.40	2.98	LS	71

a) 11°C b) 17°C c) 16°C

Table I.1.15. Micelle radius (in 10^{-9} m) of sodium dodecylsulphate in water and in aqueous NaCl at various temperatures.

$C_{12}H_{25}OSO_3Na$ concentration mol/l	NaCl conc. mol/l	Temperature 10	15	20	25	30	35	40	Method	Reference
CMC	0.03				2.22				calc	56
CMC	0.10				1.91				De	76
CMC	0.10				2.34				calc	56
CMC	0.10				2.53			2.54	LS	63
CMC	0.20				2.44				calc	56
CMC	0.20				2.55			2.55	LS	63
CMC	0.30							2.46	LS	16
CMC	0.40				2.47			2.56	LS	63
CMC	0.45				2.60			2.56	LS	63
CMC	0.50				2.69			2.58	LS	63
CMC	0.55				2.69			2.71	LS	63
CMC	0.60				2.79			2.84	LS	63
0.00867	0.60							2.5	LS	16
0.0173	0.60							2.8	LS	16
0.0347	0.60							3.3	LS	16
0.0693	0.60							3.9	LS	16
0.00867	0.80							4.1	LS	16
0.0173	0.80							6.0	LS	16
0.0260	0.80							7.0	LS	16
0.0347	0.80							7.7	LS	16

Table I.1.16. Diffusion coefficient (D_o, in 10^{-6} cm^2/s) and micelle radius (R, in 10^{-9} m) of sodium dodecylsulphate micelles in H_2O-D_2O mixtures and in NaCl solutions in D_2O.

$C_{12}H_{25}OSO_3Na$ conc. mol/l	X_{H_2O}	NaCl conc. mol/l	Do		R		Method	Ref.
			25°C	40°C	25°C	40°C		
CMC	1.0	0.2	0.975	1.40	2.47	2.46	LS	16
CMC	0.5	0.2	0.826	1.18	2.61	2.52	LS	16
CMC	0.25	0.2	0.766	1.09	2.67	2.74	LS	16
CMC	0.0	0.2	0.732	1.07	2.67	2.68	LS	16
0.00867	1.0	0.6				2.5	LS	16
0.00867	0.0	0.6				2.6	LS	16
0.0173	1.0	0.6				2.8	LS	16
0.0173	0.0	0.6				3.2	LS	16
0.0347	1.0	0.6				3.3	LS	16
0.0347	0.0	0.6				4.2	LS	16
0.0693	1.0	0.6				3.9	LS	16
0.0693	0.0	0.6				5.3	LS	16
0.00434	0.0	0.8				4.0	LS	16
0.00867	1.0	0.8				4.1	LS	16
0.00867	0.0	0.8				6.1	LS	16
0.0173	1.0	0.8				6.0	LS	16
0.0173	0.0	0.8				8.3	LS	16
0.0260	1.0	0.8				7.0	LS	16
0.0260	0.0	0.8				9.3	LS	16
0.0347	1.0	0.8				7.7	LS	16
0.0347	0.0	0.8				10.5	LS	16

REFERENCES Chapter I.1.

1) Shinoda, K., Pure Appl. Chem., **52**, 1195 (1980).
2) Lange, H., and Schwuger, M.J., Kolloid Z. & Z. Polym., **223**, 145 (1968).
3) Miyamoto, S., Bull. Chem. Soc. Japan, **33**, 371 (1960).
4) Schwuger, M.J., Kolloid Z. & Z. Polym., **233**, 979 (1969).
5) Weil, J.K., Smith, F.S., Stirton, A.J., Bistline, R.G., J. Am. Oil Chem. Soc., **40**, 538 (1963).
6) Gotte, E., Fette Seifen Anstrichmittel, **71**, 219 (1969).
7) Rico, I. and Lattes, A., J. Phys. Chem., **90**, 5870 (1986).
8) Moroi, Y., Oyama, T., and Matuura, R., J. Colloid Interface Sci., **60**, 103 (1977).
9) Moroi, Y., Sugii, R., and Matuura, R., J. Colloid Interface Sci., **98**, 184 (1984).
10) Moroi, Y., Motomura, K., and Matuura, R., J. Colloid Interface Sci., **46**, 111 (1974).
11) Huisman, H.F., Proc. Kon. Ned. Akad. Wetensch. (B), **67**, 388 (1964).
12) Evans, H.C., J. Chem. Soc., 579 (1956).
13) Ogino, K., Kakihara, T., and Abe, M., Colloid Polymer Sci., **265**, 604 (1987).
14) Kresheck, G.C. and Hargraves, W., J. Colloid Interface Sci., **48**, 481 (1974).
15) Moroi, Y., Nishikido, N., Uehara, H., and Matuura, R., J. Colloid Interface Sci., **50**, 254 (1975).
16) Chang, J.N. and Kaler, E.W., J. Phys. Chem., **89**, 2996 (1985).
17) Rassing, J., Sams, P.J., and Wyn-Jones, E., J. Chem. Soc., Faraday Trans. II, **69**, 180 (1973).
18) Lange, H. and Schwuger, M.J. in "Fettalkohole", Henkel KGaA, Düsseldorf (1981).
19) Goddard, E.D. and Benson, G.C., Can. J. Chem., **35**, 986 (1957).
20) Ramadan, M.S., Evans, D.F., and Lumry, R., J. Phys. Chem., **87**, 4538 (1983).
21) Tartar, H.V. and Lelong, A.L.M., J. Phys. Chem., **59**, 1185 (1956).
22) Mysels, K.J. and Otter, R.J., J. Colloid Science, **16**, 462 (1961).
23) Mukerjee, P., Kapauan, P., and Meyer, H.G., J. Phys. Chem., **70**, 783 (1966).
24) Aniansson, E.A.G., Wall, S.N., Almgren, M., Hoffman, H., Kielman, I., Ulbricht, W., Zana, R., Lang, J. and Tondre, C., J. Phys. Chem., **80**, 905 (1976).
25) Flockhart, B.D., J. Colloid Science, **16**, 484 (1961).
26) Goddard, E.D., Hoeve, C.A.J., and Benson, G.C., J. Phys. Chem., **61**, 593 (1957).
27) Paredes, S., Tribout, M., Ferreira, J., and Leonis, J., Colloid Polymer Sci., **254**, 637 (1976).
28) Eatough, D.J. and Rehfeld, S.J., Thermochim. Acta, **2**, 443 (1971).
29) Williams, R.J., Phillips, J.N., and Mysels, K.J., Trans. Faraday Soc., **51**, 728 (1955).

30) Klimenko, N.A., Karmazina, T.V., Yaroshenko, N.A., Bartnitskii, A.E., and Aryamova, Zh.M., Kolloidnyi Zh., **46**, 1112 (1984).

31) Quintela, P.A., Reno, R.C.S., and Kaifer, A.E., J. Phys. Chem., **91**, 3582 (1987).

32) Abu-Hamdiya, M. and Rahman, I.A., J. Phys. Chem., **91**, 1530 (1987).

33) Lianos, P. and Zana, R., J. Colloid Interface Sci., **84**, 100 (1981).

34) Rosenholm, J.B., Grigg, R.B., and Hepler, L.G., J. Chem. Thermodynamics, **18**, 1153 (1986).

35) Carrion Fité, F.J., Tenside Detergents, **22**, 225 (1985).

36) Mukerjee, P. and Mysels, K., Natl. Stand. Ref. Data Ser., Natl. Bur. Stand., No. 36 (1971).

37) Johnson, I., Olofsson, G., and Jönsson, B., J. Chem. Soc., Faraday Trans. I, **83**, 3331 (1987).

38) Moroi, Y. and Matuura, R., J. Phys. Chem., **89**, 2923 (1985).

39) Berg, R.L., Noll, L.A., and Good, W.D., ACS Symposium Series No. 91, 87 (1979).

40) Ossa, E.M. de la, and Flores, V., Tenside Detergents, **24**, 38 (1987).

41) Ramadan, M.S., Evans, D.F., Lumry, R., and Philson, S., J. Phys. Chem., **89**, 3405 (1985).

42) Miyamoto, S., Bull. Chem. Soc. Japan, **33**, 375 (1960).

43) Gunnarsson, G., Jöhnsson, B., and Wennerström, H., J. Phys. Chem., **84**, 3114 (1980).

44) Pilcher, G., Jones, M.N., Espada, L., and Skinner, H.A., J. Chem. Thermodynamics, **1**, 381 (1969).

45) Benjamin, L., J. Phys. Chem., **68**, 3575 (1964).

46) Sharma, V.K., Bhat, R., and Ahluwalia, J.C., J. Colloid Interface Sci., **115**, 396 (1987).

47) Mazer, N. and Olofsson, G., J. Phys. Chem., **86**, 4584 (1982).

48) Bergström, S. and Olofsson, G., Thermochimica Acta, **109**, 155 (1986).

49) Woolley, E.M. and Burchfield, T.E., J. Phys. Chem., **88**, 2155 (1984).

50) Stainsby, G. and Alexander, A.E., Trans. Faraday Soc., **46**, 587 (1950).

51) Elvingston, J., J. Phys. Chem., **91**, 1455 (1987).

52) Kutter, P., Schmitt-Fumian, W.W., and Bachmann, L., Proceedings of the sixth European Congress on Electron Microscopy, Jerusalem, Israel, II, 119 (1976).

53) Matijevic, E. and Pethica, B.A., Trans Faraday Soc., **54**, 587 (1958).

54) Musbally, G.M., Perron, G., and Desnoyers, J.E., J. Colloid Interface Sci., **48**, 494 (1974).

55) Rao, I.V., and Ruckenstein, E., J. Colloid Interface Sci., **113**, 375 (1986).

56) Stigter, D., J. Phys. Chem., **83**, 1670 (1979).

57) Kay, R.L. and Lee, K.S., J. Phys. Chem., **90**, 5266 (1986).

58) Moroi, Y., Humphrey-Baker, R., and Gratzel, M., J. Colloid Interface Sci., **119**, 588 (1987).

59) Malliaris, A., Progr. Colloid Polymer Sci., **73**, 161 (1987).

60) Malliaris, A., Le Moigne, J., Sturm, J., and Zana, R., J. Phys. Chem., **89**, 2709 (1985).

40

61) Hayashi, S. and Ikeda, S., J. Phys. Chem., **84**, 744 (1980).

62) Rohde, A. and Sackmann, E., J. Colloid Interface Sci., **70**, 494 (1979).

63) Corti, M. and Degiorgio, V., J. Phys. Chem., **85**, 711 (1981).

64) Croonen, Y., Geladé, E., Zegel, M. van der, Vandendriessche, H., Schrijver, F.C. de, and Almgren, M., J. Phys. Chem., **87**, 1426 (1983).

65) Tokiwa, F. and Ohki, K., J. Phys. Chem., **71**, 1343 (1967).

66) Anacker, E.W., Rush, R.M., Johnson, J.S., J. Phys. Chem., **68**, 81 (1964).

67) Coll, H., J. Phys. Chem., **74**, 520 (1970).

68) Kratohvil, J.P., Chem. Phys. Letters, **60**, 238 (1979).

69) Birdi, K.S., Dalsager, S.U., and Backlund, S., J. Chem. Soc., Faraday Trans. I, **76**, 2035 (1980).

70) Turro, N.J. and Yekta, A., J. Am. Chem. Soc., **100**, 5951 (1978).

71) Mazer, N.A., Benedek, G.B., and Carey, M.C., J. Phys. Chem., **80**, 1075 (1976).

72) Hayter, J.B. and Penfold, J., Colloid Polymer Sci., **261**, 1022 (1983).

73) Almgren, M. and Löfroth, J.E., J. Colloid Interface Sci., **81**, 486 (1981).

74) Triolo, R., Caponetti, E., and Graziano, V., J. Phys. Chem., **89**, 5743 (1985).

75) Leaist, D., J. Colloid Interface Sci., **111**, 230 (1986).

76) Douhéret, G. and Viallard, A., J. Chimie Physique, **78**, 85 (1981).

I.2 Alkanesulphonates

Table I.2.1. Krafft temperature of sodium n-alkanesulphonates

Alkane sulphonate $NaC_nH_{2n+1}SO_3$	Krafft temperature °C	Method	Reference
C_{10}	22	Co	1
	22.5	Sol/C_M	2
	22.5	Sol	5
C_{12}	33	Co	1
	31.5	Sol	5
	38	Sol	4
	34.5	Sol	9
C_{13}	35.5	Sol	4
C_{14}	42	Sol/C_M	6
	42	Co	1
	39.5	Sol	5
	48	Sol	4
C_{15}	48	Sol	4
C_{16}	50	Co	1
	47.5	Sol	5
	57	Sol	4
C_{17}	62	Sol	4
C_{18}	57	Co	1
	57	Sol	5
	70	Sol	4
C_{20}	66	Co	1
C_{22}	74	Co	1

Table I.2.2. Krafft temperature of sodium alkanesulphonates.

$$\text{NaSO}_3 \ \text{CH} < \begin{array}{l} \text{C}_9 \ \text{H}_{19} \\ \text{C}_m \ \text{H}_{2m+1} \end{array}$$

m	Krafft temp. °C	Method	Reference
0	22.5	Sol/C_M	2
2	< 0	Sol/C_M	2
4	< 0	Sol/C_M	2
6	< 0	Sol/C_M	2
8	< 0	Sol/C_M	2
9	< 0	Sol/C_M	2

Table I.2.3. Krafft temperature of metal 3-hydroxypentadecane sulphonates and metal 2-pentadecene sulphonates

Compound	Krafft temperature, °C				Ref.
	Sol		DSC		
	Na salt	Ca salt	Na salt	Ca salt	
$C_{12}H_{25}CH(OH)C_2H_4SO_3^-$	20.2	66.8	21.5	68	10
$C_{12}H_{25}CH=CHCH_2SO_3^-$	35.5	91.3	40.0		10

Table I.2.4. Krafft temperature of various substituted sodium sulphonates (Taken from reference 28).

Compound	Chain length								
	10	11	12	13	14	15	16	17	18
—2–alkene–[a]							36		54
—3–oxoalkane–[b]			28		40		51		59
—3–hydroxyalkane–[c]					21		37		51
—2–oxoalkane–[d]			25	36		48.5		60.5	
—2–hydroxyalkane–[e]	35.5	58	67.5	73		83		94	
—1–hydroxyalkane–[f]			59		73		84		93

a. $RCH = CHCH_2 SO_3 Na$

b. $RCOCH_2 CH_2 SO_3 Na$

c. $RCHOHCH_2 CH_2 SO_3 Na$

d. $RCOCH_2 SO_3 Na$

e. $RCHOHCH_2 SO_3 Na$

f. $RCHOHSO_3 Na$, according to ref. (4)

Table I.2.5. Krafft temperature of copper(II) and methylviologen(II) alkanesulphonates.

Compound	Krafft Temp. °C	Method	Reference
$Cu(C_nH_{2n+1}SO_3)_2$			
n = 10	37	Co	3
n = 12	53	Co	3
n = 14	62	Co	3
$MV^*(C_nH_{2n+1}SO_3)_2$			
n = 10	< 0	Abs	3
n = 12	10	Abs	3
n = 14	22	Abs	3
n = 16	35	Abs	3
n = 18	43	Abs	3

* MV = 1,1'-dimethyl-4,4'-bispyridinium (II)

CH_3-^+N ⬡⬡ N^+-CH_3 $(C_{14}H_{29}SO_3^-)_2$

Table I.2.6. Krafft temperature of 1,1'-(1,ω -alkanediyl)-bispyridinium tetradecanesulphonates* or $C_nBP(C_{14})_2$.

n	Krafft Temp. °C	Method	Reference
2	34.6	Sol/C_m	7
4	11.4	Sol/C_m	7
6	6.0	Sol/C_m	7
8	< 0	Sol/C_m	7
10	< 0	Sol/C_m	7
12	2.1–2.3	Sol/C_m	7
14	35.7	Sol/C_m	7

* ⬡N$^+$–(CH$_2$)$_n$–$^+$N⬡ $(C_{14}H_{29}SO_3^-)_2$

Table I.2.7. Krafft temperature of sodium dodecanesulphonate in aqueous NaCl.

Krafft temperature °C	NaCl conc. kmol/m^3	Method	Reference
34.5	0	Sol	9
40	0.1	Sol	9
42	0.2	Sol	9
45.5	0.5	Sol	9
49	1.0	Sol	9

Table I.2.8. Thermodynamic parameters for dissolution of copper(II) and methylviologen(II)* dodecanesulphonates obtained by solubility measurements (Taken from reference 33).

Surfactant	$\Delta H°$ kJ/mol	$\Delta G°$ kJ/mol	$\Delta S°$ J/mol·K
$Cu(C_{12}H_{25}SO_3)_2 \cdot 2H_2O$	127	58 (32°C)	226 (32°C)
$MV(C_{12}H_{25}SO_3)_2 \cdot 2H_2O$	75	49 (9°C)	92 (9°C)

* MV = 1,1'-dimethyl-4,4'-bispyridinium (II)

$$CH_3 - {}^+N \overbrace{\bigcirc\bigcirc} N^+ - CH_3 \ (C_{14}H_{29}SO_3{}^-)_2$$

Table I.2.9. Melting point and phase transition temperature of 1,1'-(1, ω -alkanediyl)-bispyridinium tetradecanesulphonates* or $C_nBP(C_{14})_2$ and of copper(II) and methylviologen(II) alkanesulphonates**.

Compound	Melting point °C	Method	Phase Tr. Temp. °C	Method	Reference
$C_nBP(C_{14})_2$					
n = 2	188 –	MP	100	MP	8
			128	DSC	8
n = 4	68–90	MP			8
	79	DSC			8
n = 6	92–94	MP			8
	92.6	DSC	47	MP	8
n = 8	60–81	MP	41	MP	8
	64.3	DSC			8
n =10	82–118	MP	35	MP	8
	73	DSC	55	DSC	8
n = 12	63–83	MP			8
	64	DSC			8
n = 14	134–138	MP	69	MP	8
	135	DSC			8
$Cu(C_{14})_2$	49 –	MP	88	MP	8
	147	DSC	101	DSC	8
$MV(C_{14})_2$**	74–94	MP	61	MP	8
	91.0	DSC			8

* ⬡N$^+$–$(CH_2)_n$–$^+$N⬡ $(C_{14}H_{29}SO_3{}^-)_2$

** CH_3–$^+$N⬡⬡N$^+$–CH_3 $(C_{14}H_{29}SO_3{}^-)_2$

Table I.2.10. Critical micelle concentration (CMC, in mmol/L) of sodium alkanesulphonates at various temperatures.

Alkane sulphonate	Temperature, °C												Method	Ref.
	20	25	30	35	40	45	50	55	60	65	70	80		
$C_6H_{13}SO_3^-$	425		372		319								ST	27
	458												DS	27
$C_8H_{17}SO_3^-$		140a											D_S	11
		150a		160a									Ca	12
		155b											Co	15
		155			162		177						R	18
	162		139		116								ST	27
	207												DS	27

Table I.2.10. (continued)

Alkane sulphonate	Temperature, °C												Method	Ref.
	20	25	30	35	40	45	50	55	60	65	70	80		
$C_{10}H_{21}SO_3^-$				52a									Ca	12
				41.8a									Ca	13
					40							58	Co	14
						39.8							ST	2
			38.1						43				Co	15
		41		42		45		49		55			R	18
							38.5						Co	19
	39.4d	38.8		40.1		40.6		41.6					Co	3
	50.2												DS	27
	65.6		61.5		57.5								ST	27
$C_{11}H_{23}SO_3^-$		13.5											NS	35

Table I.2.10. (continued)

Alkane sulphonate	20	25	30	35	40	45	50	55	60	65	70	80	Method	Ref.
$C_{12}H_{25}SO_3^-$				9.7									Ca	12
				10.2									Ca	13
					11							14	Co	14
				10		11	11	12	12	14			R	18
							6.9						DS	4
					9								Co	19
				10.6	10.7	11.2	11.5	12					Co	3
	11		10.5		9.2								ST	27
	11.8												DS	27
					9.3								DS	28
		9.3											ST	36
$C_{13}H_{27}SO_3^-$							3.52						DS	4
		3.49											NS	35

Table I.2.10. (continued)

Alkane sulphonate	20	25	30	35	40	45	50	55	60	65	70	80	Method	Ref.
$C_{14}H_{29}SO_3^-$	5		4.5		3.9								ST	27
					2.5				3.3			4.6	CO	14
					2.4				2.42				DS	28
												ST	17
	3.2				2.5	2.9							R	18
							2.03						DS	27
							3.3e	3.5					DS	4
						3.0							CO	3
					2.6								CO	19
$C_{15}H_{31}SO_3^-$		0.668					0.66						DS	4
						1.5							NS	35
													ST	20
$C_{15}H_{31}\text{-}8\,SO_3^-$		6.6				5.4							ST	20

Temperature, °C

Table I.2.10. (continued)

Alkane sulphonate	Temperature, °C												Method	Ref.
	20	25	30	35	40	45	50	55	60	65	70	80		
$C_{16}H_{33}SO_3^-$									0.54				ST	17
	1.8												R	18
			1.5				0.9c						ST	27
					1.37		0.45						DS	4
	1.09												DS	27
					0.74								Co	19
					0.74								DS	28
$C_{17}H_{35}SO_3^-$		0.213					0.21						DS	4
												NS	35
$C_{18}H_{37}SO_3^-$									0.19				ST	17

a) mmol/kg
b) 23°C
c) 52°C
d) 22°C
e) 51.5°C

Table I.2.11. Critical micelle concentration (CMC, in mmol/kg) of sodium dodecanesulphonate at various temperatures.

Temperature °C	CMC mmol/kg	Method	Reference
74	8.0 ± 2	Δ H	16
126	19.0 ± 3	Δ H	16
178	68.0 ± 6	Δ H	16

Table I.2.12. Critical micelle concentration (CMC, in mmol/L) of isomers of sodium dodecanesulphonates at 40 °C using conductivity (Taken from reference 34).

Sulphonate	CMC, (mmol/L)
1–dodecane	10.4
2–dodecane	14.9
3–dodecane	18.6
4–dodecane	23.2
5–dodecane	28.3
6–dodecane	36.1

Table I.2.13. Critical micelle concentration (CMC, in mmol/L) of sodium alkanesulphonates at 45 °C.

$$NaSO_3 CH < \begin{array}{c} C_9H_{19} \\ C_mH_{2m+1} \end{array}$$

m	CMC (mmol/L)	Method	Reference
0	39.8	ST	2
2	18.2	ST	2
4	6.76	ST	2
6	1.66	ST	2
8	0.24	ST	2
9	0.089	ST	2

Table I.2.14. Critical micelle concentration (CMC, in mmol/L) of various substituted sodium alkane   sulphonates at 25 °C.

Substituent		CMC (mmol/L)	Method	Reference
3–hydroxy	C_{12}	25 (24°C)	ST	36
	C_{14}	6.3	ST	36
	C_{16}	1.5	ST	36
3–methoxy	C_{12}	7.0 (24°C)	ST	36
	C_{16}	0.31	ST	36
3–ethoxy	C_{12}	4.6 (24°C)	ST	36
3–n–propoxy	C_{12}	3.5 (24°C)	ST	36
	C_{16}	0.075	ST	36
3–i–propoxy	C_{12}	3.2 (24°C)	ST	36
3–n–butoxy	C_{12}	1.4 (24°C)	ST	36
	C_{16}	0.032	ST	36
3–n–hexoxy	C_{12}	0.60 (24°C)	ST	36
3–n–octoxy	C_{12}	0.11 (24°C)	ST	36
3–(2–ethyl)hexoxy	C_{12}	0.29 (24°C)	ST	36
3–hydroxyethoxy	C_{14}	0.33	ST	36
3–hydroxyethoxy– ethoxy	C_{14}	1.1	ST	36
3–phenoxy	C_{12}	1.8 (24°C)	ST	36
	C_{14}	0.21	ST	36
3–trichlorophenoxy	C_{14}	0.015	ST	36
3–dimethylamino	C_{14}	1.0	ST	37
3–propylamino	C_{14}	0.58	ST	37
3–butylamino	C_{14}	0.18	ST	37

Table I.2.14. (continued)

Substituent		CMC (mmol/L)	Method	Reference
3–morpholino	C_{14}	0.72	ST	37
3–piperidino	C_{14}	0.45	ST	37
1–carbomethoxy	C_{11}	9.5	ST	39
	C_{13}	2.4	ST	39
	C_{15}	0.368	ST	38
		0.55	DS	38
	C_{17}	0.092	ST	38
		0.12	DS	38
1–carbethoxy	C_{15}	0.285	ST	38
	C_{17}	0.072	ST	38
1–carbo–n–propoxy	C_{15}	0.0998	ST	38
	C_{17}	0.0116	ST	38
1–carbo–i–propoxy	C_{17}	0.0233	ST	38

Table I.2.15. Critical micelle concentration (in mmol/L) of various substituted sodium sulphonates at 40 °C using dye solubilization (Taken from reference 28).

Compound	Alkyl Chain Length			
	12	14	16	18
–2–alkene[a]	13.0	2.7	0.61	0.18
–3–hydroxyalkane[b]	24.8	6.33	1.45	0.38
–3–oxoalkane[c]	28.8	6.7	1.84	0.5[d]

a. $RCH = CHCH_2SO_3Na$ c. $RCOCH_2CH_2SO_3Na$

b. $RCHOHCH_2CH_2SO_3Na$ d. at 60°C

Table I.2.16. Critical micelle concentration (CMC, in mmol/L) of 1-hydroxyalkane-2-sulphonates at 25 °C.

Alkane	CMC, mmol/L	Method	Reference
Dodecane	15.0	Sol	4
Tetradecane	3.4	Sol	4
Hexadecane	0.3	Sol	4
Octadecane	0.1	Sol	4

Table I.2.17. Critical micelle concentration (CMC, in mmol/L) of secondary sodium alkanesulphonates at 23 °C.

Alkane	CMC, mmol/L	Method	Reference
C_{11}	39	SD	21
C_{12}	14	SD	21
C_{13}	8.7	SD	21
C_{14}	4.7	SD	21
C_{15}	2.3	SD	21
C_{16}	1.08	SD	21
C_{17}	1.08	SD	21

Table I.2.18. Critical micelle concentration (CMC, in mmol/L) of disodium alkanesulphonate-1-carboxylate and disodium alkanesulphonate-1-carboxyethane at 25 °C.

Alkane sulphonate		CMC (mmol/L)	Method	Reference
1–carboxylate	C_{15}	6.6	NS	35
	C_{17}	2.5	NS	35
1–carboxy ethane sulphonate	C_{15}	7.16	Co	38
	C_{17}	2.23	Co	38

Table I.2.19. Critical micelle concentration (CMC, in mmol/L) of copper(II) alkanesulphonates at various temperatures measured using conductivity (Taken from reference 3).

Cu $(C_{10}H_{21}SO_3)_2$		Cu $(C_{12}H_{25}SO_3)_2$		Cu $(C_{14}H_{29}SO_3)_2$	
Temp. °C	CMC mmol/L	Temp. °C	CMC mmol/L	Temp. °C	CMC mmol/L
40.0	8.75	54.1	1.67	62.9	0.38
44.9	8.98	55.5	1.72	64.8	0.41
50.0	9.10	56.2	1.72	66.8	0.39
55.0	9.37	57.6	1.75		

Table I.2.20. Critical micelle concentration (CMC, in mmol/L) of methylviologen(II)* alkanesulphonates at various temperatures measured using conductivity (Taken from reference 3).

MV $(C_{10})_2$		MV $(C_{12})_2$		MV $(C_{14})_2$	
Temp. °C	CMC mmol/L	Temp. °C	CMC mmol/L	Temp. °C	CMC mmol/L
9.9	7.90	7.5	1.32	22.5	0.191
19.8	7.52	10.0	1.27	25.0	0.189
25.0	7.29	15.0	1.27	27.5	0.192
30.1	7.31	20.0	1.22	29.5	0.196
35.0	7.69	25.0	1.20	32.5	0.200
45.0	8.19	35.0	1.26	35.1	0.205
55.0	8.66	45.0	1.35	40.0	0.213
		55.0	1.55	45.0	0.233
				50.0	0.240
				55.0	0.247

MV $(C_{16})_2$		MV $(C_{18})_2$	
Temp. °C	CMC mmol/L	Temp. °C	CMC mmol/L
37.0	0.0350	50.0	0.0095
40.0	0.0376	53.0	0.0105
43.0	0.0405	56.0	0.0111
46.0	0.0429	60.0	0.0119
49.1	0.0450	63.0	0.0139

* MV = 1,1'-dimethyl-4,4'-bispyridinium (II)

$CH_3 - {}^+N$ ⬡⬡ $N^+ - CH_3 \ (C_{14}H_{29}SO_3^-)_2$

Table I.2.21. Critical micelle concentration (CMC, in mmol/L) of 1,1'-(1, ω -alkanediyl)-bispyridinium tetradecanesulphonates* or $C_nBP(C_{14})_2$ at 35 °C.

n	CMC mmol/L	Method	Reference
2	0.19	Co	8
	0.13	DSC	8
4	0.17	Co	8
	− 0.01	DSC	8
6	0.17	Co	8
	0.17	DSC	8
14	0.027	Co	8
	0.036	DSC	8

* $\bigcirc N^+-(CH_2)_n-^+N\bigcirc$ $(C_{14}H_{29}SO_3^-)_2$

Table I.2.22. Critical micelle concentration (CMC, in mmol/L) of magnesium alkanesulphonates at various temperatures.

Surfactant	Temperature °C		Method	Reference
	23	60		
$Mg(C_8H_{17}SO_3)_2$	55		Co	15
$Mg(C_{10}H_{21}SO_3)_2$		10	Co	15
$Mg(C_{12}H_{25}SO_3)_2$		1.6	Co	15

Table I.2.23. Critical micelle concentration (CMC, in mmol/L) of sodium alkanesulphonates in aqueous NaCl at 45 $^\circ$C measured using surface tension (Taken from reference 2).

$$NaSO_3-CH < \begin{array}{l} C_9H_{19} \\ C_mH_{2m+1} \end{array}$$

m	C_{NaCl} mmol/L	CMC mmol/L
0	68.4	23.4
	136	15.1
	273	10.4
	546	6.76
2	68.4	8.31
	136	5.75
	273	4.16
	546	2.63
4	68.4	2.29
	136	1.48
	273	0.954
	546	0.631
6	8.55	1.07
	17.1	0.707
	34.2	0.478
	68.4	0.301
8	8.55	0.131
	17.1	0.091
	34.2	0.061
	68.4	0.039
9	4.27	0.052
	8.55	0.036
	17.1	0.026
	68.4	0.012

Table I.2.24. Critical micelle concentration (CMC, in mmol/L), surface tension at the CMC (γ_{CMC}, in mN/m), and molar volume (V, in cm^3/mol) of sodium alkanesulphonates

Surfactant	T	CMC	γ CMC	V	Reference
	°C	mmol/L	mN/m	cm^3/mol	
$NaC_{10}H_{21}SO_3$	25	7.60	27.5	228.9	17
$NaC_{14}H_{29}SO_3$	60	2.42	28.8	296.9	17
$NaC_{16}H_{33}SO_3$	60	0.54	26.0	340.3	17
$NaC_{18}H_{37}SO_3$	60	0.19	28.6	367.2	17

Table I.2.25. Free energy of micellization (ΔG°_m, in kJ/mol) of sodium alkanesulphonates at various temperatures.

Surfactant Thermodynamic parameter	Temperature, °C									Method	Ref.
	25	30	35	40	45	50	55	60	65		
$C_8H_{17}SO_3Na$ ΔG°_m, kJ/mol	−14.75			−15.21		−15.46				Calc.	32
$C_{10}H_{21}SO_3Na$ ΔG°_m, kJ/mol	−18.02		−18.44		−18.86		−19.27		−19.48	Calc.	32
$C_{12}H_{25}SO_3Na$ ΔG°_m, kJ/mol	−21.70		−22.04		−22.50		−22.96		−23.21	Calc.	32

Table I.2.26. Enthalpy of micellization (ΔH°_m, in kJ/mol) of sodium alkanesulphonates at various temperatures.

Surfactant / Thermodynamic parameter	Temperature, °C											Method	Ref.
	20	25	30	35	40	45	50	55	60	65	70		
$C_{10}H_{21}SO_3Na$													
ΔH°_m, kJ/mol			-2.1		-5.9						-12.6	Calc.	23
				-0.586								Calc.	24
				-2.144								Calc.	12
	-4.46	-1.91		-6.43		-6.48		-6.46		-10.54		Calc.	27
												Calc.	32
$C_{12}H_{25}SO_3Na^*$													
ΔH°_m, kJ/mol					-7.5						-16.7	Calc.	23
				-4.043								Calc.	12
				-6.025								Calc.	24
	-7.26	-7.84		-7.84		-7.84		-7.84		-14.83		Calc.	27
												Calc.	32
$C_{14}H_{29}SO_3Na$													
ΔH°_m, kJ/mol	-7.95											Calc.	27

Table I.2.26. (continued)

Surfactant Thermodynamic parameter	Temperature, °C										Method	Ref.
	20	25	30	35	40	45	50	55	60	65		
$C_{16}H_{33}SO_3Na$												
$\Delta H°_m$, kJ/mol	-10.42										Calc.	27
$C_8H_{17}SO_3Na$												
$\Delta H°_m$, kJ/mol		3.817		1.084							Calc.	12
		6.884		3.964							Calc.	24
		-1.51			-1.31		-7.34				Calc.	32
	-12.09											27
$C_6H_{13}SO_3Na$												
$\Delta H°_m$, kJ/mol	-9.96										Calc.	27

* Enthalpy of dilution values at 75, 125 and 178°C given by D.G. Archer et al. (16)

Table I.2.27. Entropy of micellization (ΔS°_m, in kJ/mol) of sodium alkanesulphonates at various temperatures.

Surfactant	Temperature, °C									Method	Ref.
Thermodynamic parameter	25	30	35	40	45	50	55	60	65		
$C_8H_{17}SO_3Na$											
ΔS°_m, kJ/mol.deg	0.044			0.044		0.025				Calc.	32
$C_{10}H_{21}SO_3Na$											
ΔS°_m, kJ/mol.deg	0.054		0.039		0.039		0.039		0.039	Calc.	32
$C_{12}H_{25}SO_3Na$											
ΔS°_m, kJ/mol.deg	0.046		0.046		0.046		0.046		0.025	Calc.	32

Table I.2.28. Gibbs free energy of micellization (ΔG°_m, in kJ/mol) of sodium alkanesulphonates at 45 °C.

$$\text{Na SO}_3 \text{ CH} < \frac{C_9 H_{19}}{C_m H_{2m+1}}$$

m	ΔG°_m, kJ/mol	Method	Reference
0	− 16.91	Calc.	2
2	− 20.35	Calc.	2
4	− 24.47	Calc.	2
6	− 30.13	Calc.	2
8	− 36.54	Calc.	2
9	− 39.15	Calc.	2

Table I.2.29. Enthalpy of micellization (ΔH°_m, in kJ/mol) of copper(II), methylviologen(II)*, and 1,1' (1, ω -alkanediyl)-bispyridinium tetradecanesulphonates** at 35 °C.

Surfactant	ΔH°_m, kJ/mol	Method	Reference
$Cu(C_{14})_2$	$-$ 29	Ca	8
	$-$ 36.6	Calc.	8
$MV(C_{14})_2$	$-$ 2	Ca	8
	$-$ 5.3	Calc.	8
$C_2BP(C_{14})_2$	$-$ 13	Ca	8
	$-$ 30.9	Calc.	8
$C_4BP(C_{14})_2$	$-$ 17	Ca	8
	$-$ 1.7	Calc.	8
$C_6BP(C_{14})_2$	13	Ca	8
	7.5	Calc.	8
$C_{14}BP(C_{14})_2$	$-$ 17	Ca	8
	$-$ 31.2	Calc.	8

$*\qquad CH_3 - {}^+N$ ⬡⬡ $N^+ - CH_3 \quad (C_{14}H_{29}SO_3{}^-)_2$

$** \quad$ ⬡ $N^+ - (CH_2)_n - {}^+N$ ⬡ $\quad (C_{14}H_{29}SO_3{}^-)_2$

Table I.2.30. Apparent molar heat capacity $C_{p,\phi}$ (in J mol^{-1}K^{-1}) of sodium decanesulphonate and sodium dodecanesulphonate (0.3 mol/kg) at various temperatures at 1.86 MPa.

Surfactant	Temperature, °C			Method	Reference
	75	125	177		
$C_{10}H_{21}SO_3^-$	584	764		Ca	16
	569	832			
	571				
$C_{12}H_{25}SO_3^-$	577	724	1029	Ca	16
	570	724	1023		
	573				

Table I.2.31. Degree of ionization (α) of sodium alkanesulphonates at 45 °C.

$$Na^{+\,-}SO_3CH < \begin{array}{c} C_9H_{19} \\ C_mH_{2m+1} \end{array}$$

m	α	Method	Reference
0	0.12	In(c)−In(c′)	2
2	0.14	In(c)−In(c′)	2
4	0.17	In(c)−In(c′)	2
6	0.23	In(c)−In(c′)	2
8	0.33	In(c)−In(c′)	2
9	0.41	In(c)−In(c′)	2

Table I.2.32. Extent of counter-ion association with micelles (m/N_{agg}) of 1,1' (1, ω -alkanediyl)-bispyridinium tetradecanesulphonates* or $C_nBP(C_{14})_2$ at 35 °C.

n	(m/N_{agg})	Method	Reference
2	0.41	In(c)–In(c')	22
4	0.42	In(c)–In(c')	22
6	0.44	In(c)–In(c')	22
8	0.44	In(c)–In(c')	22
10	0.46	In(c)–In(c')	22
12	0.48	In(c)–In(c')	22
14	0.49	In(c)–In(c')	22

* $\bigcirc\!\!-N^+-(CH_2)_n-{}^+N-\!\!\bigcirc$ $(C_{14}H_{29}SO_3{}^-)_2$

Table I.2.33. Aggregation number of sodium alkanesulphonates at various temperatures.

Surfactant	Temperature, °C									Method	Ref.
	20	25	30	35	40	45	50	55	60		
$NaC_8H_{17}SO_3$		25.4[a]								LS	15
$NaC_{10}H_{21}SO_3$			40							LS	15
$NaC_{12}H_{25}SO_3$					54					LS	15
$NaC_{14}H_{29}SO_3$									80	LS	15
NaC_{15}-1-sulfonate						67				Fl	20
NaC_{15}-8-sulfonate		63				41				Fl	20

a) 23°C

Table I.2.34. Aggregation number of 1,1' (1,ω -alkanediyl)-bispyridinium tetradecanesulphonates* or $C_nBP(C_{14})_2$ at 35 $^{\circ}C$.

n	N_{agg}	Method	Reference
2	74.2	LS	22
4	54.4	LS	22
6	44.1	LS	22
8	33.8	LS	22
10	33.4	LS	22
12	35.2	LS	22
14	67.4[a]	LS	22

$$* \quad \bigodot N^+ - (CH_2)_n -^+N \bigodot \quad (C_{14}H_{29}SO_3^-)_2$$

a) 61.2 at 40°C

Table I.2.35. Aggregation number of magnesium alkanesulphonates measured using light scattering (taken from reference 15).

Surfactant	Temperature, °C	
	23	60
$Mg\ (C_8H_{17}SO_3)_2$	51	
$Mg\ (C_{10}H_{21}SO_3)_2$		103
$Mg\ (C_{12}H_{25}SO_3)_2$		107

Table I.2.36. Mean micellar mass (M_a) of sodium pentadecanesulphonate as a function of electrolyte concentration measured using light scattering (Taken from reference 31).

Electrolyte, Concentration (mmol/L)		$\overline{M}_a . 10^{-5}$
KCl	0.1	0.76
	0.2	0.79
	0.3	1.92
	0.4	2.50
	0.5	5.50
	0.6	10.00
CsCl	0.4	1.66
NaCl	0.4	0.342
LiCl	0.4	0.301

Temperature unknown

Table I.2.37. Self-diffusion coefficient (D, in $m^2 s^{-1}$) of sodium octanesulphonate in D_2O at 25 °C measured using NMR (Taken from reference 29).

Concentration mol.kg^{-1}	D^a $m^2 s^{-1}$
Infinite dilution	0.66×10^{-9}
0.5	2.2×10^{-10}
1.0	1×10^{-10}
1.5	6×10^{-11}
2.0	3.5×10^{-11}

a) Values of D obtained from graph of D against concentration

Table I.2.38. Carbon-13 chemical shifts (δ_{mo}, in ppm)[a] and chemical shift changes ($\Delta\delta$)[b] upon micellization of sodium alkanesulphonates. Numbering of the carbon atoms in the alkyl chain starts from the SO_3^- group[c]. All measurements in D_2O at 32 °C. Taken from reference 25.

Surfactant		C-1	C-2	C-3	C-4	C-5	C-6	C-7	C-8	C-9	C-10	C-11	C-12
$NaC_8H_{17}SO_3$	δ_{mo}	51.97	28.89	24.97	29.39	29.39	31.96	23.05	14.41				
	$\Delta\delta$ m	+0.21	+1.01	+0.52	+0.98	+1.11	+0.75	+0.62	+0.34				
$NaC_{10}H_{21}SO_3$	δ_{mo}	51.95	28.56	24.84	29.06	29.47	29.47	29.32	32.00	22.87	14.27		
	$\Delta\delta$ m	+0.26	+0.99	+0.47	+1.01	+0.96	+1.06	+0.88	+0.66	+0.47	+0.21		
$NaC_{12}H_{25}SO_3$	δ_{mo}	52.00	28.55	24.82	29.03	29.28	29.54	29.54	29.46	29.28	31.99	22.82	14.23
	$\Delta\delta$ m	+0.21	+0.99	+0.49	+1.04	+1.23	+0.97	+0.97	+1.18	+0.91	+0.67	+0.49	+0.25

a. Carbon-13 chemical shifts are relative to TMS; δ_{mo} monomer shift at CMC.
b. $\Delta\delta$ m = ($\delta_{micelle} - \delta_{mo}$)
c. For C-4 and C-5 of $C_8H_{17}SO_3Na$, C-5 and C-6 of $C_{10}H_{21}SO_3Na$, and for C-4 and C-9 of $C_{12}H_{25}SO_3Na$ assignments are tentative.

Table I.2.39. Carbon-13 chemical shifts (δ , in ppm)[a] and one-bond carbon-carbon coupling constants ($^1J(^{13}C^{13}C)$) for sodium octane sulphonate.

Carbon[b]	δ (ppm)	Carbon[b]	$^1J(^{13}C\ ^{13}C)$
C–1	51.93	C–8, C–7	35.4
C–2	25.02	C–7, C–6	35.1
C–3	29.24	C–6, C–5	34.7
C–4	29.67	C–3, C–2	34.2
C–5	29.77	C–2, C–1	34.2
C–6	32.27		
C–7	23.05		
C–8	14.19		

a) With respect to TMS

b) Numbering of carbon atoms starts from SO_3^-
Measurements in D_2O; temperature not known; ref. (30)

Table I.2.40. Standard thermodynamics of adsorption of sodium alkanesulphonates at the air-water interface at various temperatures.

Surfactant[a]	Temp. (°C)	ΔG°_{ads} kJ/mole	ΔH°_{ads} kJ/mole	ΔS°_{ads} J/mole K	Ref.
$NaC_{10}H_{21}SO_3$	45	26.13			2
$NaC_{12}H_{25}SO_3$	10	28.6	4.6	117	
	20	29.7	4.6	117	
	30	30.9	4.6	117	
$NaC_{14}H_{29}SO_3$	10	30.9	4.6	126	26
	20	32.1	4.6	126	
	30	33.4	4.6	126	
$NaC_{16}H_{33}SO_3$	10	34.1	8.0	149	
	20	35.4	8.1	149	
	30	37.0	8.0	149	

[a] $NaC_{10}H_{21}SO_3$: average of values obtained at various salinities; Other surfactants: all values at constant ionic strength 8×10^{-4} N.NaCl

REFERENCES Chapter I.2.

1) Saito, M., Moroi, Y., and Matuura, R., J. Colloid Interface Sci., **88**, 578 (1982).

2) Granet, R. and Piekarski, S., Colloids Surfaces, **33**, 321 (1988).

3) Moroi, Y., Sugii, R., Akine, C., and Matuura, R., J. Colloid Interface Sci., **108**, 180 (1985).

4) Weil, J.K., Smith, F.D., Stirton, A.J., and Bistline, R.G., J. Am. Oil Chem. Soc., **40**, 538 (1963).

5) Tartar, H.V., and Wright, K.A., J. Am. Chem. Soc., **61**, 539 (1939).

6) Moroi, Y., Sugii, R., and Matuura, R., J. Colloid Interface Sci., **98**, 184 (1984).

7) Moroi, Y., Matuura, R., Kuwamura, T., and Inokuma, S., J. Colloid Interface Sci., **113**, 225 (1986).

8) Moroi, Y., Matuura, R., Kuwamura, T., and Inokuma, S., Colloid Polymer Sci., **266**, 374 (1988).

9) Somasundaran, P., Ananthapadmanabhan, K.P., Celik, M.S., and Manev, E.D., Soc. Pet. Eng. Journal, 667, December (1984).

10) Tsujii, K., Saito, N., and Takeuchi, T., J. Phys. Chem., **84**, 2287 (1980).

11) Lindman, B., Puyal, M-C., Kamenka, N., Rymden, R., and Stilbs, P., J. Phys. Chem., **88**, 5048 (1984).

12) Woolley, E.M. and Burchfield, T.E., J. Phys. Chem., **89**, 714 (1985).

13) Berg, R.L., Noll, L.A., and Good, W.D., ACS Symposium Series, No. 19, 87 (1979).

14) Wright, K., Doyle Abbott, A., Sivertz, V., and Tartar, H.V., J. Am. Chem. Soc., **61**, 549 (1939).

15) Tartar, H.V. and Lelong, A.L.M., J. Phys. Chem., **59**, 1185 (1956).

16) Archer, D.G., Albert, H.J., White, D.E., and Wood, R.H., J. Colloid Interface Sci., **100**, 68 (1984).

17) Klimenko, N.A., Karmazina, T.V., Yaroshenko, N.A. Bartnitskii, A.E., and Aryamova, Zh.M., Kolloidnyi Zh., **46**, 1112 (1984).

18) Klevens, H.B., J. Phys. Chem., **52**, 130 (1948).

19) Saito, M., Moroi, Y., and Matuura, R. in "Surfactants in Solution", Mittal, K.L. and Lindman, B. (Eds.), Vol. 2, Plenum Press, New York, (1984), page 771.

20) Fanghänel, E., Ortman, W., Behrmann, K., Willscher, S., Turro, N.J., and Gould, I.R., J. Phys. Chem., **91**, 3700 (1987).

21) Quack, J.M. and Trautmann, M., Annali di Chimica, **77**, 245 (1987).

22) Moroi, Y., Matuura, R., Tanaka, M., Murata, Y., Aikawa, Y., Furutani, E., Kuwamura, T., Takahashi, H., and Inokuma, S., J. Phys. Chem., **94**, 842 (1990).

23) Stainsby, G. and Alexander, A.E., Trans. Faraday Soc., **46**, 587 (1950).

24) Hill, R.M., Ph.D. Thesis, University of Oklahoma (1982).

25) Okabayashi, H., Yoshida, T., Matsushita, K., and Terada, Y., Chemica Scripta, **20**, 117 (1982).

26) Perea-Carpio, R., Gonzalez-Caballero, F., Bruque, J.M., and Pardo, G., J. Colloid Interface Sci., **95**, 513 (1983).

27) Kovtunenko, L.I., Smirnov, N.I., and Titova, N.P., Zh. Prikl. Khimii, **48**, 323 (1975).
28) Püschel, F., Tenside, **4**, 320 (1967).
29) Lindman, B., Puyal, M-C., Kamenka, N., Rymden, R., and Stilbs, P., J. Phys. Chem., **88**, 5048 (1984).
30) Millot, C., Brondeau, J., and Canet, D., Mag. Res. Chem., **24**, 648 (1986).
31) Bagdasaryan, V.V., Shaginyan, A.A., Barkhudaryan, V.G., Aivazyan, O.M., and Aslanyan, V.M., Kolloidnyi Zhurnal, **45**, 295 (1983).
32) Yakovlev, V.D., Zaichenko, L.P., and Abramzon, A.A., Zh. Prikl. Khimii, **52**, 2471 (1979).
33) Moroi, Y., Ikeda, N., and Matuura, R., J. Colloid Interface Sci., **101**, 285 (1984).
34) Besserman, M., Ph.D. Thesis, University of Washington (1953).
35) Weil, J.K., Stirton, A.J., Smith, F.D., and Bistline, R.G., US Pat. 3228980, 11.01.66.
36) Püschel, F., and Kraatz, K.-H., IV. Intern. Tagung Grenzfl. Stoffe, Berlin, Band 1, 202 (1974).
37) Püschel, F., IV. Intern. Tagung Grenzfl. Stoffe, Berlin, Band 1, 235 (1974).
38) Weil, J.K. and Stirton, A.J., J. Phys. Chem., **60**, 899 (1956).
39) Schwuger, M.J., Fette Seifen Anstrichmittel, **72**, 565 (1970).

I.3 Alkylarenesulphonates

Table I.3.1. Krafft temperature of sodium 4-alkylbenzene-sulphonates.

COMPOUND	KRAFFT TEMPERATURE °C	METHOD	REFERENCE
1ϕC7	9	Sol/C_M	3
1ϕC8	26	Sol	8
	18.5	Sol/C_M	3
	25.5	DSC	8
	18	Sol/C_M	9
3ϕC9	< 1	Sol	2
1ϕC10	39	Sol	1
	36	Sol/C_M	3
	39	Sol/C_M	4
	47	Sol	5
	40	Cryst	5
2ϕC10	22	Sol	1
	16	Sol	6
3ϕC10	4	Sol	1
5ϕC10	< 1	Sol	1
1ϕC11	50	Sol	5
	44	Cryst	5
1ϕC12	40	Sol/C_M	4
	64	Sol	5
	52	Sol/C_M	3
	53	Cryst	5
	62.5	Sol	8
	65.0	DSC	8
	50	Sol/C_M	9
3ϕC12	14	Sol	2
1ϕC14	75	Sol	5
	45	Sol/C_M	4
	59	Cryst	5
1ϕC15	78	Sol/C_M	3
1ϕC16	49	Sol/C_M	4
1ϕC18	58	Sol/C_M	4

Table I.3.2. Krafft temperatures of 4-substituted sodium benzenesulphonates (Taken from reference 5).

R	KRAFFT TEMPERATURE °C	METHOD
$C_{11}H_{23}-O-$	75	Sol
	65	Cryst
$C_{11}H_{23}-NH-$	< 10	Sol
	< 10	Cryst
$C_{11}H_{23}-S-$	84	Sol
	74	Cryst
$C_{11}H_{23}-SO-$	32	Sol
	17.5	Cryst
$C_{11}H_{23}-SO_2-$	72	Sol
	68.3	Cryst
$C_{10}H_{21}-O-OC$	52	Sol
	42	Cryst
$C_{10}H_{21}-CO-O$	24	Sol
	12	Cryst
$C_9H_{19}-O-OC-CH_2$	< 10	Sol
	< 6	Cryst
$C_8H_{17}-O-OC-CH_2-O-$	32	Sol
	28	Cryst
$C_{10}H_{21}-CO-NH$	41	Sol
	31	Cryst
$(C_{11}H_{23}-NH-C_6H_4SO_3)_2Ca$	50	Sol
	33	Cryst

Table I.3.3. Krafft temperature of sodium 4-alkylbenzene-
sulphonates in aqueous NaCl.

COMPOUND	NaCl CONCENTRATION M	KRAFFT TEMPERATURE °C	METHOD	REFERENCE
2φC10	0.1	26	Sol	6

Table I.3.4. Krafft temperature of sodium alkylnaphtalenesulphonates measured using turbidity (Taken from reference 7)

ALKYL	KRAFFT TEMPERATURE °C
Octyl	10
Decyl	20
Dodecyl	38
Tetradecyl	52
Hexadecyl	64

Table I.3.5. Krafft temperature of sodium octylnaphtalene-
sulphonate and sodium dodecylnaphtalenesulphonate in aqueous NaCl
measured using turbidity (Taken from reference 7).

COMPOUND	NaCl wt %	KRAFFT TEMPERATURE °C
Octyl	0	10
	0.1	22
	0.5	39
Dodecyl	0	38
	0.1	54
	0.2	58
	0.3	59
	1.0	65

Table I.3.6. Krafft temperature of differently substituted sodium dodecylnaphtalenesulphonates in aqueous NaCl measured using turbidity (Taken from reference 7).

DODECYL COMPOUND	NaCl wt %	KRAFFT TEMPERATURE °C
Linear	0	38
	0.1	54
	0.2	58
	0.3	59
	1.0	65
1—methyl	0.1	32
	0.25	38
	0.5	40
	1.0	44
	2.0	48
bilinear	0.75	12
	1.0	15
	1.5	28
	2.0	35
	2.5	42

Table I.3.7. Krafft temperature of metal 4-dodecylbenzene-sulphonates[a].

Me^+	KRAFFT TEMPERATURE °C
Li	10
Na	52
K	39
Rb	19
Cs	-6[b]

a : Method : Sol/C_M

    Ref.   : (3)

b : Calculated value

Table I.3.8. Krafft temperature of calcium 4-alkylbenzene-sulphonates measured using solubility (Taken from reference 8).

COMPOUND	KRAFFT TEMPERATURE °C
Ca octylbenzenesulphonate	196.5
Ca dodecylbenzenesulphonate	269.0

Table I.3.9. Transition enthalpy at the Krafft temperature[a] for sodium and calcium alkylbenzenesulphonates (Taken from reference 8).

COMPOUND	ΔH kJ/mol	METHOD
Na 4-(octyl)benzenesulphonate	10.4	DSC
Ca 4-(octyl)benzenesulphonate	193.1	Calc.
Na 4-(dodecyl)benzenesulphonate	26.3	DSC
Ca 4-(dodecyl)benzenesulphonate	253.3	Calc.

a. See Table I.3.8.

Table I.3.10. Critical micelle concentration (CMC, in mmol/L) of sodium 4-(alkyl)benzenesulphonates at various temperatures.

ALKYL	15	20	25	30	35	40	45	50	55	60	65	70	75	80	METHOD	REF.
1-Hexyl													37.1		DS	4
1-Heptyl													20.9		DS	4
1-Heptyl			22.8[a]	24.0[a]	25.0[a]	26.2[a]	27.4[a]								CO	3
1-Octyl													14		DS	4
1-Octyl			11.4	12.0		12.4		13.3							CO	9
1-Octyl					14.7										CO	11
1-Octyl			11.1												ST	11
1-Octyl			11			12				15					CO	12
1-Octyl			11.4[a]	12.0[a]		12.7[a]		13.5[a]		14.7[a]		15.8[a]			CO	3
2-Octyl									19						DS	10
2-Ethyl-Hexyl													25.4		DS	4
1-Nonyl	10.43	10.40	10.27	10.41		10.80		11.64		12.84		14.13			DS	4
3-Nonyl													6.5		CO	2
1-Decyl								3.98		4.48		5.10			CO	1
1-Decyl													3.7		DS	4
1-Decyl													3.75[b]		IFT	5
1-Decyl								3.14							ST	11
1-Decyl								3.81							CO	11
1-Decyl							3.8[a]	4.0[a]	4.3[a]	4.5[a]	4.7[a]				CO	3
2-Decyl		4.35	4.63	4.86		4.98		5.50		6.07		6.92			CO	1
3-Decyl		6.03	6.02	6.04		6.31		6.78		7.39		8.35			CO	1
5-Decyl	8.31	8.14	8.01	8.13		8.98		9.06		10.34		11.39			CO	1

TEMPERATURE, °C

Table I.3.10. (continued)

ALKYL	TEMPERATURE, °C														METHOD	REF.
	15	20	25	30	35	40	45	50	55	60	65	70	75	80		
2-Propyl-Heptyl													8.48		DS	4
1-Undecyl										2.1[b]					IFT	5
2-Undecyl		2.45			2.53										Sol	11
2-Undecyl						1.90									Co	11
2-Undecyl															ST	11
1-Dodecyl									1.20		1.64			1.68	Co	9
1-Dodecyl										1.30			1.19		DS	4
1-Dodecyl												1.2[b]			IFT	5
1-Dodecyl										1.20					Co	11
1-Dodecyl										1.2					Co	12
1-Dodecyl					1.03[c]										ST	13
1-Dodecyl									1.3[a]	1.4[a]	1.5[a]	1.6[a]	1.8[a]		Co	3
2-Dodecyl			1.87						1.69						DS	10
2-Dodecyl			2.10												Co	10
2-Dodecyl			1.19												Co	20
3-Dodecyl	1.99	2.16		2.35		2.48		2.69		3.04			3.30		Co	2
3-Dodecyl			1.46												Co	20
4-Dodecyl			1.75												Co	15
4-Dodecyl			1.59												Co	20
6-Dodecyl			2.38	2.50		2.60									DS	4
6-Dodecyl													3.12		Co	14
2-Butyl-Octyl	2.10	2.29											3.20		DS	4

Table I.3.10. (continued)

ALKYL	15	20	25	30	35	40	45	50	55	60	65	70	75	80	METHOD	REF.
2-Tridecyl				0.71[d]											Sol	11
2-Tridecyl					0.72										Co	11
2-Tridecyl						0.62									ST	11
1-Tetradecyl													0.66		DS	4
1-Tetradecyl												0.5			IFT	5
2-Pentyl-Nonyl													3.32		DS	4
2-Pentadecyl					0.5[e]										Sol	11
2-Pentadecyl						0.31									Co	11
2-Pentadecyl						0.22									ST	11
1-Hexadecyl													0.535		DS	4
2-Heptadecyl							0.14[f]								Sol	11
2-Heptadecyl								0.13							Co	11
2-Heptadecyl								0.14							ST	11
1-Octadecyl													0.638		DS	4

TEMPERATURE, °C

a : mmol/kg
b : water/n-Heptane interface
c : CMC of undercooled, metastable solution below the Krafft point
d : 27.7°C
e : 32.6°C
f : 45.5°C

Table I.3.11. Critical micelle concentration (CMC, in mmol/L) of 4-substituted sodium benzenesulphonates at various temperatures.

R	TEMPERATURE, °C				METHOD	REF.
	60	65	70	75		
$C_{11}H_{23}-O-$			3.2		IFT	5
$C_{11}H_{23}-NH-$	3.2				IFT	5
$C_{11}H_{23}-S-$				2.8	IFT	5
$C_{11}H_{23}-SO-$	6.5				IFT	5
$C_{11}H_{23}-SO_2$			8.5		IFT	5
$C_{10}H_{21}-O-OC-$				4.1	IFT	5
$C_{10}H_{21}-CO-O-$	5.0				IFT	5
$C_9H_{19}-O-OC-CH_2$			8.5		IFT	5
$C_8H_{17}-O-OC-CH_2-O-$	28.0				IFT	5
$C_{10}H_{21}-CO-NH-$				7.3	IFT	5
$(C_{11}H_{23}-NH-C_6H_4SO_3)_2Ca$	0.7**				IFT	5

*   IFT measurements at water/n-heptane interface

** In meq/L

Table I.3.12. Critical micelle concentration (CMC, in mmol/L) of 4-(6-dodecyl)benzenesulphonates with different cations at various temperatures.

CATION	TEMPERATURE, °C						METHOD	REF.
	15	20	25	30	35	40		
Dimethyl ammonium	1.67	1.75	1.75	1.85	1.88	1.95	Co	14
Diethyl ammonium	1.36	1.43	1.50	1.54	1.58	1.59	Co	14
Dipropyl ammonium						1.18	Co	14

Table I.3.13. Critical micelle concentration (CMC, in mmol/L) of sodium alkylarenesulphonates in D_2O at various temperatures.

ALKYLARENE	TEMPERATURE, °C		METHOD	REF.
	40	50		
Dodecylorthoxylene	1.0	1.4	Co	16

Table I.3.14. Critical micelle concentration (CMC, in mmol/L) of sodium 2,5-dialkylbenzenesulphonates.

Di-ALKYL	TEMPERATURE, °C	METHOD	REF.
	25		
C_2H_5; $C_{12}H_{25}$	0.47	FL	17
	0.52	FL	18
C_3H_7; $C_{11}H_{25}$	0.72	FL	17
	0.72	FL	18
$C_{11}H_{23}$; C_3H_7	1.15	FL	17
	1.21	FL	18
C_7H_{15}; C_7H_{15}	2.1	FL	17
	2.02	FL	18
$C_{13}H_{27}$; CH_3	0.66	FL	18
$C_{12}H_{25}$; C_2H_5	0.88	FL	18

Table I.3.15. Critical micelle concentration (CMC, in mmol/kg) of sodium 4-(4-dodecyl)benzenesulphonate measured by calorimetry at elevated temperatures (Taken from reference 19).

TEMPERATURE °C	CMC (mmol/kg)
74	2.0
126	5.0
178	17.5

Table I.3.16. Critical micelle concentration (CMC, in mmol/L) of sodium 4-(alkyl)benzenesulphonates in aqueous NaCl at various temperatures.

ALKYL	NaCl CONC. M	TEMPERATURE, °C		METHOD	REF.
		25	30		
3–Nonyl	0.171		2.5	ST	22
2–Decyl	0.01	2.3		ST	21
3–Decyl	0.171		0.7	ST	22
2–Dodecyl	0.01	0.3		ST	21
4–Dodecyl	0.171		0.15	ST	22

Table I.3.17. Aggregation number of 2-(decyl)benzenesulphonate in water and in aqueous NaCl at various temperatures.

Surf.Conc. (mmol/L)	NaCl Conc. (M)	TEMPERATURE, °C						METHOD	REF.
		25	35	45	50	55	60		
CMC	0	50						FL	33
50	0	60	53	47			31[a]	FL	32
101.6	0	84	81	74			50[a]	FL	32
50	0.107	78	73	69			60[a]	FL	32
50	0.230			84			78[a]	FL	32
102	0.108	118	111	103			92[a]	FL	32
102	0.294			177			135[a]	FL	32

a : 58°C

Table I.3.18. Aggregation number of sodium 4-(alkyl)benzene-sulphonates at various temperatures.

ALKYL	Conc. mmol/L	TEMPERATURE, °C									METHOD	REF.
		5	15	20	25	30	35	40	45	60		
3−nonyl	CMC			42[a]							FL	33
1−decyl	CMC			83[a,b]							FL	33
1−decyl	50								72		FL	32
1−decyl	100								78		FL	32
2−decyl	CMC			63[a]							FL	33
3−decyl	CMC			53[a]							FL	33
5−decyl	CMC			17[a]							FL	33
5−decyl	25				33		28		24	20	FL	32
5−decyl	50				47		43		40		FL	32
5−decyl	104.4				76		75		72		FL	32
3−dodecyl	CMC					57					LS	20
3−dodecyl	CMC			64[a]							FL	33
3−dodecyl	25				70.5						FL	32
3−dodecyl	50				77						FL	32
3−dodecyl	100				89		84		78	77[c]	FL	32
4−dodecyl	CMC					24					LS	20
	3.0				32						FL,KM	15
	50				34						FL,KM	15
	70				33						FL,KM	15
	100				35						FL,KM	15
	150				70						FL,KM	15
	200				141						FL,KM	15
	300				536						FL,KM	15
6−dodecyl	CMC					21					LS	27
6−dodecyl	CMC				21						KM	24
6−dodecyl	CMC	25	22	19							KM	14
8−hexadecyl	CMC								30		est	27

a : 22°C
b : estimated
c : 58°C

Table I.3.19. Aggregation number of sodium   4-(alkyl)benzene-sulphonates calculated according   to reference 26

Alkyl	N_{agg}		
	Eq.(1)	Eq.(2)	
		(a)	(b)
hexyl	33.1	38.0	37.9
heptyl	40.3	44.8	44.8
octyl	49.0	53.0	50.3
nonyl	59.6	62.6	62.8
decyl	72.5	74.0	74.0
undecyl	88.2	87.5	−
dodecyl	107	103	103
tridecyl	131	122	−
tetradecyl	159	144	122
pentadecyl	193	170	−
hexadecyl	235	200	130
octadecyl	347	275	123
eicosyl	513	411	352.5

Eq. (1) $\log N_{agg} = 0.713 + 0.085 (n_c + 3.5)$

Eq. (2) $\log N_{agg} = 1.164 + 0.29 \log(CMC)^{-1}$

a : Calculated CMC at 75°C
b : Experimental CMC at 75°C

Table I.3.20. Aggregation number of sodium   4-(alkyl)benzene-sulphonates in aqueous NaCl at various temperatures.

ALKYL	CONCENTRATION	NaCl	TEMPERATURE, °C				METHOD	REF.
	mmol/L	mmol/L	25	35	45	60		
2-decyl	CMC	0.1	78.2				UC	6
	CMC	0.1	84^a				UC	6
5-decyl	25	0.115	74	65	59	43^b	FL	32
5-decyl	25	0.268				72^b	FL	32
5-decyl	50	0.104	81	75	69	54^b	FL	32
5-decyl	50	0.244				86^b	FL	32
5-decyl	104.4	0.104				90^b	FL	32
3-dodecyl	100	0.123				116^b	FL	32

a : Alternative method of calculation

b : 58°C

Table I.3.21. Aggregation number of sodium   4-(alkyl)benzene-sulphonates in D_2O or in D_2O/H_2O mixtures at various temperatures.

ALKYL	CONCENTRATION M	TEMPERATURE, °C 45	TEMPERATURE, °C 65	METHOD	REF.
1-dodecyl	0.0660		89[c] 90[d] 89[e] 89[f] 89[g]	SANS SANS SANS SANS SANS	28 28 28 28 28
6-dodecyl	0.055	54[a] 50[b]		SANS SANS	23 23
6-dodecyl	0.0700		44.5[c] 44.9[d] 44.9[e] 43.7[f] 43.2[g]	SANS SANS SANS SANS SANS	28 28 28 28 28
1-dodecyl	0.0198 0.0344 0.0503 0.0696 0.0905 0.0985		72[a] 93[a] 88[a] 91[a] 95[a] 98[a]	SANS SANS SANS SANS SANS SANS	28 28 28 28 28 28

a : 100% D_2O
b : 57% D_2O
c : 99% D_2O
d : 90% D_2O
e : 80% D_2O
f : 70% D_2O
g : 58.7% D_2O

Table I.3.22. Aggregation number of sodium 4-(alkyl)benzene-sulphonates in D_2O solutions of NaCl at various temperatures.

ALKYL	CONCENTRATION mmol/L	NaCl mmol/L	TEMP., °C 65	METHOD	REF.
1-dodecyl	0.0660	–	86	SANS	28
	0.0656	10	94	SANS	28
	0.0653	20	97	SANS	28
	0.0643	50	110	SANS	28
6-dodecyl	0.0700	–	44.5	SANS	28
	0.0696	12	53	SANS	28
	0.0692	22	64	SANS	28
	0.0689	29	72	SANS	28

Table I.3.23. Aggregation number of sodium 2,5-dialkylbenzene-sulphonates at 25 °C measured using fluorescence (Taken from reference 18).

R_2, R_5 =	N_{agg}
$C_{13}H_{27}$, CH_3	56
C_2H_5, $C_{12}H_{25}$	54
$C_{12}H_{25}$, C_2H_5	52
C_3H_7, $C_{11}H_{23}$	45
$C_{11}H_{23}$, C_3H_7	43
C_7H_{15}, C_7H_{15}	38

Table I.3.24. Aggregation number of 4-(6-dodecyl)benzenesulphonates with different cations at various temperatures obtained by using kinetic measurements (Taken from reference 14).

COUNTERION	TEMPERATURE, °C				
	15	25	30	35	40
$H_2N(CH_3)_2^+$	(28)	26		24	
$H_2N(C_2H_5)_2^+$	31	29		27	
$H_2N(C_3H_7)_2^+$					24

Table I.3.25. Aggregation number of sodium alkylarenesulphonates in D_2O at various temperatures calculated from conductivity data assuming a degree of dissociation (β) of 0.6 (Taken from reference 16).

COMPOUND CONCENTRATION (mole/kg)	TEMPERATURE °C	N_{agg}
Dodecylorthoxylene 0.001 < C < 0.005	40	22
Dodecylorthoxylene 0.001 < C < 0.005	50	38
3C12-o-AXS	53.6	30
3C12-o-AXS	50	30
3C12-m-ATS	25	70

Table I.3.26. Micellar diffusion coefficient (D, in 10^{-6} cm^2/s) of sodium alkylbenzenesulphonates (isomeric mixtures) in 0.01 M NaCl at 25 °C measured using Schlieren Optics (Taken from reference 6).

SURFACTANT[a]	D 10^6 cm^2/sec.
NaC$_{10}$BS—mx	1.09
NaC$_{10}$BS—in	1.10
NaC$_{10}$BS—2φ	1.04
NaC$_{11}$BS—mx	1.05
NaC$_{12}$BS—mx	0.795
NaC$_{12}$BS—in	0.800
NaC$_{13}$BS—mx	0.340
NaC$_{14}$BS—mx	0.200
NaC$_{15}$BS—mx	0.173

a : NaC$_n$BS—mx is an isomeric mixture containing 2—phenyl, 3—phenyl etc.;
NaC$_{10}$BS—2φ is the 2—phenyl isomer;
NaC$_n$BS—in is a mixture of 3—phenyl, 4—phenyl and further inner isomers. For details, see (6).

Table I.3.27. Diffusion coefficient (D, in 10^{-7} cm^2/s) of 0.07 M sodium dodecylbenzenesulphonate[a] at various NaCl concentrations and temperatures.

C_{NaCl}	D, 10^{-7} cm^2/sec. at T, °C							
M	25	35	45	55	65	75	85	95
0.1	7.77	11.8	16.0	21.3	25.3	30.4	36.2	
0.15	4.77	7.59	11.7	17.2	24.7	33.8		
0.2	3.25	5.42	8.46	13.2	18.0	25.4	34.4	41.3
0.3		2.54	4.01	6.67	10.1	14.5	20.4	28.9
0.5					2.61	7.94	11.9	15.6

a : Purified Siponate DS—10; 80% para and 20% ortho isomers
Method   : LS   Reference : (31)

Table I.3.28. Diffusion coefficient (D, in 10^{-7} cm^2/s) of sodium dodecylbenzenesulphonate[a] in 0.2 M NaCl at various surfactant concentrations and temperatures.

Surfactant concentration	D, 10^{-7} cm^2/sec. at T, °C					
M	25	35	45	55	65	95
0.018	3.23	5.49	8.48	13.2		
0.14	4.07	6.22	9.3	13.3		
0.20	4.78	7.18	10.2	13.7	18.3	38.3

a : Purified Siponate DS-10; 80% para and 20% ortho isomers

Method : LS
Ref.   : (31)

Table I.3.29. Self-diffusion coefficient (D, in 10^{-5} cm^2/s) of 0.05 mol/kg sodium 4-(octyl)benzenesulphonate in aqueous salt (0.05 mol/kg).

D 10^5 cm^2/sec	SALT 0.05 Mkg^{-1}
0.110	LiCl
0.119	NaCl
0.105	KCl
0.165	CsCl

a : probably at 33°C

Method : C
Ref.   : (9)

Table I.3.30. Diffusion coefficient (D, in 10^{-8} cm^2/s) of linear sodium alkylbenzenesulphonates (commercial mixtures) at 30 °C.

% COMPOSITION					D
C10	C11	C12	C13	C14	10^{-8} cm2 sec.−1
12.7	33.8	30.4	22.4	0.3	9.2
0.1	17.8	25.8	39.5	16.4	9.0
0.2	11.2	22.8	34.2	30.5	8.8
dodecylbenzene(propylene tetramer)					16.1

Method : AK (monomer diffusion coefficients)

Ref.   : (29)

Table I.3.31. Micellar diffusion coefficient (D, in 10^{-6} cm^2/s) of sodium alkylbenzenesulphonates[a] at 27 °C measured using a semipermeable membrane (Taken from reference 30).

ALKYL	D 10^{-6} cm^{2}s^{-1}
decyl	1.18
dodecyl	1.03
tetradecyl	0.93
hexadecyl	0.88

a : Technical grade, purified

Table I.3.32. Interfacial tension between water and heptane at the CMC of 4-substituted benzenesulphonates at various temperatures measured using a Stalagmometer (Taken from reference 5).

R	TEMP. °C	γ dyne/cm
$C_{10}H_{21}$	60	9.3
$C_{11}H_{23}$	60	8.4
$C_{12}H_{25}$	60	8.0
$C_{14}H_{29}$	60	6.9
$C_{11}H_{23}-O-$	70	9.0
$C_{11}H_{23}-NH-$	60	10.5
$C_{11}H_{23}-S-$	75	8.0
$C_{11}H_{23}-SO-$	60	12.7
$C_{11}H_{23}-SO_2-$	72	12.6
$C_{10}H_{21}-O-OC-$	60	11.4
$C_9H_{19}-O-OC-CH_2-$	60	10.0
$C_{10}H_{21}-CO-O-$	60	10.3
$C_8H_{17}-O-OC-CH_2-O-$	60	11.4
$C_{10}H_{21}-CO-NH-$	60	13.6
$(C_{11}H_{23}-NH-C_6H_4SO_3)_2Ca$	60	2.8

REFERENCES Chapter I.3.

1) Van Os, N.M., Daane, G.J., and Bolsman, T.A.B.M., J. Colloid Interface Sci., **115**, 402 (1987).
2) Van Os, N.M., Daane, G.J., and Bolsman, T.A.B.M., J. Colloid Interface Sci., **123**, 267 (1988).
3) Rouviere, J., Faucompre, B., Lindheimer, M., Partyka, S., and Brun, B., J. Chim. Phys., **80**, 309 (1983).
4) Griess, W., Fette Seifen Anstrichmittel, **57**, 24 (1955).
5) Püschel, F. and Todorov, O., Tenside, **5**, 193 (1968).
6) Tokiwa, F. and Ohki, K., Kolloid Zeitschrift u. Zeitschrift f. Polymere, **223**, 38 (1968).
7) Valint, P.L., Bock, J., Kim, M.W., Robbins, M.L., Steyn, P., and Zushma, S., 200th Am. Chem. Soc. National Meeting, Washington D.C. (1990).
8) Tsujii, K., Saito, N., and Takeuchi, T., J. Phys. Chem., **84**, 2287 (1980).
9) Kamenka, N., Chorro, M., Fabre, H., Lindman, B., Rouviere, J., and Cabos, C., Colloid Polymer Sci., **257**, 757 (1979).
10) Schick, M.J. and Fowkes, F.M., J. Phys. Chem., **61**, 1062 (1957).
11) Gershman, J.W., J. Phys. Chem., **61**, 581 (1957).
12) Paquette, R.G., Lingafelter, E.C., and Tartar, H.V., J. Am. Chem. Soc., **65**, 686 (1943).
13) La Mesa, C., J. Phys. Chem., **94**, 323 (1990).
14) Bauernschmitt, D., Hoffmann, H., and Platz, G., Ber. Bunsenges. Phys. Chem., **85**, 203 (1981).
15) Lianos, P. and Lang, J., J. Colloid Interface Sci., **96**, 222 (1983).
16) Sinton, S.W. and Huff, S.L., J. Colloid Interface Sci., **120**, 358 (1987).
17) Fanghänel, E., Willscher, S., and Ortman, W., J. Praktische Chemie, **331**, 195 (1989).
18) Fanghänel, E., Ortman, W., Behrmann, K., Willscher, S., Turro, N.J., and Gould, I.R., J. Phys. Chem., **91**, 3700 (1987).
19) Archer, D.G., Albert, H.J., White, D.E., and Wood, R.H., J. Colloid Interface Sci., **100**, 68 (1984).
20) Ludlum, D., J. Phys. Chem., **60**, 1240 (1956).
21) Tokiwa, F. and Ohki, K., J. Colloid Interface Sci., **26**, 457 (1968).
22) Scamehorn, J.F., Schechter, R.S., and Wade, W.H., J. Colloid Interface Sci., **85**, 463 (1982).
23) Triolo, R., Hayter, J.B., Magid, L.J., and Johnson, J.S., J. Chem. Phys., **79**, 1977 (1983).
24) Hoffmann, H., Ber. Bunsenges. Phys. Chem., **82**, 988 (1978).
25) Magid, L.J., Triolo, R., Johnson, J.S., and Koehler, W.C., J. Phys. Chem., **86**, 164 (1982).
26) Sowada, R., Acta Hydrochim. et Hydrobiol., **12**, 327 (1984).
27) Magid, L.J., Shaver, R.J., Gulari, E., Bedwell, B., and Alkhafaji, S., Prepr., Div. Pet. Chem., Am. Chem. Soc., **26**, 93 (1981).

28) Caponetti, E., Triolo, R., Ho, P.C., Johnson, J.S., Magid, L.J., Butler, P., and Payne, K.A., J. Colloid Interface Sci., **116**, 200 (1987).
29) Kumar, R. and Bhat, S.G.T., Tenside Detergents, **24**, 86 (1987).
30) Anand, O.N., Malik, V.P., and Kumar, V., J. Chem. Tech. Biotechnol., **33A**, 130 (1983).
31) Cheng, D.C.H., and Gulari, E., J. Colloid Interface Sci., **90**, 410 (1982).
32) Binana-Limbele, W., Van Os, N.M., Rupert, L.A.M., and Zana, R., J. Colloid Interface Sci., **141**, 157 (1991).
33) Van Os, N.M., Kok, R., and Bolsman, T.A.B.M., Tenside Detergents, **29**, 175 (1992).

PART II: CATIONIC SURFACTANTS

II.1 Alkyltrimethylammonium salts

Table II.1.1. Critical micelle concentration (CMC, in mmol/L)[a] of alkylammonium salt surfactants at various temperatures and conditions.

Compound	20	25	30	40	50	55	60	65	95	130	160	Condition	Ref
C$_{10}$N(CH$_3$)$_3$Br	65												13
		67.6						77.9	93.6	108	174		4
		60.2											9
		65											16
						72							19
						58						1 M NaBr	19
C$_{12}$N(CH$_3$)$_3$Br	5.40[b]	5.27	5.33	6.40	8.28	9.03	10.20						2
	15.3	15.4											13
	15.0[c]	13.3											1
								21.9	34.2	61.2	86.6		4
		15											9
		13.64											16
						18							21
						14						1 M NaBr	19
													19

Table II.1.1. (continued)

Compound	Temperature, °C											Condition	Ref
	20	25	30	40	50	55	60	65	95	130	160		
$C_{12}N(CH_3)_3Br$		13.40										EtOH-H_2O x_2=0.0119	21
		13.29										x_2=0.0228	21
		13.02										x_2=0.0350	21
		12.75										x_2=0.0478	21
		13.22										x_2=0.0606	21
		13.78										x_2=0.0668	21
		14.51										x_2=0.0743	21
		14.90										x_2=0.0816	21
		15.77										x_2=0.0912	21
		13.5										D_2O	9
	6.5c											n-BuOH 0.44 mol/1	1
	2.5c											n-BuOH 0.72 mol/1	1
	0.5c											n-BuOH 1.00 mol/1	1
$C_{14}N(CH_3)_3Br$	3.41d	3.32	3.41	3.60	4.10	4.30	4.49						2
	3.6												13
		3.79						6.69	9.83	21.4	39.4		4
		3.41											9

Table II.1.1. (continued)

Compound	Temperature, °C											Condition	Ref
	20	25	30	40	50	55	60	65	95	130	160		
$C_{14}N(CH_3)_3Br$		3.5											6
		3.5											16
		3.58											21
						4.3							19
		2.4										0.005 M NaBr	3
		1.49										0.010 M NaBr	3
		1.01										0.020 M NaBr	3
		0.7										0.030 M NaBr	3
		0.39										0.080 M NaBr	3
						3.9						1 M NaBr	19
	3.52											EtOH-H$_2$O x$_2$=0.0111	21
	3.40											x$_2$=0.0226	21
	3.39											x$_2$=0.0350	21
	3.26											x$_2$=0.0474	21
	3.30											x$_2$=0.0596	21
	3.83											x$_2$=0.0754	21
	4.12											x$_2$=0.0892	21

Table II.1.1. (continued)

Compound	20	25	30	40	50	55	60	65	95	130	160	Condition	Ref
$C_{14}N(CH_3)_3Br$		4.33										EtOH-H$_2$O x$_2$=0.1051	21
		3.20										D$_2$O	9
$C_{14}N(CH_3)_3OH$		4.5											6
$C_{14}N(CH_3)_3NO_3$		2.7											6
$C_{16}N(CH_3)_3Br$	0.9												13
	0.97[c]												1
		0.824	0.870	0.949	1.050		1.170						2
		0.955						1.55	2.61	6.12	13.03		4
		0.8											6
		1.00											9
		0.96											12
		0.92											16
		0.8				1.2							17
		0.84											21
													19
		0.538										0.001 M NaBr	3
		0.259										0.0025 M NaBr	3

Temperature, °C

Table II.1.1. (continued)

Compound	Temperature, °C											Condition	Ref
	20	25	30	40	50	55	60	65	95	130	160		
$C_{16}N(CH_3)_3Br$		0.182										0.005 M NaBr	3
		0.103										0.010 M NaBr	3
		0.082										0.020 M NaBr	3
						$\underline{0.9}$						1 M NaBr	19
		0.82										0.001 M NaCl	12
		0.49										0.010 M NaCl	12
		0.23										0.100 M NaCl	12
		0.81										EtOH-H_2O x_2=0.0111	21
		0.80										x_2=0.0227	21
		0.75										x_2=0.0346	21
		0.69										x_2=0.0469	21
		0.76										x_2=0.0601	21
		0.85										x_2=0.0748	21
		1.03										x_2=0.0893	21
		1.10										x_2=0.1047	21
	0.30[c]											n-BuOH 0.44 mol/1	1
	0.10[c]											n-BuOH 0.72 mol/1	1

Table II.1.1. (continued)

116

Compound	Temperature, °C											Condition	Ref
	20	25	30	40	50	55	60	65	95	130	160		
$C_{16}N(CH_3)_3Br$		0.82										D_2O	9
$C_{16}N(CH_3)_3Cl$		1.4											6
		1.3											17
		1.34											21
		1.26										EtOH-H_2O x_2=0.0111	21
		1.16										x_2=0.0224	21
		1.08										x_2=0.0348	21
		0.95										x_2=0.0467	21
		0.94										x_2=0.0594	21
		1.04										x_2=0.0751	21
		1.13										x_2=0.0879	21
$C_{16}N(CH_3)_3OH$		2.3–3.4											6
		1.8											17
$C_{16}N(CH_3)_3NO_3$		0.81										EtOH-H_2O x_2=0.0114	21
		0.80											21
		0.82										x_2=0.0230	21
		0.81										x_2=0.0307	21

Table II.1.1. (continued)

Compound	\multicolumn{11}{Temperature, °C}											Condition	Ref
	20	25	30	40	50	55	60	65	95	130	160		
$C_{16}N(CH_3)_3NO_3$		0.80										EtOH-H$_2$O x_2=0.0426	21
		0.83										x_2=0.0542	21
		0.93										x_2=0.0669	21
		1.27										x_2=0.0795	21
$(C_{16}N(CH_3)_3)_2SO_4$		0.6											6
$(C_{16}N(CH_3)_3)_2CO_3$		0.8											6
$C_{16}N(CD_3)_3Br$		0.91										D$_2$O	9
		0.74											9
$C_{18}N(CH_3)_3Br$			0.292	0.292	0.299	0.306	0.313						2

a CMC are given in mmol/l. Underlined values are in mmol/kg

b At 18°C

c At 21 °C

d At 19 °C

Table II.1.2. Residual charge of a micelle of alkylammonium surfactants at various temperatures.

Compound	Temperature, °C							Ref
	20	25	30	40	50	55	60	
$C_8N(CH_3)_3Br$		0.36						11
$C_{10}N(CH_3)_3Br$		0.27						11
		0.27						5
$C_{12}N(CH_3)_3Br$	0.34[a]	0.36	0.35	0.39	0.40	0.40	0.42	2
		0.23						5
		0.23						11
		0.24						11
$C_{14}N(CH_3)_3Br$	0.24[b]	0.26	0.29	0.29	0.30	0.29	0.28	2
		0.20						5
		0.27						6
		0.20						11
		0.22						11
$C_{14}N(CH_3)_3OH$		0.47						6
$C_{12}C_2N(CH_3)_2Br$		0.28						11
$C_{12}C_3N(CH_3)_2Br$		0.32						11
$C_{12}C_4N(CH_3)_2Br$		0.44						11
$C_{12}C_8N(CH_3)_2Br$		0.62						11

Table II.1.2. (continued)

Compound	Temperature, °C							Ref
	20	25	30	40	50	55	60	
$C_{12}C_{10}N(CH_3)_2Br$		0.74						11
$C_{14}N(C_2)_3Br$		0.35						11
$C_{14}N(C_2)_3Br$		0.42						11
$C_{14}N(C_4)_3Br$		0.48						11
$C_{16}N(CH_3)_3Br$		0.21	0.23	0.21	0.22		0.24	2
		0.16						5
		0.24						6
		0.16						11
$C_{16}N(CH_3)_3OH$		0.48						6
$C_{16}N(CH_3)_3Cl$		0.36						6
$C_{16}N(CH_3)_3NO_3$		0.36						6
$(C_{16}N(CH_3)_3)_2SO_4$		0.37						6

[a] At 18 °C

[b] At 19.5 °C

Table II.1.3. Aggregation number of alkylammonium salt surfactants at various temperatures and conditions.

Compound	Condition	Surfactant conc. [mM]	Temperature [°C]	N	Remark	Ref
$C_{10}N(CH_3)_3Br$		65	20	39		13
		100	20	38		13
		200	20	41		13
$C_{12}N(CH_3)_3Br$		15.3	20	54		13
		50	20	53		13
		100	20	57		13
		100	23	65		17
$C_{12}N(CH_3)_3Cl$		31	23	47		17
		100	23	51		17
	0.020 M NaCl	31	23	48		17
	0.072 M NaCl	31	23	53		17
	0.155 M NaCl	31	23	56		17
	0.310 M NaCl	31	23	63		17
	0.520 M NaCl	31	23	65		17

Table II.1.3. (continued)

Compound	Condition	Surfactant conc. [mM]	Temperature [°C]	N	Remark	Ref
$C_{12}N(CH_3)_3OH$		100	23	20		17
$C_{14}N(CH_3)_3Br$		3.6	20	72		13
		7	20	63		13
		10	20	70		13
		10	23	68		17
		50	20	65		13
		50	20	68		13
		100	20	65		13
		100	23	97		17
		140	20	73		13
		150	20	80		13
		250	20	75		13
$C_{14}N(CH_3)_3Cl$		100	23	66		17
			23	69		17
		84.8	24	65		17
		84.8	30	57		17

Table II.1.3. (continued)

Compound	Condition	Surfactant conc. [mM]	Temperature [°C]	N	Remark	Ref
$C_{14}N(CH_3)_3Cl$		84.8	40	58		17
		84.8	50	54		17
		400		58		18
	0.037 M hexane	400		89	$N_{sol} = 8$	18
	0.07 M hexane	400		94	$N_{sol} = 17$	18
	0.094 M hexane	400		100	$N_{sol} = 24$	18
	0.034 M heptane	400		90	$N_{sol} = 7$	18
	0.065 M heptane	400		99	$N_{sol} = 16$	18
	0.09 M heptane	400		107	$N_{sol} = 24$	18
	0.031 M octane	400		92	$N_{sol} = 7$	18
	0.061 M octane	400		104	$N_{sol} = 16$	18
	0.086 M octane	400		114	$N_{sol} = 25$	18
	0.027 M nonane	400		91	$N_{sol} = 6$	18
	0.054 M nonane	400		105	$N_{sol} = 14$	18
	0.067 M nonane	400		110	$N_{sol} = 19$	18
	0.023 M decane	400		91	$N_{sol} = 5$	18
	0.048 M decane	400		106	$N_{sol} = 13$	18

Table II.1.3. (continued)

Compound	Condition	Surfactant conc. [mM]	Temperature [°C]	N	Remark	Ref
$C_{14}N(CH_3)_3OH$		64	23	42		17
$C_{16}N(CH_3)_3Br$		0.9	20	92		13
		10	20	88		13
		31	23	104		17
		50	20	90		13
$C_{16}N(CH_3)_3Cl$		31	23	89		17
			23	81		17
		400		120		18
	0.050 M NaCl	31	23	107		17
	0.083 M NaCl	31	23	127		17
	0.104 M NaCl	31	23	136		17
	0.155 M NaCl	31	23	176		17
	0.032 M hexane	400		119	$N_{sol} = 10$	18
	0.061 M hexane	400		125	$N_{sol} = 19$	18
	0.084 M hexane	400		135	$N_{sol} = 29$	18
	0.109 M hexane	400		142	$N_{sol} = 39$	18
	0.028 M heptane	400		124	$N_{sol} = 9$	18

Table II.1.3. (continued)

Compound	Condition	Surfactant conc. [mM]	Temperature [°C]	N	Remark	Ref
$C_{16}N(CH_3)_3Cl$	0.065 M heptane	400		139	$N_{sol} = 23$	18
	0.102 M heptane	400		151	$N_{sol} = 38$	18
	0.025 M octane	400		128	$N_{sol} = 8$	18
	0.071 M octane	400		154	$N_{sol} = 27$	18
	0.095 M octane	400		160	$N_{sol} = 38$	18
	0.022 M nonane	400		129	$N_{sol} = 7$	18
	0.032 M nonane	400		135	$N_{sol} = 11$	18
	0.075 M nonane	400		162	$N_{sol} = 30$	18
	0.02 M decane	400		130	$N_{sol} = 6$	18
	0.039 M decane	400		142	$N_{sol} = 14$	18
	0.06 M decane	400		157	$N_{sol} = 23$	18
	0.078 M decane	400		171	$N_{sol} = 33$	18
$C_{16}N(CH_3)_3OH$		50	23	46		17

Table II.1.4. Thermodynamic parameters of micellization [a,b] of alkylammonium salt surfactants at various temperatures.

$\Delta X°$	Compound	5	10	20	25	30	40	50	55	60	Model[c]	Ref.
$\Delta G°$	$C_{12}N(CH_3)_3$ Br			44.71[d]	45.92	46.63	47.22	47.34	47.59	47.63	PS	2
				37.09[d]	37.64	38.47	39.23	37.89	38.06	37.64	MA	2
	$C_{14}N(CH_3)_3$ Br			47.17[e]	48.18	48.85	50.18	51.19	51.69	52.19	PS	2
				41.53[e]	41.90	41.78	42.91	43.49	44.20	44.87	MA	2
	$C_{16}N(CH_3)_3$ Br				55.04	55.70	57.13	58.38		59.64	PS	2
					49.26	49.31	51.10	51.94		52.84	MA	2
$\Delta H°$	$C_{10}N(CH_3)_3$ Br		5.6		0.2		-4.7		-8.3		cal	5
	$C_{12}N(CH_3)_3$ Br			1.3[d]	-8.0	-14.2	-28.0	-43.9	-43.9	-61.1	PS	2
				0.8[d]	-6.7	-11.7	-22.6	-35.1	-41.8	-48.9	MA	2
					-2.3		-8.7		-14.2		cal	5
	$C_{14}N(CH_3)_3$ Br			0.1[e]	-3.0	-6.0	-12.7	-20.2	-24.2	-28.4	PS	2
				0.1[e]	-2.6	-5.1	-10.9	-17.1	-20.6	-24.4	MA	2
		2.8			-4.9		-12.4		-19.9		cal	5
	$C_{16}N(CH_3)_3$ Br				-16.8	-17.0	-17.2	-17.4	-17.5	-17.5	PS	2
					-15.0	-15.0	-15.4	-15.5	-15.4	-15.4	MA	2

Table II.1.4. (continued)

$\Delta X°$	Compound	\multicolumn{9}{c}{Temperature, °C}	Model[c]	Ref.
		5	10	20	25	30	40	50	55	60		
$\Delta H°$	$C_{16}N(CH_3)_3$ Br				-9.8		-18.1		-27.0		cal	5
					-9.2						cal	12
					-12.0	-13.6	-18.5	-23.8			cal	15
					-8.8[f]						cal	12
					-5.9[g]						cal	12
					-3.4[h]						cal	12
$\Delta S°$	$C_{12}N(CH_3)_3$ Br			158[d]	127	107	61	10	-5	-13	PS	2
				131[d]	104	88	49	8	-3	-10	MA	2
	$C_{14}N(CH_3)_3$ Br			161[e]	152	141	120	96	84	72	PS	2
				142[e]	132	121	102	82	72	61	MA	2
	$C_{16}N(CH_3)_3$ Br				128	128	128	127		127	PS	2
					115	113	114	113		112	MA	2

a. Units: $\Delta G°$ in kJ/mol; $\Delta H°$ in kJ/mol; $\Delta S°$ in J/mol·deg

b. Obtained from the temperature dependence of the CMC by using conductivity, unless noted otherwise.

c. <u>PS</u>: phase separation model; <u>MA</u>: mass action model; <u>cal</u>: direct microcalorimetric determination

d. At 18°C   f. In 0.001 M NaCl   h. In 0.100 M NaCl

e. At 19°C   g. In 0.010 M NaCl

Table II.1.5. Krafft temperature of hexadecyltrimethylammonium bromide in various solvents.

Compound	Condition	T_{Krafft} [°C]	Ref.
$C_{16}N(CH_3)_3$ Br	water formamide	26 43	8 8

Table II.1.6. Thermodynamic parameters of adsorption of hexadecyl-trimethylammonium bromide at different interfaces at various temperatures.

Compound: $C_{16}N(CH_3)_3$ Br

$\Delta G°$ [kJ/mol]

Interface	temperature [°C]					Ref.
	30	40	50	60	70	
air − water	−32.5	−33.4	−33.2	−33.5	−33.8	10
octane − water	−34.7	−35.4	−36.2	−36.8	−37.4	10

$\Delta H°$ [kJ/mol]

Interface	temperature [°C]					Ref.
	30	40	50	60	70	
air − water	−21.0	−21.0	−21.0	−21.0	−21.0	10
octane − water	−12.0	−12.0	−12.0	−12.0	−12.0	10

$\Delta S°$ [J/mol*deg]

Interface	temperature [°C]					Ref.
	30	40	50	60	70	
air − water	37.9	39.9	37.7	37.5	37.6	10
octane − water	74.9	74.9	74.9	74.4	74.0	10

Table II.1.7. Apparent molar heat capacity (Φ_c, in J mol^{-1}K^{-1}) of aqueous hexadecyltrimethylammonium bromide at various temperatures and surfactant concentrations.

Compound: C$_{16}$N(CH$_3$)$_3$ Br			
m	Φ_c	t	Ref.
[mol/kg]	[J/mol K]	[°C]	
0.010000	837	40	20
0.012227	824	40	20
0.015132	784	40	20
0.009357	992	55	20
0.010413	946	55	20
0.012435	892	55	20
0.014989	874	55	20
0.049702	813	55	20

Table II.1.8. Apparent molar volume (Φ_v, in cm^3 mol^{-1}) of aqueous hexadecyltrimethylammonium bromide at various temperatures and surfactant concentrations.

Compound: C$_{16}$N(CH$_3$)$_3$ Br			
m	Φ_v	t	Ref.
[mol/kg]	[cm^3/mol]	[°C]	
0.010000	364.4	40	20
0.012227	364.9	40	20
0.015132	364.7	40	20
0.009357	366.7	55	20
0.010413	369.6	55	20
0.012435	369.5	55	20
0.014989	368.1	55	20
0.049702	369.1	55	20

Table II.1.9. Density of aqueous decyltrimethylammonium bromide solutions at various surfactant concentrations.

Compound: $C_{10}N(CH_3)_3$ Br			
m [mol/kg]	d [g/cm^3]	t [°C]	Ref.
0.005913	0.9972	25	14
0.01053	0.9973	25	14
0.03007	0.9978	25	14
0.1009	0.9993	25	14
0.3000	1.0026	25	14
0.5630	1.0066	25	14
0.5655	1.0067	25	14

Table II.1.10. Refractive index increment of alkyltrimethylammonium bromides.

Compound	t [°C]	dn/dc	Ref.
$C_{12}N(CH_3)_3$ Br	25	0.147	2
$C_{14}N(CH_3)_3$ Br	25	0.143	2
$C_{16}N(CH_3)_3$ Br	30	0.148	2

Table II.1.11. Apparent micellar molecular mass (M_{agg}, in 10^3 D) of alkyltrimethylammonium bromide aggregates.

Compound	t [°C]	M_{agg} [10^3 D]	Ref.
$C_{12}N(CH_3)_3$ Br	25	20.9	2
$C_{14}N(CH_3)_3$ Br	25	27.3	2
$C_{16}N(CH_3)_3$ Br	30	33.3	2

Table II.1.12. Self-diffusion coefficient (D, in 10^{-9} m^2/s) of hexadecyltrimethylammonium bromide and of hexadecyltrimethyl-ammonium chloride at various concentrations in D$_2$O at 33 °C measured using radio-active tracer method (Taken from reference 7).

Compound : C$_{16}$N(CH$_3$)$_3$ Br

Condition : D$_2$O at 33°C

m	D
[mol/kg]	$[10^{-9}$ m^2/s]
0.000503	0.550
0.000639	0.550
0.000823	0.527
0.00115	0.406
0.00148	0.305
0.00200	0.245
0.00247	0.212
0.00326	0.168
0.00328	0.156
0.00334	0.165
0.00404	0.145
0.00415	0.138
0.00484	0.123
0.00578	0.107
0.00585	0.110
0.00702	0.098
0.00763	0.091
0.00780	0.098
0.00988	0.084
0.0127	0.076

Table II.1.12. (continued)

0.0154	0.065
0.0160	0.071
0.0200	0.069
0.0252	0.059
0.0265	0.067
0.0384	0.063
0.0574	0.061
0.0790	0.0557
0.0967	0.056
0.102	0.0505
0.102	0.053
0.122	0.054
0.132	0.050
0.148	0.049
0.148	0.0505
0.186	0.0480
0.217	0.042
0.226	0.041
0.242	0.040
0.275	0.025
0.297	0.0345
0.316	0.019
0.344	0.021
0.372	0.016
0.389	0.013
0.425	0.007

Table II.1.12. (continued)

Compound : $C_{16}N(CH_3)_3$ Cl

Condition : D_2O at 33°C   Ref. 7

m [mol/kg]	D $[10^{-9}\ m^2/s]$
0.000492	0.538
0.000637	0.519
0.000748	0.560
0.00101	0.498
0.00130	0.430
0.00171	0.387
0.00210	0.306
0.00248	0.269
0.00273	0.239
0.00278	0.279
0.00312	0.212
0.00315	0.217
0.00336	0.217
0.00398	0.192
0.00410	0.181
0.00500	0.153
0.00520	0.149
0.00668	0.127
0.00674	0.127
0.00902	0.114

Table II.1.12. (continued)

0.0115	0.104
0.0153	0.084
0.0158	0.086
0.0184	0.080
0.0214	0.0774
0.0218	0.078
0.025	0.069
0.0280	0.071
0.042	0.061
0.052	0.061
0.0750	0.058
0.077	0.058
0.090	0.054
0.0923	0.054
0.093	0.054
0.117	0.054
0.121	0.055
0.130	0.050
0.174	0.048
0.202	0.045
0.226	0.043
0.237	0.046
0.268	0.043
0.323	0.041
0.352	0.037
0.504	0.031

Table II.1.13. Molecular dimensions of alkylammonium bromide surfactants.

Compound	l_c [a]	a_o [b]	v_c [c]	Ref.
$C_{10}N(CH_3)_3$ Br	14.15	64	297.8	22
$C_{12}N(CH_3)_3$ Br	16.68	63	351.9	22
$C_{14}N(CH_3)_3$ Br	19.21	65	406.0	22
$C_{16}N(CH_3)_3$ Br	21.74	64	460.1	22
$2C_{10}N(CH_3)_2$ Br	14.15	68	565.5	22
$C_{10}C_{12}N(CH_3)_2$ Br	16.68	68	649.6	22
$2C_{12}N(CH_3)_2$ Br	16.68	68	703.7	22
$C_8C_{16}N(CH_3)_2$ Br	21.74	68	703.7	22
$2C_{14}N(CH_3)_2$ Br	19.21	68	811.9	22
$2C_{16}N(CH_3)_2$ Br	21.74	68	920.1	22

a. Length of the longest alkylchain in Å

b. Head group area in $Å^2$

c. Volume of the hydrocarbon part of the surfactant in $Å^3$

REFERENCES Chapter II.1.

1) Bijsterbosch, B.H., J. Colloid Interface Sci., **47**, 186 (1974).
2) Barry, B.W. and Russell, G.F.J., J. Colloid Interface Sci., **40**, 174 (1972).
3) Barry, B.W., Morrisson, J.C., and Russell, G.F.J., J. Colloid Interface Sci., **33**, 554 (1970).
4) Evans, D.F., Allen, M., Ninham, B.W., and Fouda, A., J. Sol. Chem., 13, **87** (1984).
5) Bashford, M.T. and Wooley, E.M., J. Phys. Chem., **89**, 3173 (1985).
6) Sepulveda, L. and Cortes, J., J. Phys. Chem., **89**, 5322 (1985).
7) Lindman, B., Puyal, M.C., Kamenka, N., Rymden, R., and Stilbs, P., J. Phys. Chem., **88**, 5048 (1984).
8) Rico, I. and Lattes, A., J. Phys. Chem., **90**, 5870 (1986).
9) Berr, S.S., Caponetti, E., Johnson Jr., J.S., Jones, R.R.M., and Magid, L.J., J. Phys. Chem., **90**, 5766 (1986).
10) Zadymova, N.M. and Marliina, Z.I., Coll. J. USSR, **48**, 304 (1986).
11) Zana, R., J. Colloid Interface Sci., **70**, 330 (1980).
12) Paredes, S., Tribaut, M., Ferreira, J., and Leonis, J., Colloid Polym. Sci., **254**, 637 (1976).
13) Lianos, P. and Zana, R., J. Colloid Interface Sci., **84**, 100 (1981).
14) Archer, D.J., J. Sol. Chem., 15, 581 (1986).
15) Bergström, S. and Olofsson, G., Thermochimica Acta, **109**, 155 (1986).
16) Levièvre, J., Haddad-Fahed, O., and Gabonaud, R., J. Colloid Interface Sci., **82**, 2301 (1986).
17) Roelants, E. and De Schrijver, F.C., Langmuir, **3**, 209 (1987).
18) Malliaris, A., J. Phys. Chem., **91**, 6511 (1987).
19) Dearden, L.V. and Woolley, E.M., J. Phys. Chem., **91**, 2004 (1987).
20) Dearden, L.V. and Woolley, E.M., J. Phys. Chem., **91**, 4123 (1987).
21) Cipiciani, A., Onori, G., and Savelli, G., Chem. Phys. Lett., **143**, 505 (1988).
22) Warr, G.G., Sen, R., Evans, D.F., and Trend, J.E., J. Phys. Chem., **92**, 774 (1988).

II.2 Alkylpyridinium salts

Table II.2.1. Critical micelle concentration (CMC, in mol/L) of alkylpyridinium salts in water, and in the presence of other surfactants and/or electrolytes at various temperatures.

Cationic	Additive[a]	20	25	30	35	40	45	Meth.	Ref.
C_6-pyridinium Br^-	$C_{10}OSO_3Na$		$4.61*10^{-3}$					ST	1
	+ 0.1 M NaBr			$4.8*10^{-3}$				ST	2
	$C_{12}OSO_3Na$		$1.02*10^{-3}$	$4.4*10^{-3}$				ST	2
	$C_7F_{15}COONa$		$1.69*10^{-3}$					ST	1
	+ 0.1 M NaBr			$2.4*10^{-3}$				ST	2
				$2.0*10^{-3}$				ST	2
	0.05 M KBr			$1.93*10^{-1}$				LS	3
				$1.89*10^{-1}$				LS	3
C_8-pyridinium Br^-	C_8OSO_3Na		$6.35*10^{-3}$	$6.3*10^{-3}$				ST	1
	+ 0.1 M NaBr			$7.4*10^{-3}$				ST	2
	$C_8OSO_3Na(1:2)$			$9.6*10^{-3}$				ST	2
	+ 0.1 M NaBr			$10.8*10^{-3}$				ST	2
	$C_{10}OSO_3Na$		$1.64*10^{-3}$					ST	1

Table II.2.1. (continued)

Cationic	Additive [a]	Temperature, °C 20	25	30	35	40	45	Meth.	Ref.
C_{10}-N$^+$(pyridinium) Cl$^-$			8.682×10^{-2}					ST	4
C_{11}-N$^+$(pyridinium) Br$^-$				19.5×10^{-3}				ST	5
C_{12}-N$^+$(pyridinium) Cl$^-$				4.2×10^{-2}				LS	3
				3.1×10^{-2}				LS	3
	0.05 M KBr	1.887×10^{-2}	1.52×10^{-2}					ST	4
	0.01 M NaCl		1.24×10^{-2}					Pot	6
	0.1 M NaCl		4.5×10^{-3}					Pot	6
	0.15 M NaCl		4.0×10^{-3}					Pot	6
				1.40×10^{-2}				ST	7
								ST	5
C_{12}-N$^+$(pyridinium) Br$^-$		1.15×10^{-2} [b]						Co	8
		1.12×10^{-2} [c]						Co	8
		1.10×10^{-2} [d]						Co	8
		1.12×10^{-2}	1.14×10^{-2}	1.18×10^{-2}	1.22×10^{-2}	1.28×10^{-2}	1.35×10^{-2}	Co	8

Table II.2.1. (continued)

Cationic	Additive [a]	20	25	30	35	40	45	Meth.	Ref.
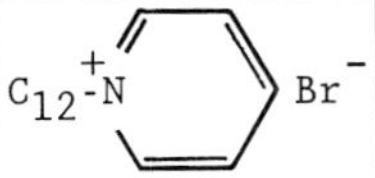 C_{12}-N⁺ pyridine Br⁻							$1.40*10^{-2}$ e	Co	8
							$1.48*10^{-2}$ f	Co	8
							$1.54*10^{-2}$ g	Co	8
							$1.63*10^{-2}$ h	Co	8
							$1.72*10^{-2}$ i	Co	8
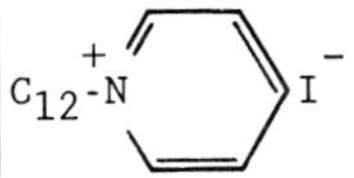 C_{12}-N⁺ pyridine I⁻					$6.6*10^{-3}$			-	9
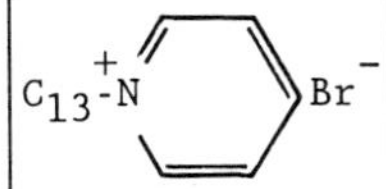 C_{13}-N⁺ pyridine Br⁻				$4.57*10^{-3}$				ST	5
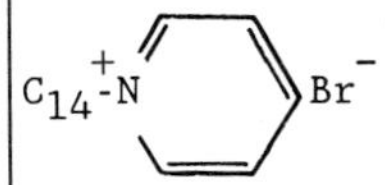 C_{14}-N⁺ pyridine Br⁻				$2.65*10^{-3}$				ST	5
				$4.1*10^{-3}$				LS	3
	0.025 M KBr			$2.0*10^{-3}$				LS	3
	0.05 M KBr			$1.5*10^{-3}$				LS	3

Temperature, °C

Table II.2.1. (continued)

Cationic	Additive [a]	20	25	30	35	40	45	Meth.	Ref.
C_{16}–N$^+$⬡ Cl$^-$	0.01754 M NaCl			1.2×10^{-4} j				LS	10
	0.05843 M NaCl			8.7×10^{-5} j				LS	10
	0.4382 M NaCl			4.8×10^{-5} j				LS	10
	0.7304 M NaCl			4.1×10^{-5} j				LS	10
			6.3×10^{-4}					Pot	6
	0.01 M NaCl		1.6×10^{-4}					Pot	6
	0.1 M NaCl		3.8×10^{-5}					Pot	6
		9.0×10^{-4} k						ST	11
		9.0×10^{-4}						ST	11
	0.03 M NaCl			9.2×10^{-5}				ST	12
	+ 0.0006 M KOH			7.8×10^{-5}				ST	12
			9.2×10^{-4} l					ST	13
C_{16}–N$^+$⬡ Br$^-$		6.6×10^{-3}	6.7×10^{-3} m					ST	11
				6.2×10^{-4}				ST	5

Temperature, °C

Table II.2.1. (continued)

Cationic	Additive[a]	45	40	35	30	25	20	Meth.	Ref.
C_{16}-N$^+$ (pyridinium) I$^-$			$4.2*10^{-4}$ [n]					$2.5*10^{-4}$	ST	13
C_{16}-N$^+$ (pyridinium) NO$_3^-$							$6.2*10^{-4}$ [o]	ST	11
C_{16}-N$^+$ (pyridinium) BF$_4^-$		$4.0*10^{-4}$ [p]						$5.0*10^{-4}$	ST	11
C_{16}-N$^+$ (pyridinium) HSO$_4^-$							$2.7*10^{-4}$	ST	11
C_{16}-N$^+$ (pyridinium) H$_2$PO$_4^-$							$7.3*10^{-4}$	ST	11
C_{16}-N$^+$ (pyridinium) Tos$^-$				$2.1*10^{-4}$ [q]				$1.6*10^{-4}$	ST	11
C_{12}-N$^+$-Me I$^-$ (pyridinium)						$2.45*10^{-3}$		Co	14
	1.0 mM NaCl					$2.40*10^{-3}$		Co	14
	2.0 mM NaCl					$2.35*10^{-3}$		Co	14

Temperature, °C

Table II.2.1. (continued)

Cationic	Additive[a]	Temperature, °C 20	25	30	35	40	45	Meth.	Ref.
C_{12}–N-Me$^+$ I$^-$ (pyridinium)	3.0 mM NaCl		$2.35*10^{-3}$					Co	14
	4.0 mM NaCl		$2.35*10^{-3}$					Co	14
	5.0 mM NaCl		$2.35*10^{-3}$					Co	14
	5.9 mM NaCl		$2.25*10^{-3}$					Co	14
	6.9 mM NaCl		$2.25*10^{-3}$					Co	14
	7.9 mM NaCl		$2.35*10^{-3}$					Co	14
	8.8 mM NaCl		$2.35*10^{-3}$					Co	14
	9.8 mM NaCl		$2.25*10^{-3}$					Co	14
	1.0 mM NaBr		$2.25*10^{-3}$					Co	14
	2.0 mM NaBr		$2.25*10^{-3}$					Co	14
	3.0 mM NaBr		$2.20*10^{-3}$					Co	14
	4.0 mM NaBr		$2.10*10^{-3}$					Co	14
	5.0 mM NaBr		$2.10*10^{-3}$					Co	14
	5.9 mM NaBr		$2.00*10^{-3}$					Co	14
	6.9 mM NaBr		$2.00*10^{-3}$					Co	14

Table II.2.1. (continued)

Cationic	Additive[a]	Temperature, °C						Meth.	Ref.
		20	25	30	35	40	45		
C_{12}—pyridinium $\overset{+}{N}$-Me I^-	7.9 mM NaBr		$2.10*10^{-3}$					Co	14
	8.8 mM NaBr		$2.10*10^{-3}$					Co	14
	9.8 mM NaBr		$2.10*10^{-3}$					Co	14
	1.0 mM NaI		$2.15*10^{-3}$					Co	14
	2.0 mM NaI		$1.70*10^{-3}$					Co	14
	3.0 mM NaI		$1.50*10^{-3}$					Co	14
	4.0 mM NaI		$1.30*10^{-3}$					Co	14
	5.0 mM NaI		$1.20*10^{-3}$					Co	14
	5.9 mM NaI		$1.05*10^{-3}$					Co	14
	6.9 mM NaI		$1.00*10^{-3}$					Co	14
	7.9 mM NaI		$0.80*10^{-3}$					Co	14
	8.8 mM NaI		$0.80*10^{-3}$					Co	14
	9.8 mM NaI		$0.75*10^{-3}$					Co	14
	1.0 mM $NaNO_3$		$2.20*10^{-3}$					Co	14
	2.0 mM $NaNO_3$		$2.00*10^{-3}$					Co	14

146

Table II.2.1. (continued)

Cationic	Additive [a]	\multicolumn{6}{c}{Temperature, °C}	Meth.	Ref.
		20	25	30	35	40	45		
C_{12} pyridinium N-Me I$^-$	3.0 mM NaNO$_3$		$2.00*10^{-3}$					Co	14
	4.0 mM NaNO$_3$		$1.95*10^{-3}$					Co	14
	5.0 mM NaNO$_3$		$1.90*10^{-3}$					Co	14
	5.9 mM NaNO$_3$		$1.90*10^{-3}$					Co	14
	6.9 mM NaNO$_3$		$1.85*10^{-3}$					Co	14
	7.9 mM NaNO$_3$		$1.85*10^{-3}$					Co	14
	8.8 mM NaNO$_3$		$1.90*10^{-3}$					Co	14
	9.8 mM NaNO$_3$		$1.75*10^{-3}$					Co	14
	1.0 mM Na Tos		$2.00*10^{-3}$					Co	14
	2.0 mM Na Tos		$1.90*10^{-3}$					Co	14
	3.0 mM Na Tos		$1.45*10^{-3}$					Co	14
	4.0 mM Na Tos		$1.20*10^{-3}$					Co	14
	5.0 mM Na Tos		$1.05*10^{-3}$					Co	14
	5.9 mM Na Tos		$1.05*10^{-3}$					Co	14
	6.9 mM Na Tos		$0.85*10^{-3}$					Co	14

Table II.2.1. (continued)

Cationic	Additive [a]	20	25	30	35	40	45	Meth.	Ref.
C_{12} pyridinium $N\text{-Me}$ I^-	7.9 mM NaTos		$0.85*10^{-3}$					Co	14
	8.8 mM Na Tos		$0.95*10^{-3}$					Co	14
	9.8 mM Na Tos		$0.65*10^{-3}$					Co	14
$(CH_3)_3\text{-}C\text{-}(CH_2)_8\text{-}$ pyridinium $N\text{-Me}$ I^-			$4.90*10^{-3}$					Co	15
$C_{10}\text{-}CH(CH_3)$ pyridinium $N\text{-Me}$ I^-				$3.93*10^{-3}$				Co	16
$C_6\text{-}C{\equiv}C\text{-}C_4$ pyridinium $N\text{-Me}$ I^-			$4.17*10^{-3}$					Co	15
			$1.3*10^{-2}$					Co	15
$(CH_3CH_2)_2CH\text{-}(CH_2)_7\text{-}$ pyridinium $N\text{-Me}$ I^-				$3.76*10^{-3}$				Co	16

Table II.2.1. (continued)

Cationic	Additive [a]	20	25	30	35	40	45	Meth.	Ref.
C$_{12}$ N-Et I$^-$				$2.21*10^{-3}$				Co	16
C$_{12}$ N-Pr I$^-$				$1.91*10^{-3}$				Co	16
C$_{12}$ N-CH(CH$_3$)(CH$_3$) I$^-$				$1.93*10^{-3}$				Co	16
C$_{12}$ N-Bu I$^-$				$1.54*10^{-3}$				Co	16

Temperature, °C

a. The first page of this table gives the CMCs of (1:1) molar mixtures of N-hexylpyridinium bromide with sodium decylsulphate, sodium dodecylsulphate, or sodium pentadecafluorooctanoate in water; and of (1:1) molar mixtures of N-hexylpyridinium bromide with sodium decylsulphate, or sodium pentadecafluorooctanoate in 0.1 M NaBr.

Likewise, the first page gives the CMCs of N-octylpyridinium bromide in water and in 0.05 M KBr; and of (1:1) molar mixtures with sodium octylsulphate in water and in 0.1 M NaBr, and of (1:1) molar mixtures with sodium decylsulphate in water; and of a (1:2) molar mixture with sodium octylsulphate in water and in 0.1 M NaBr.

The remaining pages give the CMCs of other alkylpyridinium salts in water or in aqueous electrolyte solutions.

NaTos = sodium tosylate

b. 5 °C
c. 10 °C
d. 15 °C
e. 50 °C
f. 55 °C
g. 60 °C
h. 65 °C
i. 70 °C
j. 31 °C
k. 11.3 °C
l. 27.5 °C
m. 24.6 °C
n. 39.4 °C
o. 17.1 °C
p. 49.6 °C
q. 36.0 °C

Table II.2.2. Critical micelle concentration (CMC, in mole fraction scale) of alkylpyridinium salts in ethylammonium nitrate (EAN) at 50 °C and in ethylene glycol (EG) at 27.5 °C.

Cationic	Solvent	Temperature °C	CMC[a]	Ref.
C_{14}–N⁺⟨⟩ Br⁻	EAN	50	$6.49*10^{-3}$	17
C_{16}–N⁺⟨⟩ Cl⁻	EG	27.5	$2.3*10^{-1b}$	13
C_{16}–N⁺⟨⟩ Br⁻	EAN	50	$1.69*10^{-3}$	17
C_{18}–N⁺⟨⟩ Br⁻	EAN	50	$5.69*10^{-4}$	17

a. mole fraction scale
b. mol/l

Table II.2.3. Critical rod concentration (CRC, in mmol/kg) of alkylpyridinium salts in D_2O at 30 °C.

Cationic	CRC mmol/kg	Ref.
C_{12}—pyridinium—$\overset{+}{N}$—Me I^-	45	16
C_{10}—CH(CH$_3$)—pyridinium—$\overset{+}{N}$—Me I^-	> 440	16
$(CH_3)_3C$—$(CH_2)_8$—pyridinium—$\overset{+}{N}$—Me I^-	25	16
$(CH_3CH_2)_2CH$—$(CH_2)_7$—pyridinium—$\overset{+}{N}$—Me I^-	30	16
C_{12}—pyridinium—$\overset{+}{N}$—Et I^-	37	16
C_{12}—pyridinium—$\overset{+}{N}$—Pr I^-	28	16
C_{12}—pyridinium—$\overset{+}{N}$—CH(CH$_3$)$_2$ I^-	30	16

Table II.2.4. Krafft temperature of alkylpyridinium salts

Cationic	Krafft temperature °C	Ref.
$C_{16}-N^+$ ⟨pyridine⟩ Cl^-	11.3	11
$C_{16}-N^+$ ⟨pyridine⟩ Br^-	24.6	11
$C_{16}-N^+$ ⟨pyridine⟩ I^-	39.4	11
$C_{16}-N^+$ ⟨pyridine⟩ NO_3^-	17.1	11
$C_{16}-N^+$ ⟨pyridine⟩ BF_4^-	49.6	11
$C_{16}-N^+$ ⟨pyridine⟩ HSO_4^-	< 10	11
$C_{16}-N^+$ ⟨pyridine⟩ $H_2PO_4^-$	< 10	11
$C_{16}-N^+$ ⟨pyridine⟩ Tos^-	36.0	11

Table II.2.5. Aggregation number (N) of alkylpyridinium salts in electrolyte solutions at 30 °C

Cationic	Additive	N	Ref.
 C_8-N⁺ ⬡ Br⁻		24	3
	0.05 M KBr	23	3
 C_{11}-N⁺ ⬡ Br⁻		42	3
	0.05 M KBr	61	3
 C_{14}-N⁺ ⬡ Br⁻		79	3
	0.025 M KBr	113	3
	0.05 M KBr	156	3
 C_{16}-N⁺ ⬡ Cl⁻	0.01754 M NaCl	95	10*
	0.05843 M NaCl	135	10*
	0.4382 M NaCl	135	10*
	0.7304 M NaCl	137	10*
 C_{18}-N⁺ ⬡ Br⁻	0.05 M KBr	510	3

* at 31 °C

Method: LS

Table II.2.6. Degree of micellar dissociation (β) of alkylpyridinium iodide in water and in aqueous NaI at 25 °C

Cationic	Additive	β	Ref.
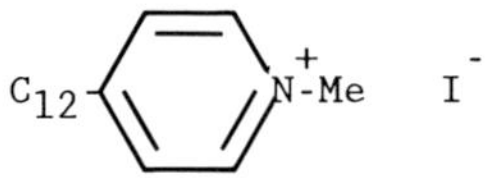 C_{12} — pyridinium N-Me I⁻		0.83	15
	2.0 mM NaI	0.87	14
	4.0 mM NaI	0.89	14
	6.0 mM NaI	0.91	14
	8.0 mM NaI	0.91	14
	10.0 mM NaI	0.94	14
	75 mM NaI	0.98	14
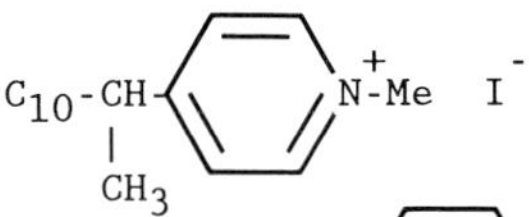 C_{10}—CH(CH₃)— pyridinium N-Me I⁻		0.80	15
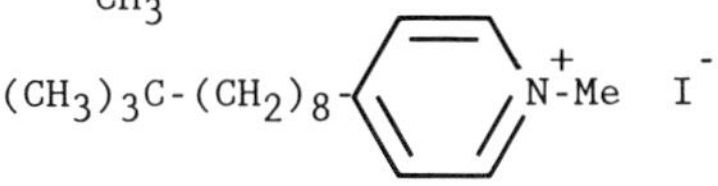 $(CH_3)_3C\text{-}(CH_2)_8$— pyridinium N-Me I⁻		0.84	15
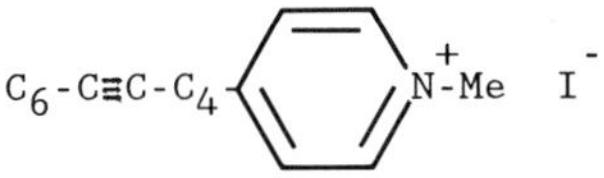 $C_6\text{-}C{\equiv}C\text{-}C_4$— pyridinium N-Me I⁻		0.78	15

Method: Co

Table II.2.7. Charge transfer band of alkylpyridinium iodides in micellar aggregates at 25 °C

Cationic	λ_{max} nm	Ref.
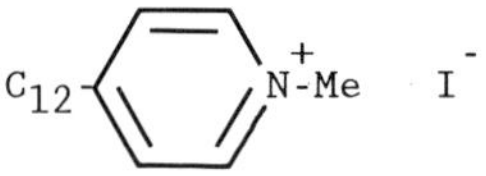 C_{12}— pyridinium N-Me I⁻	286	15
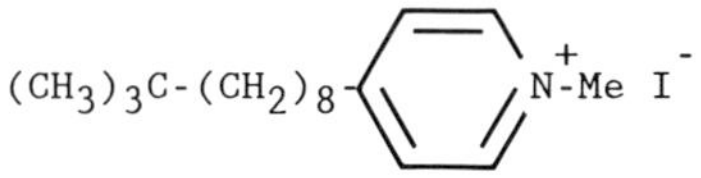 $(CH_3)_3C\text{-}(CH_2)_8$— pyridinium N-Me I⁻	292.5	15

Method: Abs

Table II.2.8. Surface tension (γ, in mN/m) of alkylpyridinium salts in water, and in the presence of other surfactants and/or electrolytes at various temperatures.

Cationic	Additive	Conc. M	Temp., °C	γ mN·m^{-1}	Ref.
C_6-N$^+$ pyridinium Br$^-$	$C_{10}OSO_3Na$	CMC	25	29.8	1
		CMC	30	29.7	2
	+ 0.1 M NaBr	CMC	30	29.2	2
	$C_{12}OSO_3Na$	CMC	25	31.4	1
	$C_7F_{15}COONa$	CMC	25	15.2	1
		CMC	30	14.8	2
	+ 0.1 M NaBr	CMC	30	14.8	2
C_8-N$^+$ pyridinium Br$^-$		CMC	30	44.8	2
	C_8OSO_3Na	CMC	25	23.8	1
		CMC	30	23.6	2
	+ 0.1 M NaBr	CMC	30	23.6	2
	C_8OSO_3Na (1:2)	CMC	30	23.7	2
	+ 0.1 M NaBr	CMC	30	23.7	2
	$C_{10}OSO_3Na$	CMC	35	24.2	1
C_{10}-N$^+$ pyridinium Cl$^-$		CMC	25	34.5	4
C_{12}-N$^+$ pyridinium Cl$^-$		CMC	20	36.2	4

Table II.2.9. Surface tension (γ, in mN/m) of alkylpyridinium salts in ethylammonium nitrate at 50 °C (Taken from reference 17)

Cationic	Conc. M	γ mN.m^{-1}
C_{14}-N Br (pyridinium)	0.0	46.6
	0.0102	42.4
	0.0197	40.1
	0.0284	38.4
	0.0366	37.0
	0.0442	35.8
	0.0514	35.4
	0.0581	34.8
	0.0643	34.4
	0.0702	34.2
	0.0758	33.8
	0.0811	34.2
	0.118	34.2
C_{16}-N Br (pyridinium)	0.0	46.3
	0.0030	34.0
	0.0056	32.1
	0.0078	32.0
	0.0098	30.6
	0.0114	30.4
	0.0129	29.7
	0.0142	29.4
	0.0154	29.0
	0.0165	28.5

Table II.2.9. (continued)

Cationic	Conc. M	γ mN.m^{-1}
C$_{16}$-N⁺ ⟩Br⁻	0.0174	28.0
	0.0183	27.5
	0.0205	27.5
C$_{18}$-N⁺ ⟩Br⁻	0.0	46.6
	0.0075	44.6
	0.0144	41.7
	0.0208	39.8
	0.0268	37.3
	0.0324	36.4
	0.0377	35.0
	0.0426	34.2
	0.0472	33.3
	0.0515	32.9
	0.0556	31.6
	0.0594	31.5
	0.0631	31.2
	0.0865	31.4

Table II.2.10. Absorbed amount (Γ) of alkylpyridinium bromide at the air-water interface[a] in the presence of other surfactants at various temperatures.

Cationic	Temp., °C	Γ_{Total} 10^{-10} mol.cm^{-2}	Γ^{+} 10^{-10} mol.cm^{-2}	Γ^{-} 10^{-10} mol.cm^{-2}
C_8-N$^+$ ⬡ Br$^-$ C_8OSO_3Na (1:1)	15	5.32		
	20	5.26		
	25	5.21		
	30	5.08		
		4.99	2.38	2.61
	35	4.94		
	40	4.88		
C_8-N$^+$ ⬡ Br$^-$ C_8OSO_3Na (2:1)	30	5.11	2.28	2.83
C_6-N$^+$ ⬡ Br$^-$ $C_{10}OSO_3Na$ (1:1)	30	4.71	1.42	3.29
C_6-N$^+$ ⬡ Br$^-$ $C_7F_{15}COONa$ (1:1)	30	4.96	1.70	3.26

a. In 0.1 M NaBr, total surfactant concentration $4.00 * 10^{-3}$ M; Ref. 2. Equimolar (1:1) surfactant mixtures; or (2:1) molar surfactant mixtures

Table II.2.11. Diffusion coefficient (D, in 10^{-6} cm^2/s) of a (1:1) mixture of N-octylpyridinium bromide and sodium octylsulphate at various temperatures.

Cationic	Temperature °C	D 10^{-6} cm^2 s^{-1}
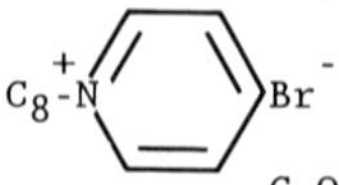 C_8-N⁺〈benzene〉Br⁻ C_8OSO_3Na (1:1)	15	1.62
	20	1.84
	25	2.16
	30	2.72
	30	2.83 [a]
	35	3.32
	40	3.25

a. in 0.1 M NaBr solution

Reference 2

Method: AK

Table II.2.12. Rate constant of monomer exchange between micelle and solution state for N-dodecylpyridinium iodide at 35 °C

Cationic	k_f^{+} $1.mol^{-1}$ s^{-1}	k_b^{++} s^{-1}
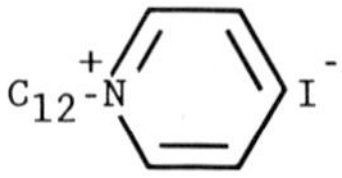 C_{12}-N⁺〈benzene〉I⁻	$1.6 * 10^6$	$1.6 * 10^3$

+   Rate constant associated with the formation of micelles

++ Rate constant associated with the dissociation of micelles

Reference 9

method: KM

Table II.2.13. Gibbs free energy (ΔG°, in kJ/mol), enthalpy (ΔH°, in kJ/mol), entropy (ΔS°, in J mol⁻¹K⁻¹), and heat capacity (ΔC$_p$°, in J mol⁻¹K⁻¹) of micellization of alkylpyridinium salts in water, in electrolyte solutions, and in aqueous urea at various temperatures.

Cationic	Temp. K	ΔG° kJ.mol⁻¹	ΔH° kJ.mol⁻¹	ΔS° J.mol⁻¹.K⁻¹	ΔC$_p$° J.mol⁻¹.K⁻¹	Ref.
C$_{10}$-N⁺ Cl⁻	298	-28.1				18
C$_{12}$-N⁺ Cl⁻	278	-32.0	8.70	146		18
	298	-34.7				18
	298	-34.2				18
	298	-20.6	1.88	75.5	-490	19
	303	-20.6	-1.55	62.9		19
	308	-20.6	-2.93	57.4		19
	310	-21.4	-3.35			18
	318	-36.5	-8.7	87.8		18
	338	-38.2	-15.2	66.9		18
	298[a]	-33.7				18
	298[b]	-34.0				18

Table II.2.13. (continued)

Cationic	Temp. K	ΔG° kJ.mol^{-1}	ΔH° kJ.mol^{-1}	ΔS° J.mol^{-1}.K^{-1}	ΔC_p° J.mol^{-1}.K^{-1}	Ref.
C_{12}-N$^+$ (pyridinium) Cl$^-$	298^c	-33.7				18
C_{12}-N$^+$ (pyridinium) Br$^-$	278	-19.62	5.56	90.4		18
	283	-20.0	2.5	79.5		18
	288	-20.4	0.08	71.1		18
	293	-20.8	-2.3	62.8		18
	294		-2.05	61.7		19
	298	-21.1	-4.06	56.9		18
			-3.47	56.1	-385	19
	303	-21.3	-5.48	52.3		18
			-5.56	48.3		19
	308	-21.6	-6.57	48.5		18
			-7.32	41.8		19
	313	-21.8	-7.41	46.0		18
	318	-22.0	-7.95	44.3		18

Table II.2.13. (continued)

Cationic	Temp. K	ΔG^{o} kJ.mol⁻¹	ΔH^{o} kJ.mol⁻¹	ΔS^{o} J.mol⁻¹.K⁻¹	ΔC_p^{o} J.mol⁻¹.K⁻¹	Ref.
C₁₂-N⁺ (pyridinium) Br⁻	323	-22.3	-8.45	42.7		18
	328	-22.5	-8.83	41.6		18
	333	-22.7	-9.25	40.3		18
	338	-22.9	-9.79	38.6		18
	343	-23.1	-10.63	36.3		18
	298[d]	-36.4				18
	298[e]	-37.4				18
	298[f]	-36.9				18
	298[g]	-36.2				18
	298[h]	-36.7				18
C₁₂-N⁺ (pyridinium) I⁻	298	-24.2	-9.3	39.4		18
			-13.5			18
	303	-24.2	-14.5	31.8		18

Table II.2.13. (continued)

Cationic	Temp. K	ΔG^o kJ.mol^{-1}	ΔH^o kJ.mol^{-1}	ΔS^o J.mol^{-1}.K^{-1}	$\Delta C_p^{\,o}$ J.mol^{-1}.K^{-1}	Ref.
C_{12}-N^{+} pyridinium, I^-	308	-24.3	-16.7	24.7		18
	310	-42.6	-12.43	39.33		18
	298^i	-42.6				18
	298^j	-44.6				18
	298^k	-25.2	-11.97	44.5		18
	303^k	-25.3	-13.93	37.66		18
	308^k	-25.5	-15.86	31.30		18
	298^l	-23.7	-12.6	37.1		18
	298^m	-23.1	-12.6	35.4		18
	298^n	-22.6	-12.9	32.6		18
	298°	-22.1	-13.1	30.5		18
	303°	-22.5	-14.9	25.1		18
	308°	-22.8	-17.6	16.8		18

Table II.2.13. (continued)

Cationic	Temp. K	ΔG° kJ.mol^{-1}	ΔH° kJ.mol^{-1}	ΔS° J.mol^{-1}.K^{-1}	ΔC_p° J.mol^{-1}.K^{-1}	Ref.
C_{12}-N$^+$ ⬡ OMe Cl$^-$	278	-20.0	13	117		18
	283	-20.5	9.2	105		18
	288	-21.0	5.9	92		18
	293	-21.5	3.0	84		18
	298	-21.8	0.6	75		18
	303	-22.2	-1.5	67		18
	308	-22.5	-3.4	63		18
	313	-22.8	-5.0	54		18
	318	-23.1	-6.7	50		18
	323	-23.4	-8.8	46		18
	328	-23.6	-10.9	39		18
	333	-23.7	-13.8	30		18
C_{12}-N$^+$ ⬡ OMe Br$^-$	278	-22.0	11	121		18
	283	-22.5	5.0	96		18

Table II.2.13. (continued)

Cationic	Temp. K	$\Delta G°$ kJ.mol^{-1}	$\Delta H°$ kJ.mol^{-1}	$\Delta S°$ J.mol^{-1}.K^{-1}	$\Delta C_p°$ J.mol^{-1}.K^{-1}	Ref.
C_{12}-N$^+$ pyridinium—OMe Br$^-$	288	-22.9	0.3	80		18
	293	-23.3	-3.3	67		18
	298	-23.6	-6.3	59		18
	303	-23.9	-8.4	50		18
	308	-24.1	-11	42		18
	313	-24.3	-14	34		18
	318	-24.5	-18	22		18
	323	-24.5	-23	4.6		18
	328	-24.5	-31	-18.4		18
C_{12}O pyridinium N-Me Br$^-$	288	-27.2	0.67	96		18
	293	-27.6	-3.2	84		18
	298	-28.0	-7.1	71		18
	303	-28.3	-11	59		18
	308	-28.6	-14	50		18

Table II.2.13. (continued)

Cationic	Temp. K	ΔG° kJ.mol^{-1}	ΔH° kJ.mol^{-1}	ΔS° J.mol^{-1}.K^{-1}	ΔC_p° J.mol^{-1}.K^{-1}	Ref.
C$_{12}$O—⟨pyridinium⟩—N$^+$-Me Br$^-$	313	-28.8	-17	38		18
	318	-29.0	-20	29		18
	323	-29.1	-22	22		18
	328	-29.2	-24	15		18
	333	-29.3	-26	10		18
C$_{14}$-N$^+$ Br$^-$	298				-669	18
	318				-586	18
	338				-460	18
	358				-251	18
C$_{14}$-N$^+$—OCH$_3$ Br$^-$	293	-30.2	-8.4	75		18
	298	-30.5	-13	59		18
	303	-30.8	-15	50		18
	308	-31.0	-16	50		18
	313	-31.3	-17	46		18

Table II.2.13. (continued)

Cationic	Temp. K	ΔG° kJ.mol^{-1}	ΔH° kJ.mol^{-1}	ΔS° J.mol^{-1}.K^{-1}	ΔC_p° J.mol^{-1}.K^{-1}	Ref.
C$_{14}$-N$^+$ —OCH$_3$ Br$^-$	318	-31.5	-18	42		18
	323	-31.7	-21	34		18
	328	-31.8	-27	15		18
	333	-31.8	-38	18		18
C$_{16}$-N$^+$ Cl$^-$	298	-51.0				18
	320		-5.4	73.6		18
C$_{16}$-N$^+$ Br$^-$	313		-24.1	15.48		18

a. In 0.02 M KCl b. In 0.05 M KCl c. In 0.08 M KCl d. In 0.08 M KCl e. In 0.04 M KBr
f. In 0.06 M KBr g. In 0.08 M KBr h. In 0.10 M KBr i. In 0.0025 M KI j. In 0.005 M KI
k. In 0.01 M KI l. In 0.5 M urea m. In 1.0 M urea n. In 1.5 M urea o. In 2.0 M urea.

Table II.2.14. Enthalpy (ΔH, in J/mol) and entropy (ΔS, in J mol^{-1}K^{-1}) of solubilization of apolar components in N-hexadecyl-pyridinium chloride micelles at 35 °C.

Cationic	Conc. M	Solute	ΔH J.mol^{-1}	ΔS J.mol^{-1} K^{-1}	Ref.
C$_{16}$-N$^+$ ⬡ Cl$^-$	0.1	Cyclohexane	2200	-2.2	20
	0.5	Cyclohexane	2400	-0.1	20
	0.1	Benzene	1480	2.3	20
	0.5	Benzene	2100	4.4	20
	0.1	tBuOH	-9400	-19	20

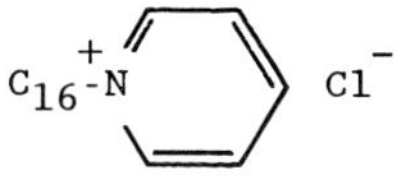

Method: VP

Table II.2.15. Regular solution theory interaction parameter (w/RT, in J/mol) of N-hexadecylpyridinium chloride with two other surfactants at 30 °C.

Surfactant	w/RT J.mol^{-1}	Ref.
C$_9$-⬡-O(CH$_2$CH$_2$O)$_{10}$-H	-5.36	12
C$_{16}$N(CH$_3$)$_3$$^+$ Cl$^-$	-0.84	12

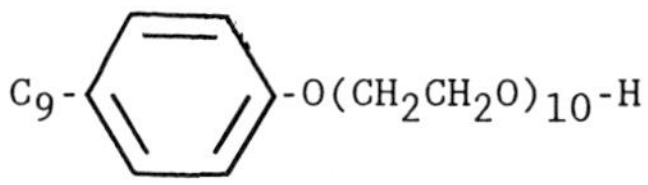

Method: calc

Table II.2.16. Natural logarithm of the partition coefficient (ln K) for the transfer of apolar components from the gas phase into an 0.1 M N-hexadecylpyridinium chloride solution at 25 °C.

Solute	ln K^a	Ref.
Benzene	6.14	24
Toluene	7.07	24
Phenol	15.15	24
p-Cresol	16.18	24
Pentane	2.66	24
Hexane	4.18	24
Cyclohexane	4.97	24
1-Butanol	11.01	24
1-Pentanol	11.66	24

a. Partition coefficient, K, is molarity of organic solute within the micelle, divided by molarity in the gas phase, extrapolated to infinite dilution.

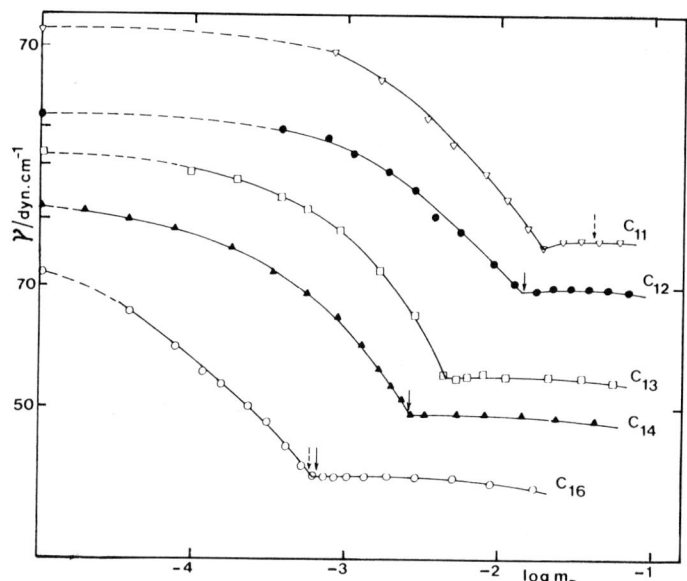

Figure II.2.1. Surface tension (γ , in dyne/cm) of alkylpyridinium bromides (C$_{11}$, C$_{13}$, C$_{14}$, and C$_{16}$) and chloride (C$_{12}$) as a function of surfactant concentration (m$_D$). Arrows indicate literature values for the CMC. (© American Chemical Society, 1984. Reproduced with permission. Taken from reference 5).

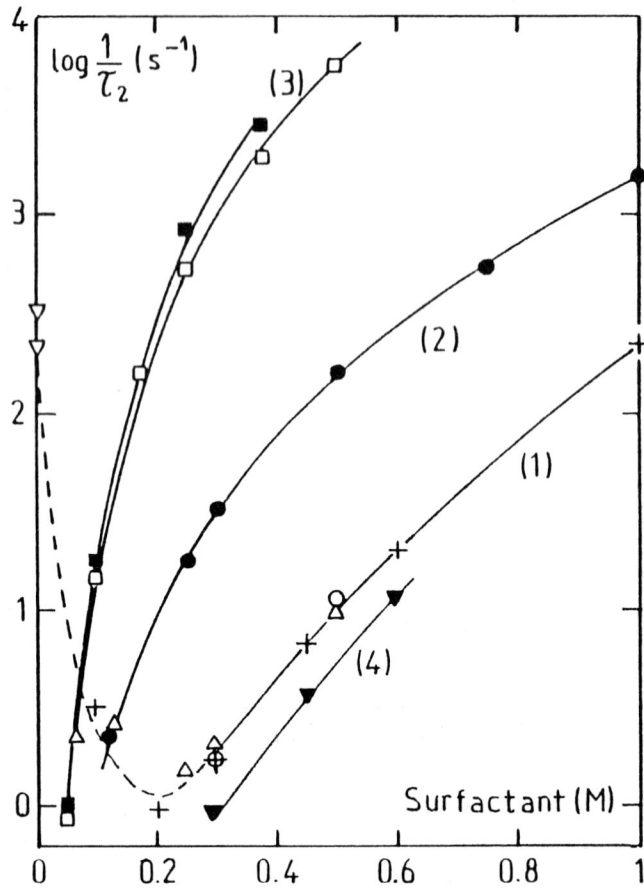

Figure II.2.2. Variation in the slow relaxation rate (1/τ_2, in s-1) with surfactant concentration for solutions of : (1) N-hexadecyl-pyridinium chloride at 25 °C, (+) without dye, (o) with eosine, (Δ) on the absorption band of the surfactant, (∇) on the absorption band of eosine: (2) N-tetra-decylpyridinium chloride at 25 °C ($\bullet$): (3) N-tetradecylpyridinium bromide at 25 °C (▢) and at 30 °C (■): (4) hexadecyltrimethylammonium chloride at 25 °C (▼). (© American Chemical Society, 1986. Reproduced with permission. Taken from reference 21).

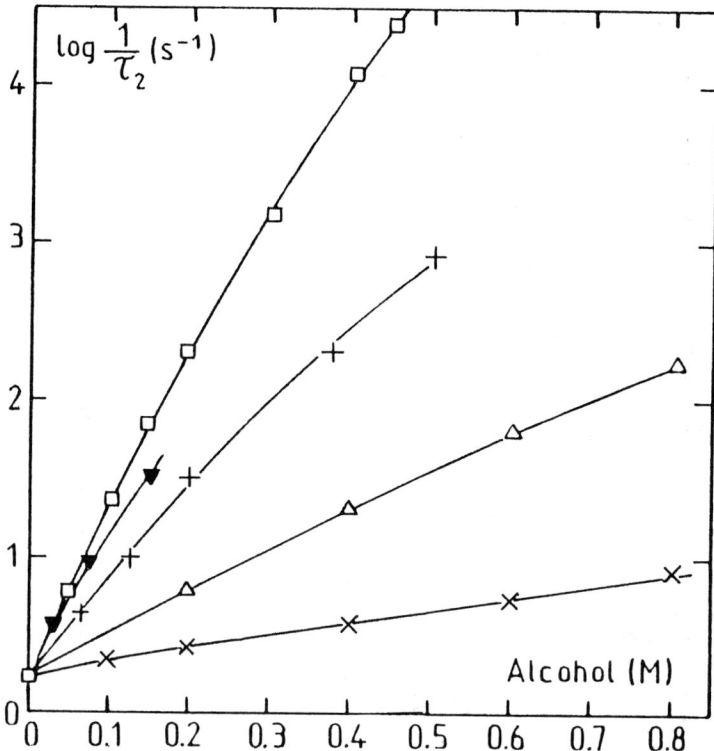

Figure II.2.3. Effect of concentration of added alcohol on the slow relaxation rate (1/τ $_2$, in s-1) for a 0.3 M N-hexadecylpyridinium chloride solution at 25 ℃: (x) ethanol;(Δ) 1-propanol; (+) 1-butanol; (∇) benzyl alcohol; (▢) 1-pentanol. (© American Chemical Society, 1986. Reproduced with permission. Taken from reference 21).

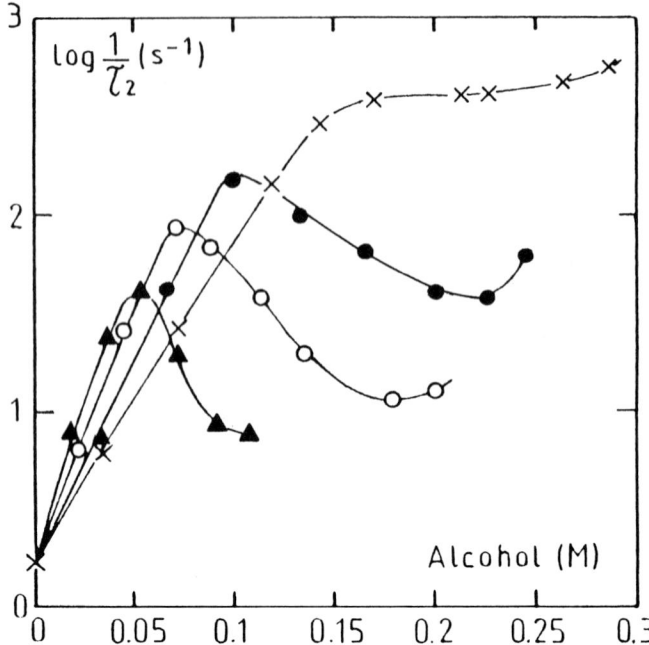

Figure II.2.4. Effect of concentration of added alcohol on the slow relaxation rate (1/τ $_2$, in s^{-1}) for a 0.3 M N-hexadecylpyridinium chloride solution at 25 °C: (x) 1-hexanol;(●) 1-heptanol; (o) 1-octanol; (▲) 1-decanol. (© American Chemical Society, 1986. Reproduced with permission. Taken from reference 21).

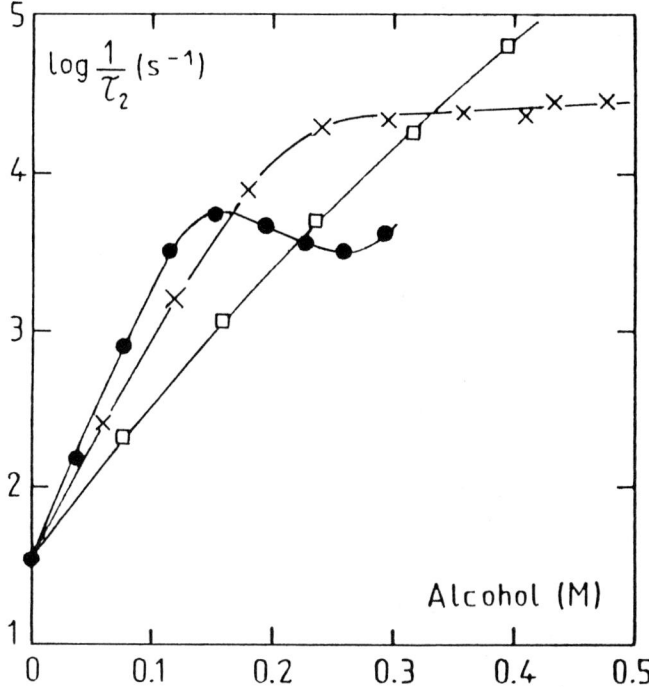

Figure II.2.5. Effect of concentration of added alcohol on the slow relaxation rate ($1/\tau_2$, in s^{-1}) for a 0.3 M N-tetradecylpyridinium chloride solution at 25 °C: (□) 1-pentanol; (x) 1-hexanol; (●) 1-heptanol. (© American Chemical Society, 1986. Reproduced with permission. Taken from reference 21).

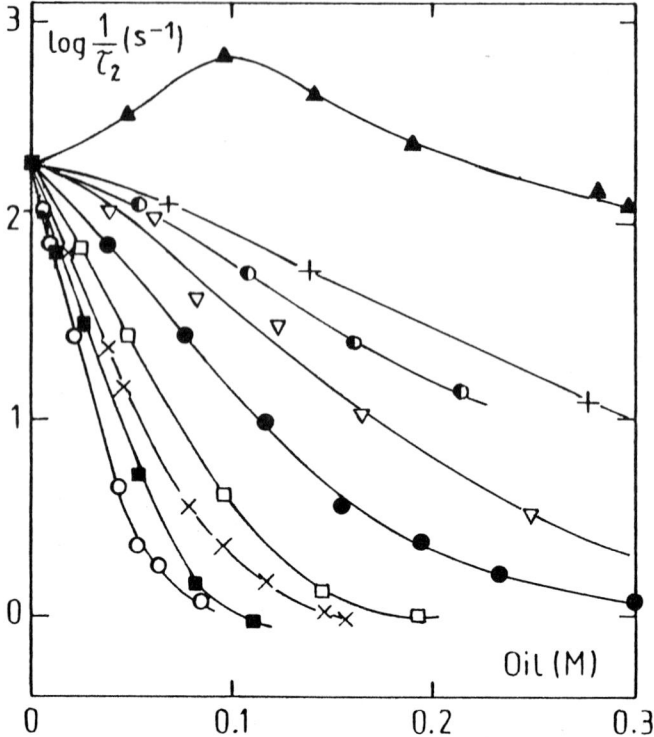

Figure II.2.6. Effect of concentration of added oil on the slow relaxation rate ($1/\tau_2$, in s^{-1}) for the mixed micellar solution of 0.3 M N-hexadecylpyridinium chloride and 0.2 M 1-pentanol at 25 °C: (∇) n-hexane; (●) n-heptane; (☐) n-octane; (x) n-nonane; (■) n-decane; (o) n-dodecane; (▼) n-tetradecane; (+) cyclohexane; (▲) toluene; (◑) n-butylbenzene. (© American Chemical Society, 1986. Reproduced with permission. Taken from reference 21).

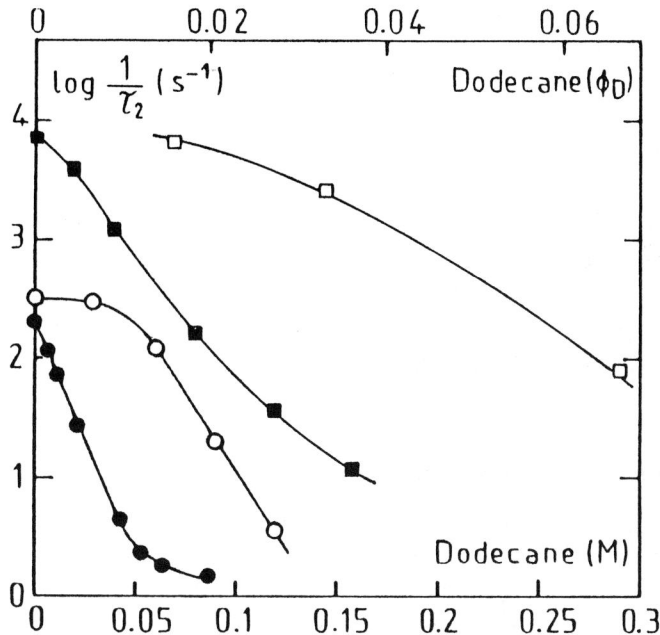

Figure II.2.7. Effect of counter-ion nature at 25 ºC. Variation of the slow relaxation rate (1/τ_2, in s^{-1}) upon the addition of dodecane to:
(●) 0.3 M N-hexadecylpyridinium chloride + 0.2 M 1-pentanol;
(o) 0.3 M N-hexadecylpyridinium bromide + 0.2 M 1-pentanol;
(■) 0.4 M N-hexadecylpyridinium chloride + 0.4 M 1-pentanol;
(▢) 0.4 M N-hexadecylpyridinium bromide + 0.4 M 1-pentanol.
Φ_D is the volume fraction of dodecane.

(© American Chemical Society, 1986. Reproduced with permission. Taken from reference 21).

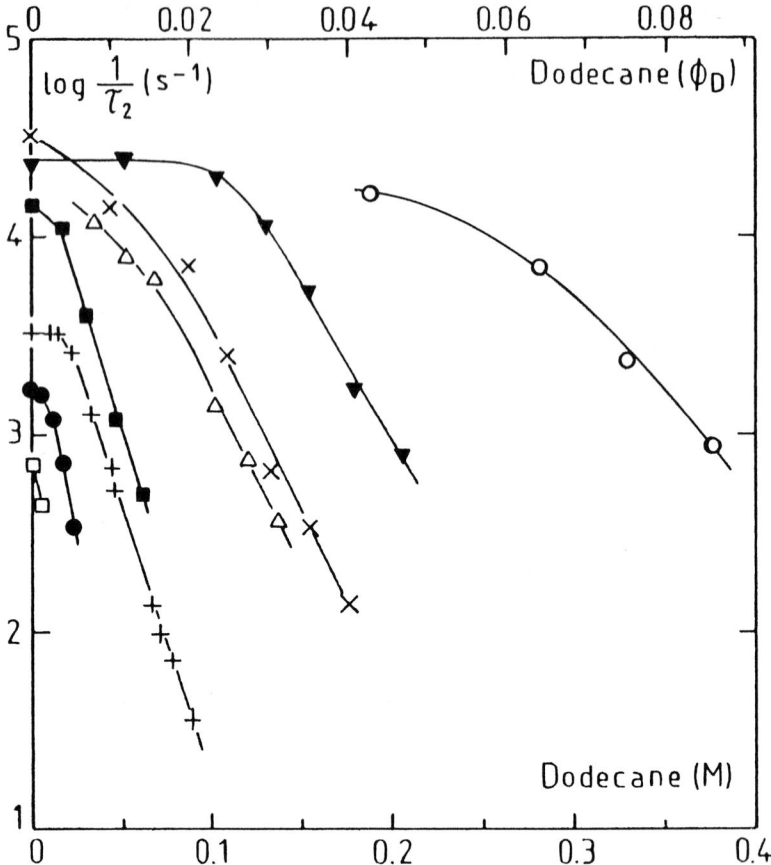

Figure II.2.8. Effect of surfactant and alcohol concentration. Variation of the slow relaxation rate (1/τ_2, in s-1) upon the addition of dodecane to:

(□)   0.1 M N-tetradecylpyridinium bromide + 0.2 M 1-pentanol at 25 oC;
(●)   0.3 M N-tetradecylpyridinium bromide + 0.1 M 1-pentanol at 15 oC;
(+)   0.3 M N-tetradecylpyridinium bromide + 0.2 M 1-pentanol at 15 oC;
(■)   0.3 M N-tetradecylpyridinium bromide + 0.2 M 1-pentanol at 25 oC;
(Δ)   0.4 M N-tetradecylpyridinium bromide + 0.3 M 1-pentanol at 15 oC;
(∇)   0.4 M N-tetradecylpyridinium bromide + 0.4 M 1-pentanol at 15 oC;
(o)   0.4 M N-tetradecylpyridinium bromide + 0.5 M 1-pentanol at 15 oC;
(x)   0.5 M N-tetradecylpyridinium bromide + 0.3 M 1-pentanol at 15 oC.

(© American Chemical Society, 1986. Reproduced with permission. Taken from reference 21).

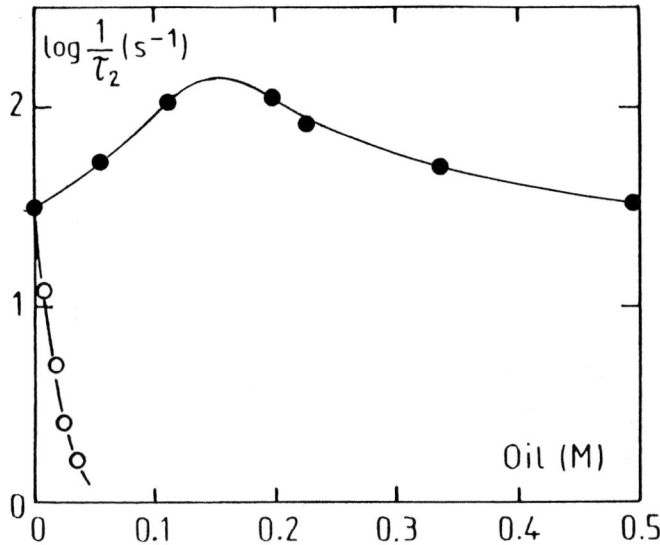

Figure II.2.9. Effect of concentration of added toluene (●) and dodecane (o) on the slow relaxation rate (1/τ_2, in s^{-1}) for 0.3 M N-hexadecylpyridinium chloride + 0.2 M 1-pentanol at 25 ºC.
(© American Chemical Society, 1986. Reproduced with permission. Taken from reference 21).

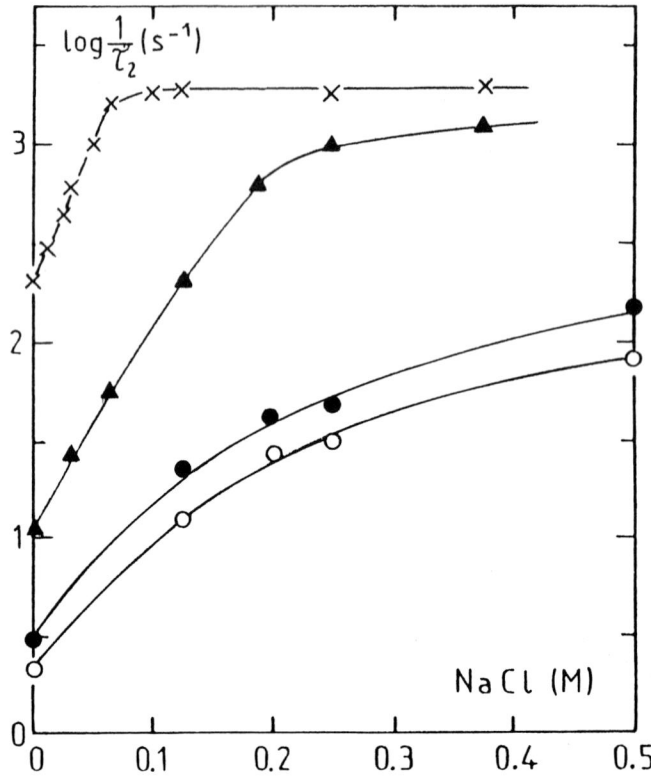

Figure II.2.10. Effect of NaCl. Variation of the slow relaxation rate ($1/\tau_2$, in s^{-1}) upon the addition of NaCl to:
(o) 0.3 M N-hexadecylpyridinium chloride at 25 ºC;
(●) 0.3 M N-hexadecylpyridinium chloride at 30 ºC;
(x) 0.3 M N-hexadecylpyridinium chloride + 0.2 M 1-pentanol at 25 ºC;
(▲) 0.3 M N-hexadecylpyridinium chloride + 0.2 M 1-pentanol + 0.0412 M decane at 25 ºC.

(© American Chemical Society, 1986. Reproduced with permission. Taken from reference 21).

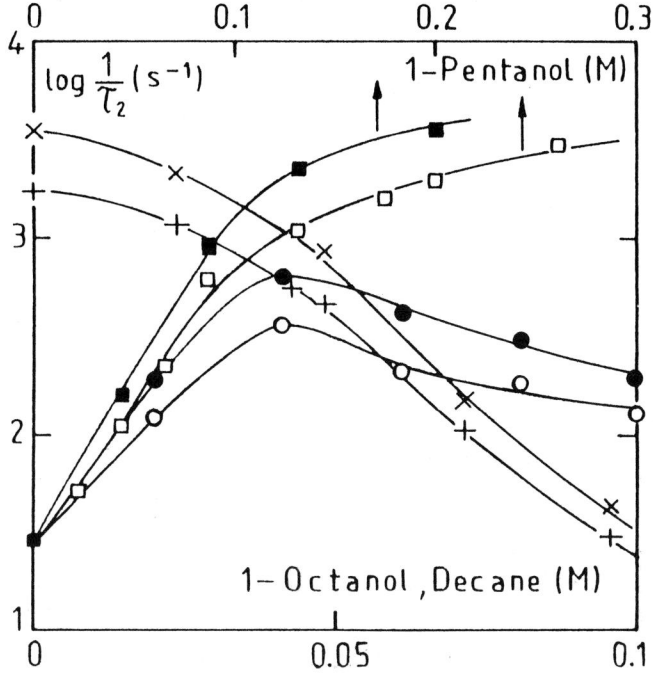

Figure II.2.11. Variation of the slow relaxation rate ($1/\tau_2$, in s^{-1}) with the concentration of added 1-pentanol (□) and 1-octanol (o) to 0.3 M N-hexadecylpyridinium chloride + 0.2 M NaCl and of added decane to 0.3 M N-hexadecylpyridinium chloride + 0.2 M 1-pentanol + 0.2 M NaCl (+) at 25 ºC. The symbols ■, ●, and x show the changes for the same systems at 30 ºC.

(© American Chemical Society, 1986. Reproduced with permission. Taken from reference 21).

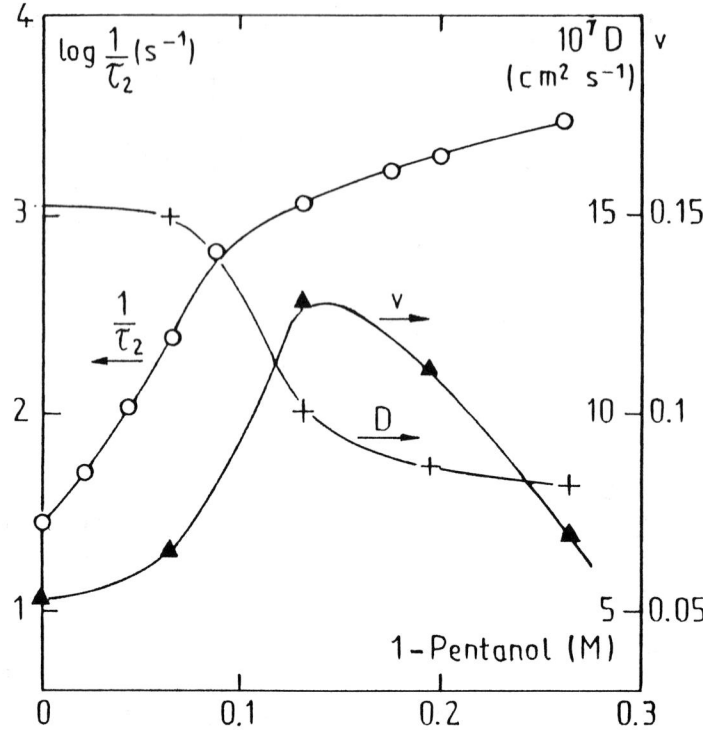

Figure II.2.12. Variation of the slow relaxation rate ($1/\tau_2$, (o) in s^{-1}), micellar collective diffusion coefficient (D, in 10^{-7} cm2/s), and variance of the autocorrelation function decay v (▲) with concentration of added 1-pentanol to 0.3 M N-hexadecyl-pyridinium chloride + 0.2 M NaCl at 25 oC. (© American Chemical Society, 1986. Reproduced with permission. Taken from reference 21).

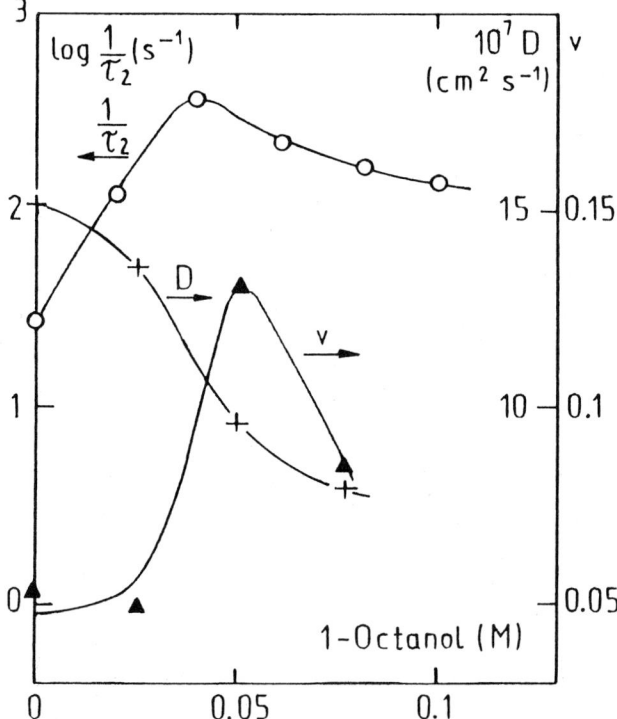

Figure II.2.13. Variation of the slow relaxation rate ($1/\tau_2$, (o) in s-1), micellar collective diffusion coefficient (D, in 10-7 cm2/s), and variance of the autocorrelation function decay v (▲) with concentration of added 1-octanol to 0.3 M N-hexadecyl-pyridinium chloride + 0.2 M NaCl at 25 oC. (© American Chemical Society, 1986. Reproduced with permission. Taken from reference 21).

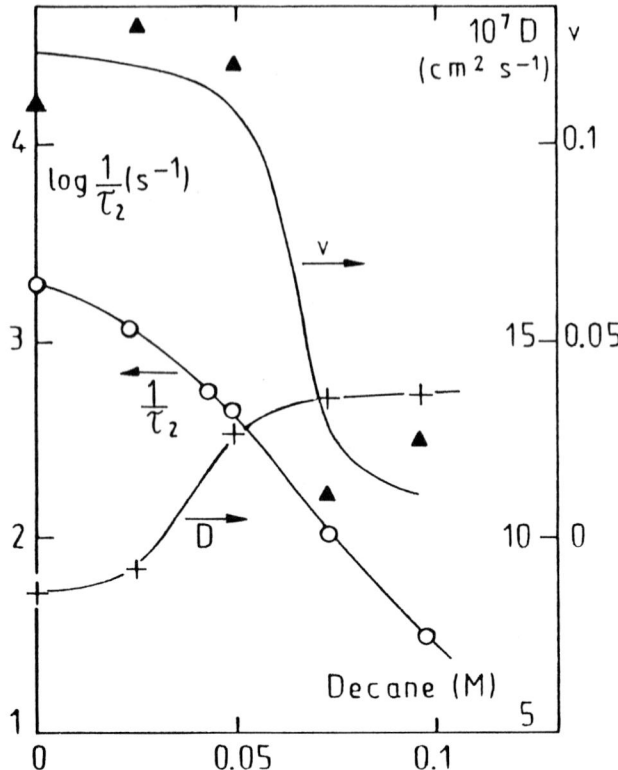

Figure II.2.14. Variation of the slow relaxation rate ($1/\tau_2$, (o) in s⁻¹), micellar collective diffusion coefficient (D, in 10⁻⁷ cm²/s), and variance of the autocorrelation function decay v (▲) with concentration of added decane to 0.3 M N-hexadecyl-pyridinium chloride + 0.2 M 1-pentanol + 0.2 M NaCl at 25 oC.

(© American Chemical Society, 1986. Reproduced with permission. Taken from reference 21).

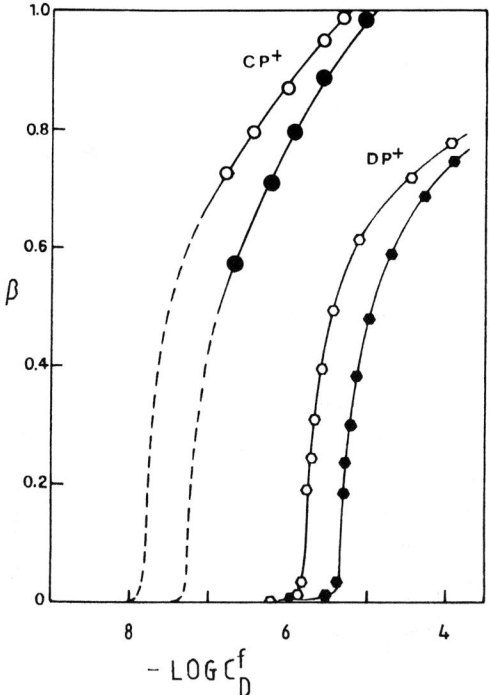

Figure II.2.15. Binding isotherm of N-hexadecylpyridinium and N-dodecylpyridinium cations to poly(styrenesulphonate) anion at 25 ºC in the presence of 0.01 M NaCl (○,◇) and 0.1 M NaCl (●,◆) (© American Chemical Society, 1988. Reproduced with permission. Taken from reference 6).

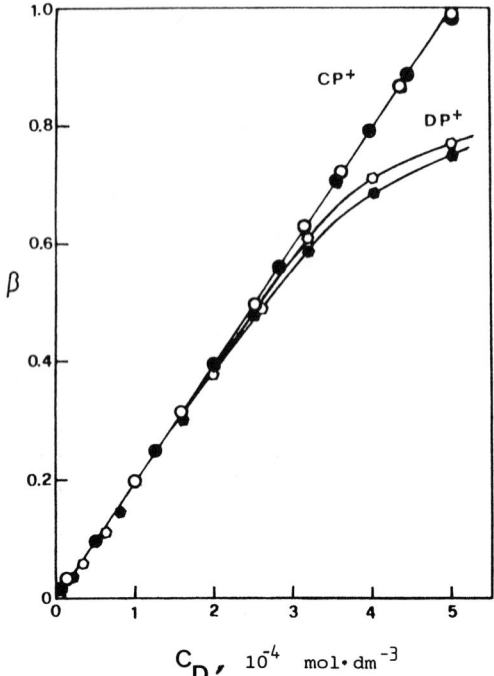

Figure II.2.16. Amount of binding (β) as a function of the total surfactant concentration (C_D, in 10^{-4} mol/dm³) at 25 °C. Symbols are as in Figure II.2.15. (© American Chemical Society, 1988. Reproduced with permission. Taken from reference 6).

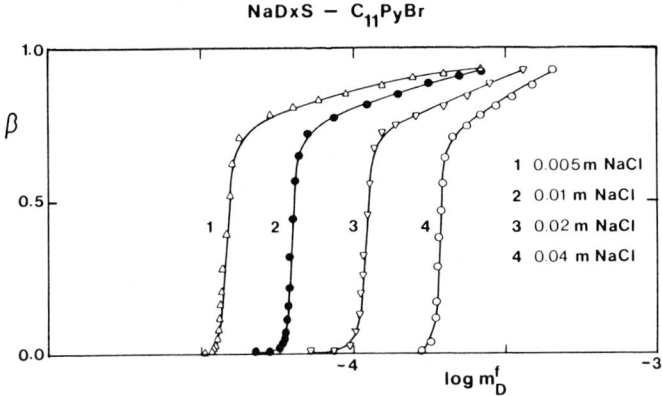

Figure II.2.17. Binding isotherm for sodium dextransulphate (NaDxS) (5*10⁻⁴ mol/kg) - N-undecylpyridinium chloride - NaCl at 30 ºC.
(© American Chemical Society, 1984. Reproduced with permission. Taken from reference 5).

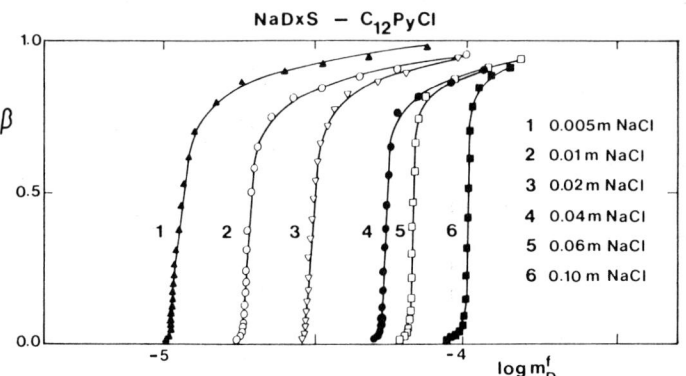

Figure II.2.18. Binding isotherm for sodium dextransulphate (NaDxS) (5*10⁻⁴ mol/kg) - N-dodecylpyridinium chloride - NaCl at 30 ºC.
(© American Chemical Society, 1984. Reproduced with permission. Taken from reference 5).

186

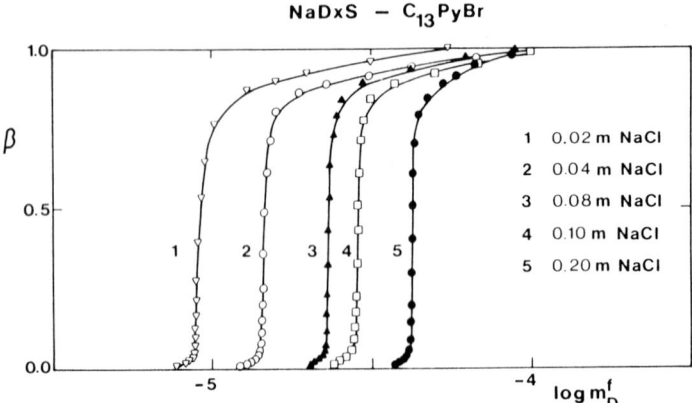

Figure II.2.19. Binding isotherm for sodium dextransulphate (NaDxS) (5*10⁻⁴ mol/kg) - N-tridecylpyridinium bromide - NaCl at 30 °C.
(© American Chemical Society, 1984. Reproduced with permission. Taken from reference 5).

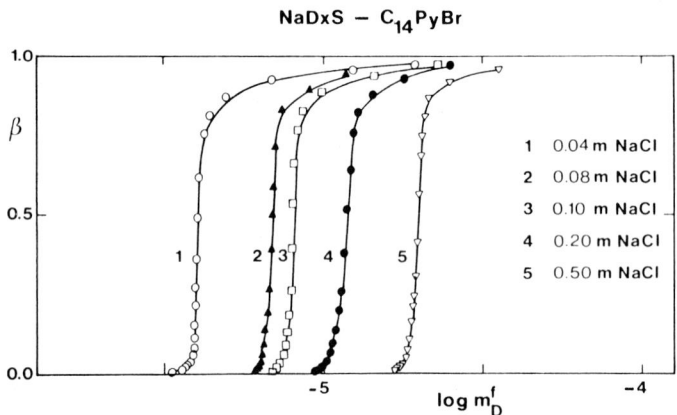

Figure II.2.20. Binding isotherm for sodium dextransulphate (NaDxS) (5*10⁻⁴ mol/kg) - N-tetradecylpyridinium bromide - NaCl at 30 °C.
(© American Chemical Society, 1984. Reproduced with permission. Taken from reference 5).

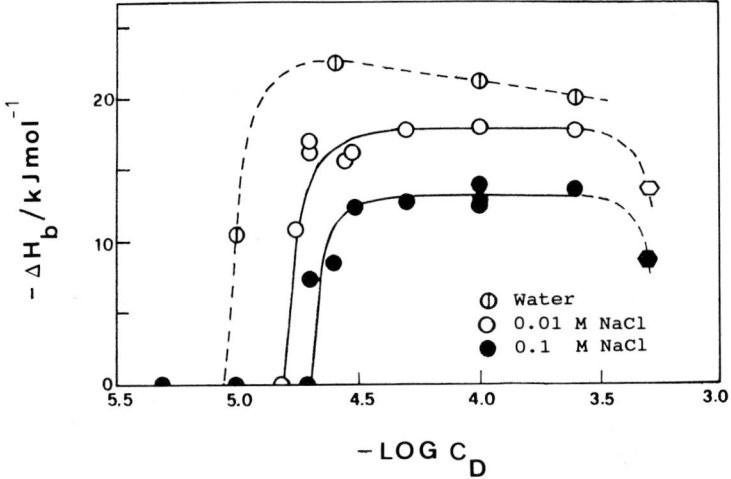

Figure II.2.21. Enthalpy of binding of N-hexadecylpyridinium
cations to poly(styrenesulphonate) anion at 25 °C in water, in 0.1 M
NaCl, and in 0.01 M NaCl. The values of ΔH_b are calculated per mole
of bound   surfactant cations. Cloudy final solutions : ◯ .
(© American Chemical Society, 1988. Reproduced with permission.
Taken from reference 6).

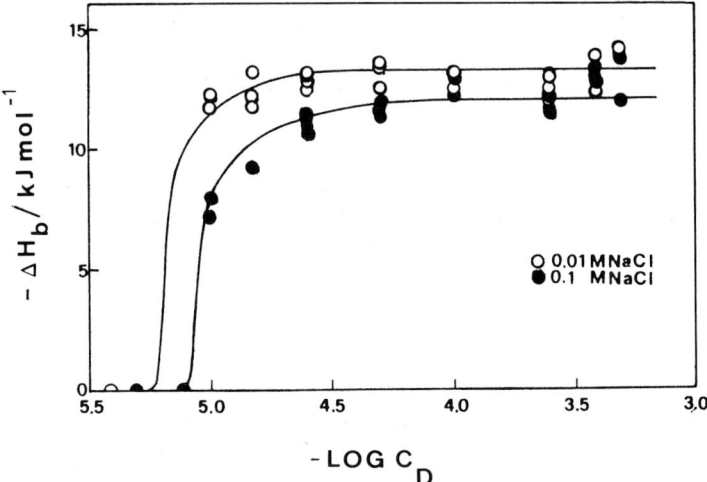

Figure II.2.22. Enthalpy of binding of N-dodecylpyridinium cations
to poly(styrenesulphonate) anion at 25 °C in water, in 0.1 M NaCl,
and in 0.01 M NaCl. The values of ΔH_b are calculated per mole of
bound   surfactant cations.
(© American Chemical Society, 1988. Reproduced with permission.
Taken from reference 6).

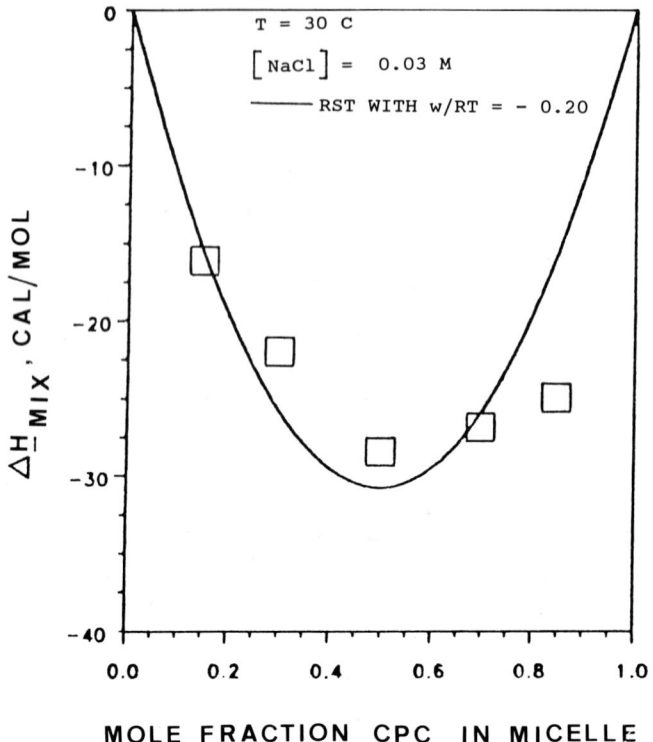

Figure II.2.23. Enthalpy of mixing of N-hexadecylpyridinium chloride and hexadecyltrimethylammonium chloride. (© American Chemical Society, 1988. Reproduced with permission. Taken from reference 12).

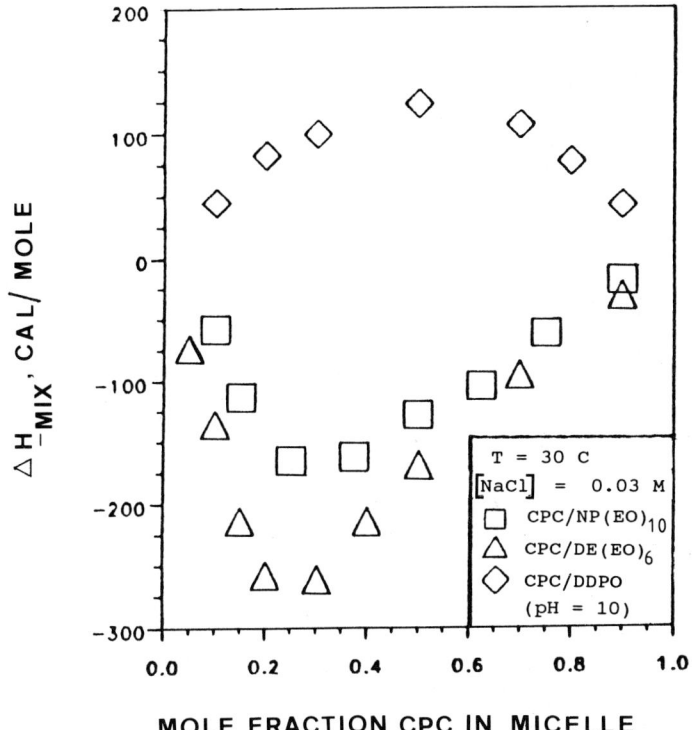

Figure II.2.24. Enthalpy of mixing of N-hexadecylpyridinium chloride (CPC) and nonylphenol(ethylene oxide)$_{10}$ ether (NP(EO)$_{10}$), of N-hexadecylpyridinium chloride and dodecylhexaoxyethylene glycol ether (DE(EO)$_6$), and of N-hexadecylpyridinium chloride and dodecyldimethyl phosphoxide (DDPO). (© American Chemical Society, 1988. Reproduced with permission. Taken from reference 12).

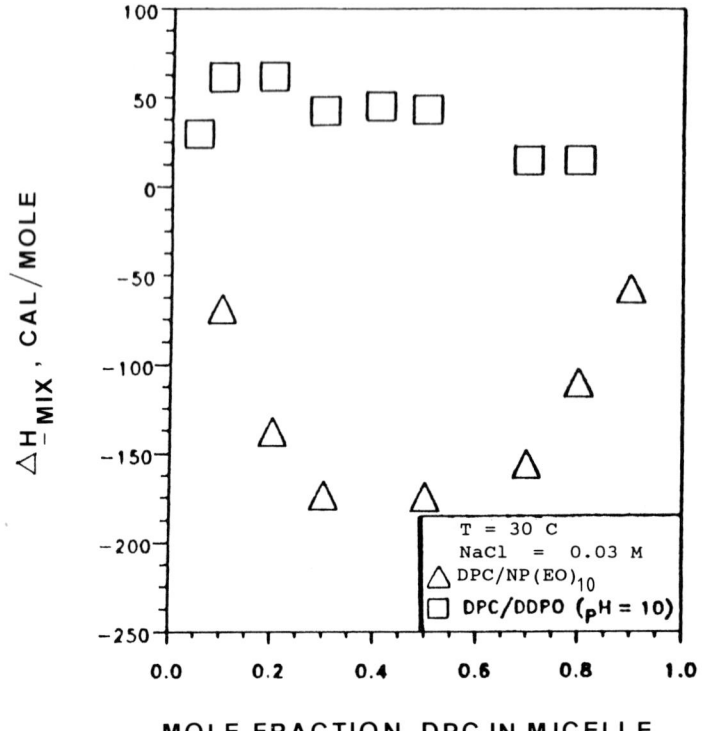

Figure II.2.25. Enthalpy of mixing of N-dodecylpyridinium chloride (DPC) and nonylphenol(ethylene oxide)$_{10}$ ether (NP(EO)$_{10}$), and of N-dodecylpyridinium chloride and dodecyldimethyl phosphoxide (DDPO). (© American Chemical Society, 1988. Reproduced with permission. Taken from reference 12).

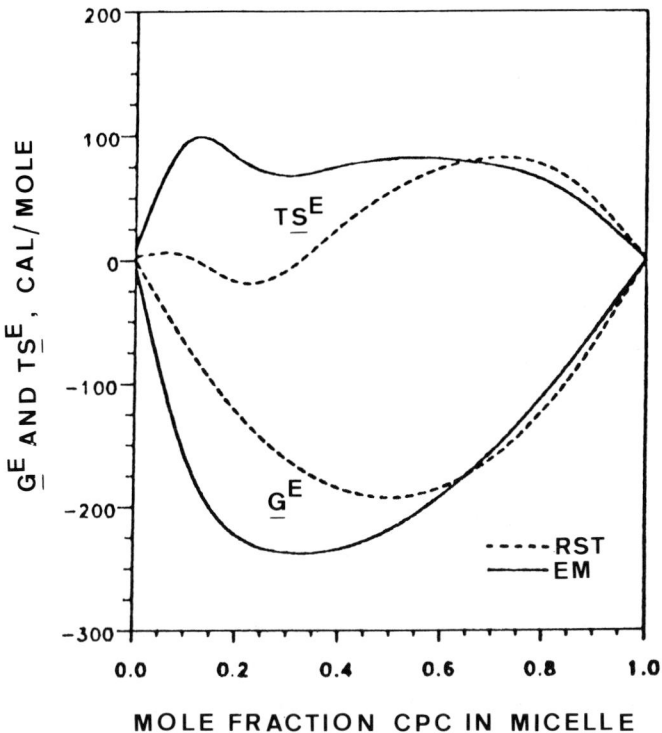

Figure II.2.26. Excess Gibbs free energy and entropy of mixing for N-hexadecylpyridinium chloride and nonylphenol(ethylene oxide)$_{10}$ ether calculated from the electrostatic model (EM) and from regular solution theory (RST).

(© American Chemical Society, 1988. Reproduced with permission. Taken from reference 12).

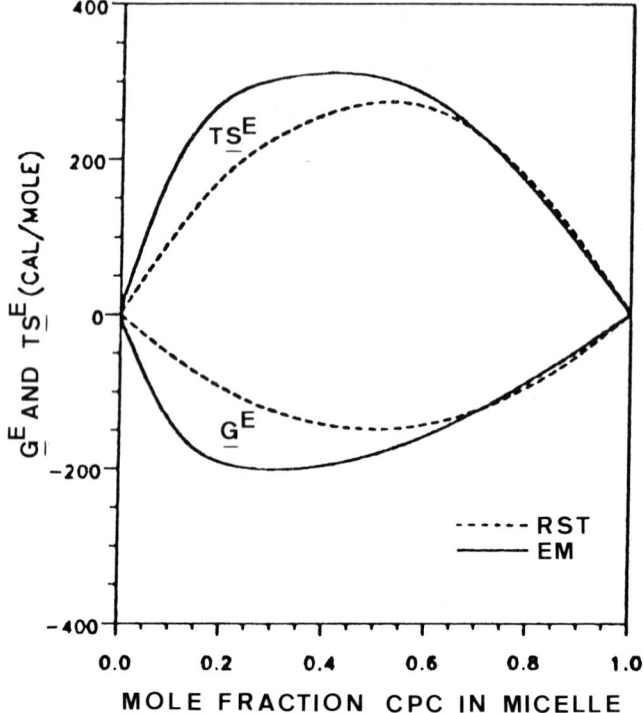

Figure II.2.27. Excess Gibbs free energy and entropy of mixing for N-hexadecylpyridinium chloride and dodecyldimethyl phosphoxide calculated from the electrostatic model (EM) and from regular solution theory (RST).

(© American Chemical Society, 1988. Reproduced with permission. Taken from reference 12).

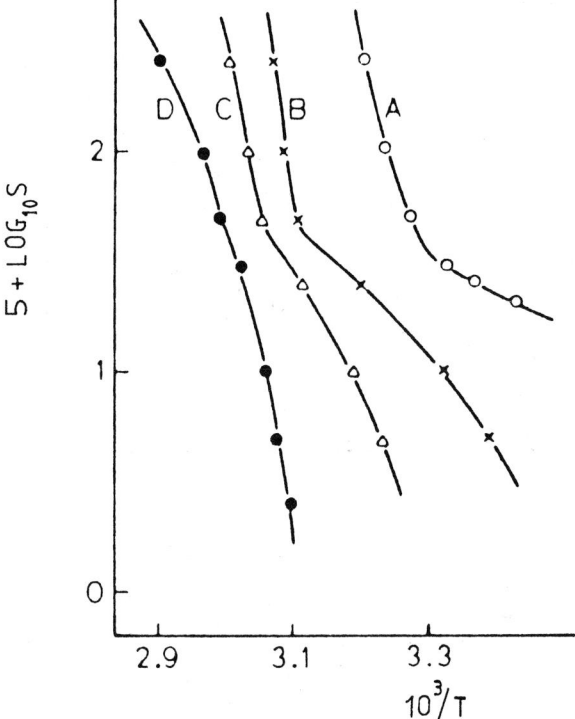

Figure II.2.28. Solubility of alkylpyridinium iodides :
(●) N-octadecyl-pyridinium,
(Δ) N-hexadecylpyridinium,
(x) N-tetradecyl-pyridinium, and
(o) N-dodecylpyridinium in xylene as a function of the reciprocal
temperature.

(© Springer Verlag, 1980. Reproduced with permission.
Taken from reference 22).

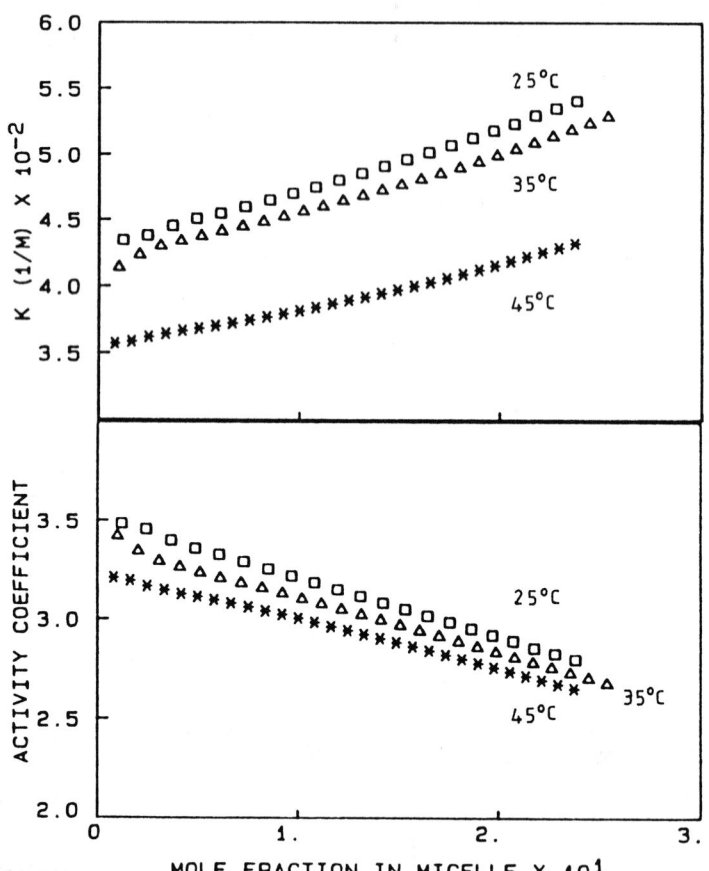

Figure II.2.29. Dependence of solubilization constant (K) and activity coefficient on the mole fraction in the micelle (X_H) for cyclohexane in solutions of N-hexadecylpyridinium chloride at various temperatures. Surfactant concentration approximately 0.1 M.

(© Academic Press, Inc., 1989. Reproduced with permission. Taken from reference 20).

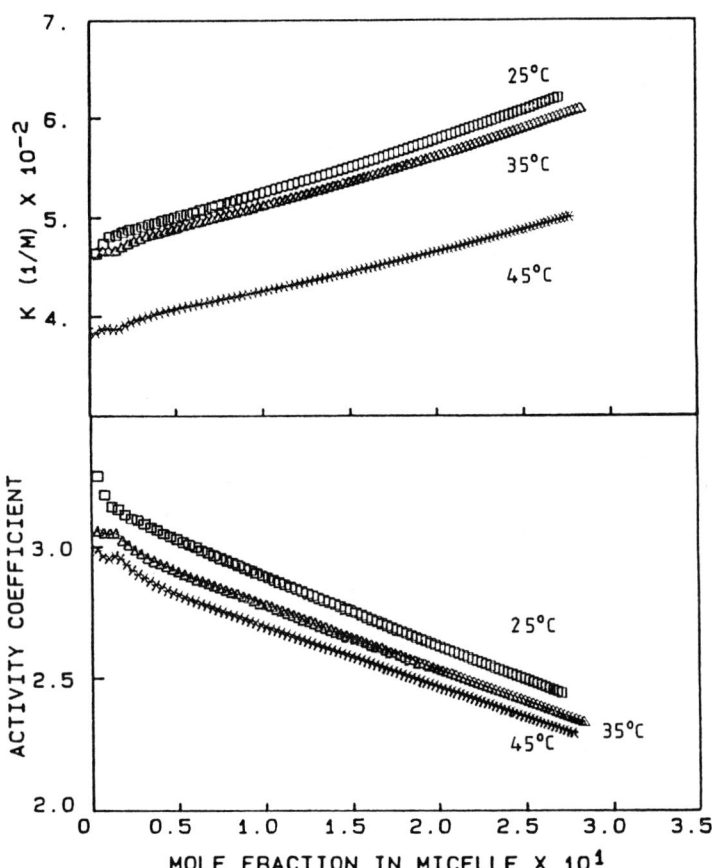

CYCLOHEXANE INTO 0.5 M CPC

Figure II.2.30. Dependence of solubilization constant (K) and activity coefficient on the mole fraction in the micelle (X_H) for cyclohexane in solutions of N-hexadecylpyridinium chloride at various temperatures. Surfactant concentration approximately 0.5 M.

(© Academic Press, Inc., 1989. Reproduced with permission. Taken from reference 20).

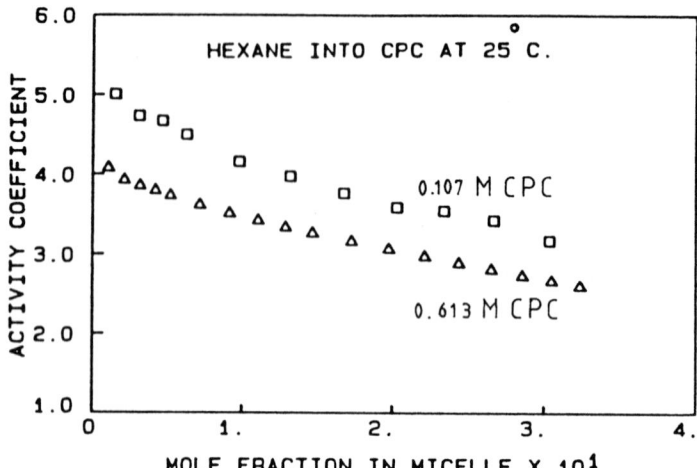

Figure II.2.31. Dependence of activity coefficient on the mole fraction in the micelle (X_H) for hexane in solutions of N-hexa-decylpyridinium chloride at 25 ºC. (© Academic Press, Inc., 1989. Reproduced with permission. Taken from reference 20).

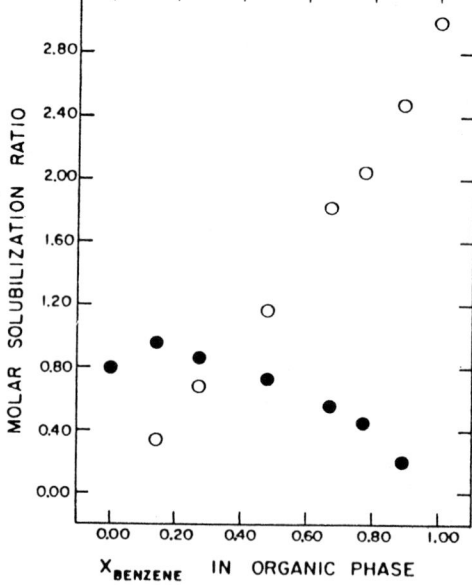

Figure II.2.32. Molecular solubilization ratio of benzene and hexane in 0.1 M N-hexadecylpyridinium chloride solution as a function of the composition of the bulk solubilizate: o hexane; o benzene. (© Academic Press, Inc., 1984. Reproduced with permission. Taken from reference 23).

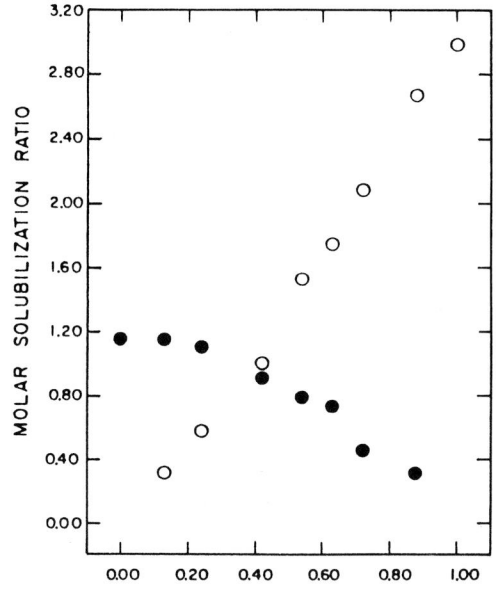

Figure II.2.33. Molecular solubilization ratio of benzene and cyclohexane in 0.1 M N-hexadecylpyridinium chloride solution as a function of the composition of the bulk solubilizate: ● cyclohexane; ○ benzene. (©Academic Press, Inc., 1984. Reproduced with permission. Taken from reference 23).

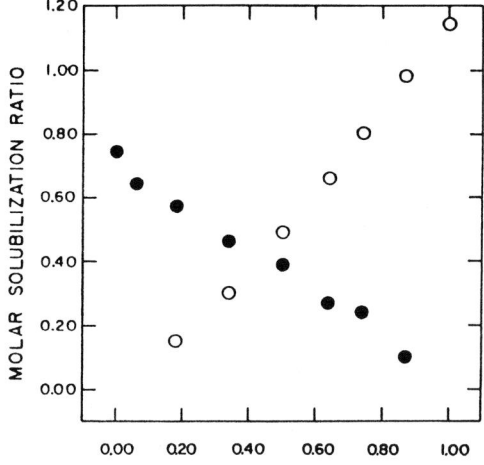

Figure II.2.34. Molecular solubilization ratio of hexane and cyclohexane in 0.1 M N-hexadecylpyridinium chloride solution as a function of the composition of the bulk solubilizate: ● hexane; ○ cyclohexane (©Academic Press, Inc., 1984. Reproduced with permission. Taken from reference 23).

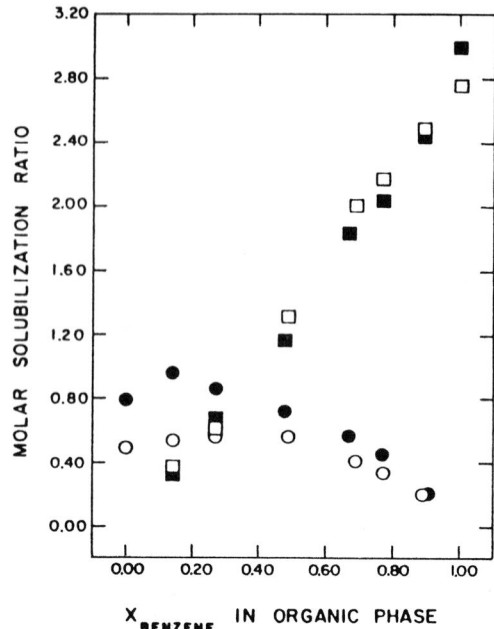

Figure II.2.35. Influence of concentration on the molar solubiliza-tion ratio of benzene and hexane in N-hexadecylpyridinium chloride solutions as a function of the composition of the bulk solubilizate phase. Circle: hexane; square: benzene; filled symbol: 0.1 M $C_{16}PyCl$; open symbol: 0.5 M $C_{16}PyCl$. (© Academic Press, Inc., 1984. Reproduced with permission. Taken from reference 23).

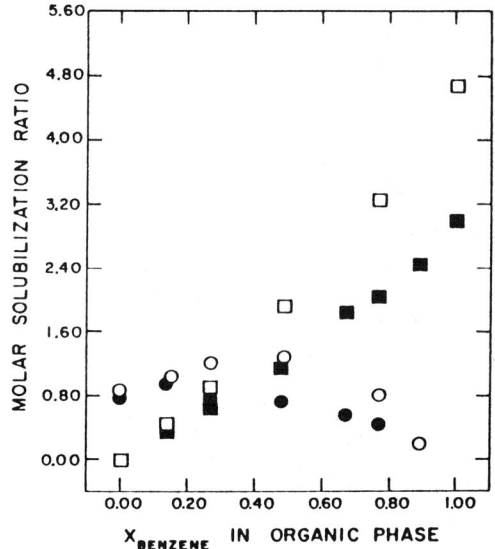

Figure II.2.36. Influence of electrolyte concentration on the molar solubilization ratio of benzene and hexane in 0.1 M N-hexadecyl-pyridinium chloride solutions as a function of the composition of the bulk solubilizate phase. Circle: hexane; square: benzene; filled symbol: no NaCl added; open symbol: 0.1 M NaCl. (© Academic Press, Inc., 1984. Reproduced with permission. Taken from reference 23).

REFERENCES Chapter II.2.

1) Yu, Z.I. and Zhao, G-X., J. Colloid Interface Sci., **130**, 414 (1989).
2) Zhang, L-H. and Zhao, G-X., J. Colloid Interface Sci., **127**, 353 (1989).
3) Trap, H.J.L. and Hermans, J.J., Proc. Kon. Ned. Acad. Wetensch. (B), **58**, 97 (1955).
4) Klimenko, N.A., Karmazina, T.V., Yaroshenko, N.A., Bartnitskii, A.E., and Aryamova, Zh.M., Kolloidnyi Zhurnal, **46**, 1112 (1984).
5) Malovikova, A., Hayakawa, K., and Kwak, J.C.T., J. Phys. Chem., **88**, 1930 (1984).
6) Skerjanc, J., Kogej, K., and Vesnaer, G., J. Phys. Chem., **92**, 6382 (1988).
7) Stellner, K.L., Amante, J.C., Scamehorn, J.F., and Harwell, J.H., J. Colloid Interface Sci., **123**, 186 (1988).
8) Adderson, J.E. and Taylor, H., J. Colloid Science, **19**, 495 (1964).
9) Sams, P.J., Wyn-Jones, E., and Rassing, J., Chem. Phys. Lett., **13**, 233 (1972).
10) Anacker, E.W., J. Phys. Chem., **62**, 41 (1958).
11) Heckman, K., Schwarz, R., and Strnad, J., J. Colloid Interface Sci., **120**, 114 (1987).
12) Rathman, J.F. and Scamehorn, J.F., Langmuir, **4**, 474 (1988).
13) Ray, A., J. Am. Chem. Soc., **91**, 6511 (1969).
14) Sudhölter, E.J.R. and Engberts, J.B.F.N., J. Phys. Chem., **83**, 1854 (1979).
15) Nusselder, J.J.H., de Groot, T.J., Trimbos, M., and Engberts, J.B.F.N., J. Org. Chem., **53**, 2423 (1988).
16) Engberts, J.B.F.N. and Nusselder, J.J.H., Pure Appl. Chem., **62**, 47 (1990).
17) Evans, D.F., Yamauchi, A., Roman, R., and Casassa, E.Z., J. Colloid Interface Sci., **88**, 89 (1982).
18) Stenius, P., Backlund, S., and Ekwall, P. in "Thermodynamics and Transport Properties of Organic Salts", Franzosina P. and Sanesi M. (Eds.), Pergamon Press (1980).
19) Kresheck, G.C. and Hargraves, W.A., J. Colloid Interface Sci., **48**, 481 (1974).
20) Smith, G.A., Christian, S.D., Tucker, E.E., and Scamehorn, J.F., J. Colloid Interface Sci., **130**, 254 (1989).
21) Lang, J. and Zana, R., J. Phys. Chem., **90**, 5258 (1986).
22) Eicke, H.F. in "Topics in Current Chemistry", Vol. 87, Springer Verlag, Berlin (1980).
23) Chaiko, M.A., Nagarajan, R., and Ruckenstein, E., J. Colloid Interface Sci., **99**, 168 (1984).
24) Smith, G.A., Christian, S.D., Tucker, E.E., and Scamehorn, J.F., Langmuir, **3**, 598 (1987).

PART III: NONIONIC SURFACTANTS

III.1 Alkylpolyoxyethylene glycol ethers

Table III.1.1. Melting point (MP, in oC), boiling point (BP, in oC at /mbar), refractive index (n_D^{50}), and density (d_4^{50}) of alkylpolyoxyethylene glycol ethers (C_nE_m).

Nonionic	MP (°C)	BP (°C/mbar)	n_D^{50}	d_4^{50}	Ref
C_2E_0		78.5	1.3493	0.764	76
C_2E_1		135.5	1.3962	0.903	76
C_2E_2		91/13	1.4150	0.963	76
C_2E_3		133/13	1.4258	0.995	76
C_2E_4		170/13	1.4325	1.018	76
C_2E_5		202/13	1.4371	1.031	76
C_2E_6		190/1.33	1.4405	1.041	76
C_2E_7		211/1.33	1.4430	1.049	76
C_2E_8		192/0.03	1.4456	1.055	76
C_2E_9		207/0.03	1.4463	1.059	76
C_2E_{10}		220/0.03	1.4473	1.062	76
C_6E_0	−51.6	65/13	1.4062	0.797	76
C_6E_1	−50.1	96/13	1.4170	0.865	76
C_6E_2	−40.2	135/13	1.4253	0.911	76
C_6E_3	−34.5	175/13	1.4211	0.943	76
C_6E_4	−18.1	206/13	1.4353	0.966	76
C_6E_6	1.2	208/1,2	1.4412	0.992	76

Table III.1.1. (continued)

Nonionic	MP (°C)	BP (°C/mbar)	n_D^{50}	d_4^{50}	Ref
C_8E_0	−16	196	1.4304[a]		76
C_8E_1	−16.2	99/3	1.4236	0.861	76
C_8E_2	−12.4	139/3	1.4296	0.898	76
C_8E_3	−10.4	171/3	1.4337	0.929	76
C_8E_4	−2.0	150/0.006	1.4370	0.951	76
C_8E_5	1.0	176/0.001	1.4398	0.947	76
C_8E_6	−3.4	194/0.001	1.4416	0.982	76
	7.5		1.4528[a]		67
$C_{10}E_0$	7	236	1.4375[a]		76
$C_{10}E_1$	5.5	78/0.24			76
$C_{10}E_2$	5.4	106/0.24			76
$C_{10}E_3$	6.8		1.4521[a]		68
$C_{10}E_4$	9.9	164/0.20	1.4392	0.934	76
$C_{10}E_6$	16.7	204/0.26	1.4433	0.968	76
	17.4		1.4521[a]		68
$C_{10}E_8$	26.3		1.4460	0.988	76
$C_{10}E_{10}$			1.4485	1.003	76

Table III.1.1. (continued)

Nonionic	MP (°C)	BP (°C/mbar)	n_D^{50}	d_4^{50}	Ref
$C_{12}E_0$	23	140/16	1.4320	0.815	76
$C_{12}E_1$	20.8	174/16	1.4335	0.852	76
$C_{12}E_2$	19.1	205/16	1.4367	0.883	76
$C_{12}E_3$	17.6	234/16	1.4391	0.908	76
$C_{12}E_4$	20.5	152/0.013	1.4413	0.928	76
$C_{12}E_5$	25.0	174/0.013	1.4431	0.943	76
$C_{12}E_6$	28.0	195/0.013	1.4446	0.955	76
$C_{12}E_7$		214/0.013	1.4460	0.967	76
$C_{12}E_8$		232/0.013	1.4471	0.976	76
$C_{12}E_9$		247/0.013	1.4480	0.985	76
$C_{12}E_{10}$		260/0.013	1.4489	0.992	76
$C_{12}E_{11}$		272/0.013	1.4496	0.999	76
$C_{12}E_{12}$		281/0.013	1.4503	1.006	76
$C_{14}E_0$	38.3	167/20			76
$C_{14}E_1$	31.7	132/0.03	1.4330[b]		76
$C_{14}E_2$	28.5	146/0.03	1.4352[b]		76
$C_{14}E_3$	25.2	156/0.024	1.4373[b]		76
$C_{14}E_4$	28.5	183/0.024	1.4390[b]		76
$C_{14}E_6$	35	206/0.03			76

Table III.1.1. (continued)

Nonionic	MP (°C)	BP (°C/mbar)	n_D^{50}	d_4^{50}	Ref
$C_{16}E_0$	49.3	189.5/20		0.818	76
$C_{16}E_1$	42.9	151/1.33	1.4355^b		76
$C_{16}E_2$	37.0	154/0.03	1.4373^b		76
$C_{16}E_3$	33.6	192/04	1.4390^b		76
$C_{16}E_4$	35.2	193/0.013	1.4407^b		76
$C_{16}E_5$	37.6	233/0.4			76
$C_{16}E_6$	36.4	234/0.07			76
$C_{16}E_7$	39				76
$C_{16}E_8$	41.5				76
$C_{16}E_9$	43				76
$C_{16}E_{12}$	45.5				76
$C_{16}E_{15}$	47				76
$C_{16}E_{21}$	49				76
$C_{18}E_0$	59	210/20	1.4390^b		76
$C_{18}E_1$	52		1.4381^b		76
$C_{18}E_2$	45	175/0.13	1.4393^b		76
$C_{18}E_3$	42	187/0.024	1.4407^b		76
$C_{18}E_4$	40.8	214/0.07	1.4416^b		76

a n_D^{25}

b n_D^{60}

Table III.1.2. Clouding temperature (T_c, in °C) and critical concentration (W_c, in wt%) of alkylpolyoxyethylene glycol ethers (C_nE_m).

Nonionic	W_c (Wt %)	T_c (°C)	Ref.
C_4E_1	27	44.5	61, 62
	29.5	48.7	66
C_5E_2	10	36	61, 62
C_6E_2	6.6	0	61, 62
C_6E_3		35	61
	10	39.6	62
	13.5	45.4	72
	13	44.7	73
		37	74, 79
C_6E_4	15	60	61, 62
		67.5	79, 74
C_6E_5		75	73, 79
C_6E_6		83	76
C_8E_3		5–8	17
	4	8	62
		6	61
C_8E_4		43	17
C_8E_4	5	35.5	62, 79, 76
		34	61
	7		70
	6.9	39.6	72
	7	40.3	73
C_8E_5		60.9	22
		60	31, 73
		55	76, 79
	7.5	55	62, 61
C_8E_6		75	15, 76
		68	74, 24
		71	15, 17
		76	67, 17
C_8E_8		~ 96	17
C_8E_{12}		~ 106	17
C_9E_4		32	75
C_9E_5		55	75
C_9E_6		75	75

Table III.1.2. (continued)

Nonionic	W_c (Wt %)	T_c (°C)	Ref.
$C_{10}E_3$		< 0	17, 68
$C_{10}E_4$		21	63
		20	61
	2.1		70
		18	74
		20.5	75
		19	76
$C_{10}E_5$		44	63, 17
		36	79, 74
		45.5	75
	3.5	40.5	72
		42	61, 63
$C_{10}E_6$		60	15, 64, 61, 74
		59–63	73
		61	74, 24
		57	74
		63	75
		61.5	76
		59	17
$C_{10}E_8$		84.5	75, 76
$C_{10}E_{10}$		95	76
$C_{11}E_4$		10.5	75
$C_{11}E_5$		37	75
$C_{11}E_6$		57.5	75
$C_{11}E_8$		82	75
$C_{12}E_3$		< 0	17
$C_{12}E_4$		5	17
	0.8		70
	0.8	3.6	4
		4	73, 17
		7.0	79, 74, 76, 25
		7.5	35
		9	75
$C_{12}E_5$		26	61
	1	26.5	4, 17
$C_{12}E_5$		31	35, 76, 74
		31.5	25
		25	74, 79
		30.5	75

Table III.1.2. (continued)

Nonionic	W_c (Wt %)	T_c (°C)	Ref.
$C_{12}E_6$	1.5	50	17, 31, 61
		53.0	25
		48	17, 74
	1.25	50.4	4, 73
	1.25		71
	1.65		7
	2.25		70
	2.2	48.0	72
		52	74
		55	74
		51.6	76
$C_{12}E_7$		67.2	76
		65	73
$C_{12}E_8$		80	17, 61
	3.5	75.3	70, 4
	3.2	75.5	70, 73
		77	73, 17
		78	74
		79	74, 76
		82	74
		79.5	75
	3	75	31
$C_{12}E_9$		87.5	74
		88	74
$C_{12}E_{10}$		94	74
		95	74
		98	74
		94.8	76
$C_{12}E_{11}$		100.3	76, 74
$C_{12}E_{12}$		>100	76, 74
		98	17
$C_{13}E_5$		27	75
$C_{13}E_6$		42	75
$C_{13}E_8$		72.5	75

Table III.1.2. (continued)

Nonionic	W_c (Wt %)	T_c (°C)	Ref.
$C_{14}E_3$		> 20	61, 17
$C_{14}E_5$		20	75
$C_{14}E_6$		42	24, 17,74, 61
		40	75
		45	12, 76
$C_{14}E_7$	1.9	57.6	73
$C_{14}E_8$		70.5	75
$C_{15}E_6$		37.5	75
$C_{15}E_8$		66	75
$C_{16}E_3$		< 20	17
$C_{16}E_4$		< 20	17
$C_{16}E_6$		37	17
	0.64		70
		38	73
$C_{16}E_6$	0.64		70
		32	74, 76
		35	24, 74
$C_{16}E_7$		53	76
		54	74
		55	74
$C_{16}E_8$	0.5	67	31, 73
		63	17
$C_{16}E_9$		75	76, 74
$C_{16}E_{12}$		92	17
$C_{18}E_8$	1.2		70

Table III.1.3. Clouding temperature (T_c, in °C) of alkylpolyoxy-ethylene glycol ethers (C_nE_m) in D_2O.

Nonionic	T_c (°C)	Ref.
C_8E_5	58	31
$C_{12}E_8$	71.5	31

Table III.1.4. Clouding temperature (T_c, in °C) of octylpenta-oxyethylene glycol ether (C_8E_5) at various NaCl concentrations (Taken from reference 22).

[NaCl] (mol/dm^3)	T_c (°C)
0	60.9
0.5	52.0
0.9	47.0
1.3	42.5

Table III.1.5. Critical micelle concentration (CMC, in mol/L) of alkylpolyoxyethylene glycol ethers (C_nE_m) at various temperatures and with various additives.

Nonionic	Additive	Temperature, °C						
		5	10	15	20	25	30	35
C_4E_6					0.8		0.76	
C_6E_3				$10.7*10^{-2}$		$10.0*10^{-2}$		$7.8*10^{-2}$
C_6E_4					$9.0*10^{-2}$			
C_6E_5					$9.0*10^{-2}$ $9.3*10^{-2}$			
C_6E_6					$7.4*10^{-2}$	$6.5*10^{-2}$		
C_8E_1						$4.9*10^{-3}$		
C_8E_3				$9.3*10^{-3}$		$7.5*10^{-3}$		
C_8E_4						$7.1*10^{-3a}$		
C_8E_4	+ 50 mol % SDS					$3.5*10^{-3}$		
C_8E_6				$11.9*10^{-3}$	$11.3*10^{-3}$	$9.9*10^{-3}$	$8.9*10^{-3}$	$7.7*10^{-3}$
C_8E_6	+ 40 mol % SDS					$3.33*10^{-3}$		
C_8E_9				$1.6*10^{-2}$		$1.3*10^{-2}$		$1.1*10^{-2}$
C_8E_{12}	+ 50 mol % SDS					$3.03*10^{-3}$		
C_9E_8						$3.0*10^{-3b}$		
$C_{10}E_3$		$11.2*10^{-4c}$		$7.3*10^{-4}$		$6.0*10^{-4}$		$5.6*10^{-4}$
$C_{10}E_4$			$9.8*10^{-4}$	$9.0*10^{-4}$	$6.4*10^{-4}$	$6.8*10^{-4}$		

Table III.1.5. (continued)

Nonionic		40	45	50	55	Method	Ref.
C_4E_6		0.71				ST	41,42,38
C_6E_3						ST	36
C_6E_4						ST,R	77
C_6E_5						ST	77
						RI	77
C_6E_6		$5.2*110^{-2}$				ST	38,41,42
C_8E_1						ST	77
C_8E_3						ST	36
C_8E_4	+ 50 mol % SDS					PFG NMR	20
C_8E_4			$6.7*10^{-3}$			ST	83
C_8E_6	+ 40 mol % SDS	$7.2*10^{-3}$				ST	36
						ST	77
C_8E_6						ST	83
C_8E_9						ST	36
C_8E_{12}	+ 50 mol % SDS					ST	83
C_9E_8						ST	59
$C_{10}E_3$						ST	36
$C_{10}E_4$						ST	77

214

Table III.1.5. (continued)

Nonionic	Additive	Temperature, °C						
		5	10	15	20	25	30	35
$C_{10}E_4$							7.3×10^{-4}	
$C_{10}E_5$		14.1×10^{-4}	11.8×10^{-4}	9.7×10^{-4}	9.0×10^{-4}; 8.6×10^{-4}; 9.5×10^{-4}; 9.25×10^{-4}	8.1×10^{-4}	7.6×10^{-4}	7.2×10^{-4}
$C_{10}E_6$				11.4×10^{-4}	9.2×10^{-4}; 9.5×10^{-4}; 9.6×10^{-4}	9.00×10^{-4}		6.6×10^{-4}
$C_{10}E_8$				1.4×10^{-3}	1.2×10^{-3}	1.0×10^{-3}	9.3×10^{-4}	
$C_{10}E_8$	0 mol % NaC[d]					1.02×10^{-3}		
	25 mol % NaC					1.09×10^{-3}		
	50 mol % NaC					1.25×10^{-3}		
	75 mol % NaC					2.58×10^{-3}		
	90 mol % NaC					4.52×10^{-3}		
	100 mol % NaC					10.3×10^{-3}		
$C_{10}E_9$				1.4×10^{-3}	1.3×10^{-3}		1.1×10^{-3}	
$C_{11}E_8$				4.0×10^{-4}	3.5×10^{-4}	3.0×10^{-4}	2.7×10^{-4}	
$C_{12}E_2$	in benzene					7.6×10^{-3}[b]		

215

Table III.1.5. (continued)

Nonionic		40	45	50	55	Method	Ref.
$C_{10}E_4$						ST	77
$C_{10}E_5$		$6.8*10^{-4}$				ST	77
						ST	77
						IS	77
						R	77
$C_{10}E_6$			$6.4*10^{-4}$			ST	77,36
						IS	77
						R	77
$C_{10}E_8$		$7.6*10^{-4}$				ST	59, 13
$C_{10}E_8$	0 mol % NaC					Pyr	23
	25 mol % NaC					Pyr	23
	50 mol % NaC					Pyr	23
	75 mol % NaC					Pyr	23
	90 mol % NaC					Pyr	23
	100 mol % NaC					Pyr	23
$C_{10}E_9$						ST	36
$C_{11}E_8$		$2.3*10^{-4}$				ST	59,13
$C_{12}E_2$	in benzene						45

Table III.1.5. (continued)

Nonionic	Additive	Temperature, °C						
		5	10	15	20	25	30	35
$C_{12}E_4$						4.6×10^{-5}; 6.4×10^{-5}		
$C_{12}E_5$			7.86×10^{-5}[e]; 9.0×10^{-5}		4.0×10^{-4}; 3.5×10^{-4}; 5.0×10^{-4}	5.78×10^{-5}; 6.4×10^{-5}		
$C_{12}E_6$				10.8×10^{-5}	8.2×10^{-5}; 7.8×10^{-5}; 10.0×10^{-5}	8.7×10^{-5}; 8.5×10^{-5}; 8.7×10^{-5}		7.2×10^{-5}
$C_{12}E_6$		13.5×10^{-5}	10.16×10^{-5}; 11.9×10^{-5}	9.89×10^{-5}	6.0×10^{-5}; 7.51×10^{-5}	6.45×10^{-5}; 6.99×10^{-5}	6.69×10^{-5}	
	5% v/v Ethanol	12.3×10^{-4}	11.5×10^{-5}	11.0×10^{-5}	10.5×10^{-5}	9.43×10^{-5}		
	10% v/v Ethanol	13.1×10^{-4}	12.1×10^{-5}	11.3×10^{-5}	11.3×10^{-5}	11.6×10^{-5}		
	15% v/v Ethanol	13.0×10^{-4}	12.0×10^{-4}	10.7×10^{-5}	11.9×10^{-5}	13.8×10^{-5}		
$C_{12}E_7$						8.0×10^{-5}[i]		

Table III.1.5. (continued)

Nonionic	40	45	50	55	Method	Ref.
$C_{12}E_4$					Cal	4
$C_{12}E_5$			4.77×10^{-5}[f,g]		ST	4
					ST	33
					ST	34
					IS	77
					R	77
$C_{12}E_6$					ST	77,36
					ST	77
					IS	77
					R	77
			5.17×10^{-5}[gh]		ST	51
					Pyr	65,4,7
					Cal	1
					GF	33
$C_{12}E_6$ 5 % v/v Ethanol					GF	37
$C_{12}E_6$ 10 % v/v Ethanol					GF	37
$C_{12}E_6$ 15 % v/v Ethanol					GF	37
$C_{12}E_7$					ST	77
				2×10^{-5}		38,43

Table III.1.5. (continued)

Nonionic	Additive	10	15	20	25	30
$C_{12}E_7$	+ 0 M Urea	$8.0*10^{-5}$			$5.0*10^{-5}$	
	+ 3 M Urea	$12.0*10^{-5}$			$6.3*10^{-5}$	
	+ 6 M Urea	$20.8*10^{-5}$			$12.5*10^{-5}$	
$C_{12}E_8$		$14.36*10^{-5}$ $15.6*10^{-5}$	$9.7*10^{-5}$	$8.3*10^{-5}$	$7.1*10^{-5}$ $10.9*10^{-5}$ $9.04*10^{-5}$ $10.9*10^{-5}$	$6.9 10^{-5}$
$C_{12}E_8$	50 mol % $C_{18}N(CH_3)_3Cl$ in 2.4 mM NaCl				$4.25*10^{-5}$	
	50 mol % $C_{20}N(CH_3)_3Cl$				$3.14*10^{-5}$	
	50 mol % SDS in 0.5M NaCl				$4.57*10^{-5}$	
$C_{12}E_9$					$10.0*10^{-5}i$	
$C_{12}E_{12}$					$14.0*10^{-5}i$	
$C_{12}E_{14}$					$5.5*10^{-5}$	
$C_{12}E_{23}$					$6.0*10^{-5}$	

Temperature, °C

Table III.1.5. (continued)

Nonionic	40	45	50	55	Method	Ref.
$C_{12}E_7$ + 0 M Urea		$2.8*10^{-5}$				77,81
+ 3 M Urea		$3.4*10^{-5}$				77,81
+ 6 M Urea		$5.6*10^{-5}$				77,81
$C_{12}E_8$	$5.8*10^{-5}$				ST	59,13
	$7.30*10^{-5}$				CAL	4
			$6.74*10^{-5}$ g,j		ST	33
						34
$C_{12}E_8$ 50 mol % $C_{18}N(CH_3)_3Cl$ in 2.4 mM NaCl					ST	83
50 mol % $C_{20}N(CH_3)_3Cl$					ST	83
50 mol % SDS in 0.5M NaCl					ST	83
$C_{12}E_9$					ST	77
$C_{12}E_{12}$					ST	77
$C_{12}E_{14}$				$2.5*10^{-5}$		43
$C_{12}E_{23}$				$3.0*10^{-5}$		43

220

Table III.1.5. (continued)

Nonionic	Additive	Temperature, °C					
		5	10	15	20	25	30
$C_{12}E_{23}$	0 mol % k Ethanol					$9.1*10^{-5}$	
	1.6 mol % Ethanol					$9.9*10^{-5}$	
	3.3 mol % Ethanol					$12.4*10^{-5}$	
	5.2 mol % Ethanol					$16.5*10^{-5}$	
	7.0 mol % Ethanol					$24.0*10^{-5}$	
	1.1 mol % Dioxane					$10.7*10^{-5}$	
	2.3 mol % Dioxane					$16.5*10^{-5}$	
	3.6 mol % Dioxane					$21.5*10^{-5}$	
	5.0 mol % Dioxane					$29.7*10^{-5}$	
$C_{12}E_{30}$	0 M Urea		$9.0*10^{-5}$			$8.0*10^{-5}$	
	3 M Urea		$36.0*10^{-5}$			$16.0*10^{-5}$	
	6 M Urea		$72.0*10^{-5}$			$25.0*10^{-5}$	
$C_{12}E_{31}$						$8.0*10^{-5}$	
$C_{13}E_6$	In benzene					$2.6*10^{-3}i$	
$C_{13}E_8$				$3.2*10^{-5}$	$2.8*10^{-5}$	$2.7*10^{-5}$ $11.2*10^{-5}$	$2.5*10^{-5}$

Table III.1.5. (continued)

Nonionic		40	45	50	55	Method	Ref.
$C_{12}E_{23}$	0 mol % Ethanol						77
	1.6 mol % Ethanol						77
	3.3 mol % Ethanol						77
	5.2 mol % Ethanol						77
	7.0 mol % Ethanol						77
	1.1 mol % Dioxane						77
	2.3 mol % Dioxane						77
	3.6 mol % Dioxane						77
	5.0 mol % Dioxane						77
$C_{12}E_{30}$	0 M Urea		$4.8*10^{-5}$				77,81
	3 M Urea		$6.7*10^{-5}$				77,81
	6 M Urea		$9.5*10^{-5}$				77,81
$C_{12}E_{31}$					$4.0*10^{-5}$		38,43
$C_{13}E_6$	in benzene	$2.0*10^{-5}$					45
$C_{13}E_8$						LS	59 77
$C_{13}E_{10}$						LS	77

Table III.1.5. (continued)

Nonionic	Additive	Temperature,°C					
		5	10	15	20	25	30
$C_{13}E_{10}$						$11.7*10^{-5}$	
$C_{13}E_{12}$						$20.2*10^{-5}$	
$C_{14}E_6$						$1.04*10^{-5}$	
$C_{14}E_8$				$1.1*10^{-5}$	$9.8*10^{-6}$	$9.0*10^{-6}$	$8.0*10^{-6}$
$C_{15}E_8$				$4.1*10^{-6}$	$3.7*10^{-6}$	$3.5*10^{-6}$	$3.2*10^{-6}$
$C_{16}E_6$						$1.0*10^{-6}$ $1.7*10^{-6}$	
$C_{16}E_7$						$1.7*110^{-6}$ $2.1*10^{-6}$	
$C_{16}E_9$					$3.6*10^{-5}$ $3.9*10^{-5}$ $3.1*10^{-5}$		
$C_{16}E_{12}$						$2.3*10^{-6}$	
$C_{16}E_{15}$						$3.1*10^{-6}$	
$C_{16}E_{21}$						$3.9*10^{-6}$	
$C_{16}E_{30}$	0 M Urea		$2.0*10^{-5}$			$1.1*10^{-5}$	
	3 M Urea		$3.2*10^{-5}$			$1.6*10^{-5}$	
	6 M Urea		$4.0*10^{-5}$			$2.0*10^{-5}$	

Table III.1.5. (continued)

Nonionic		$40\,^{\circ}C$	Method	Ref.
$C_{13}E_{12}$			LS	77
$C_{14}E_6$			ST	77
$C_{14}E_8$		$7.2*10^{-6}$	ST	59
$C_{15}E_8$		$3.0*10^{-6}$	ST	59
$C_{16}E_6$			ST	77
			ST	77
$C_{16}E_7$			ST	38,44,77
			ST	38,44
$C_{16}E_9$			ST	77
			IS	77
			IS	77
$C_{16}E_{12}$			ST	38,44,77
$C_{16}E_{15}$			ST	38,44,77
$C_{16}E_{21}$			ST	38,44,77
$C_{16}E_{30}$	0 M Urea			77,81
	3 M Urea			77,81
	6 M Urea			77,81

a unit mole/kg; b temp. not specified; c at 6.8°C; d NaC is sodium cholate; e at 11.75°C; f at 47.2°C; g minimal CMC as function of temp; h at 48.3°C; i at 23°C; j at 51.7°C; k with respect to solvent.

Table III.1.6. Relation between the critical micelle concentration (CMC) and the number of carbon atoms (n) in the hydrophobe and the number of ethylene oxide units (m) in the hydrophile of alkylpoly-oxyethylene glycol ethers.

n	m	temp (°C)	Relation	Ref
12	–	23	$\log CMC = -4.4 + 0.046\ m$	38
16	–	25	$\log CMC = -5.9 + 0.024\ m$	38
–	6	25	$\log CMC = 1.82 - 0.49\ n$	38
–	8	15	$\log CMC = 2.18 - 0.51\ n$	13
–	8	20	$\log CMC = 2.07 - 0.51\ n$	13
–	8	25	$\log CMC = 1.89 - 0.50\ n$	13
–	8	30	$\log CMC = 1.86 - 0.50\ n$	13
–	8	40	$\log CMC = 1.66 - 0.48\ n$	13

Table III.1.7. Gibbs free energy of micellization (ΔG°_m, in kJ/mol) of alkylpolyoxyethylene glycol ethers (C_nE_m) at various temperatures and with various additives

Nonionic	Additive	Temperature (°C)				Method	Ref.
		20	25	30	35		
C_6E_3			-15.6			ST	36
C_8E_3			-22.1			ST	36
C_8E_6			-21.3			ST	36
C_8E_9			-20.8			ST	36
$C_{10}E_3$			-28.3			ST	36
$C_{10}E_6$			-27.3			ST	36
$C_{10}E_8$			-27.0			ST	13
$C_{10}E_9$			-26.4			ST	36
$C_{11}E_8$			-30.0			ST	13
$C_{12}E_2$			-25.6				14
$C_{12}E_3$			-24.4				14
$C_{12}E_4$			-23.9				14
$C_{12}E_5$			-23.9				14
			-24.19			CA	33

Table III.1.7. (continued)

Nonionic	Additive	Temperature (°C)						Method	Ref.
		5	10	15	20	25	30		
$C_{12}E_6$						-23.92		CAL	33
						-33.1		ST	36
$C_{12}E_6$	0% v/v Ethanol	-29.5	-30.7	-31.7	-32.9	-33.6	– 34.3	GF	37
	0% v/v Ethanol	-28.7[b]						GF	37
	5% v/v Ethanol	-30.0	-30.7	-31.3	-32.0	-32.8		GF	37
	5% v/v Ethanol	-28.7[b]						GF	37
	10% v/v Ethanol	-29.8	-30.5	-31.2	-31.8	-32.2		GF	37
	10% v/v Ethanol	-28.9[b]						GF	37
	15% v/v Ethanol	-29.7	-30.4	-31.2	-31.5	-31.7		GF	37
	15% v/v Ethanol	-28.9[b]						GF	37
$C_{12}E_6$	0% w/w $HCONH_2$					-33.0			46
	55% w/w $HCONH_2$					-25.2			46
	100% w/w $HCONH_2$					-16.6			46
$C_{12}E_7$						-23.3			14
						-22.6			14
$C_{12}E_8$						23.08		CA	33

Table III.1.7. (continued)

Nonionic	Additive	Temperature (°C)	Method	Ref.
		25		
$C_{12}E_8$	0% w/w $HCONH_2$ 55% w/w $HCONH_2$ 100% w/w $HCONH_2$	-32.0 -24.3 -16.2		46 46 46
$C_{12}E_8$		-33.6		13
$C_{13}E_8$		-36.0		13
$C_{14}E_8$		-38.7		13
$C_{15}E_8$		-41.1		13
$C_{16}E_{30}$		226		21

a at 11.75°C;

b at 0°C

Table III.1.8. Enthalpy of micellization (ΔH, in kJ/mol) of alkyl-polyoxyethylene glycol ethers (C_nE_m) at various temperatures and with various additives.

Nonionic	Additive	Temperature (°C)					Method	Ref.
		10	15	20	25	30		
C_6E_3					13.8		ST	36
C_8E_3					15.5		ST	36
C_8E_6					18.0		ST	36
					14.6			21
C_8E_9					13.8			21
					15.5		ST	36
$C_{10}E_3$					9.96			21
					9.62		ST	36
$C_{10}E_6$					17.6		ST	36
					20.3			21
$C_{10}E_8$					18.5		ST	13
$C_{10}E_9$					9.0			21
					6.7		ST	36
$C_{11}E_8$					16.7		ST	13
$C_{12}E_2$					4.2			14
$C_{12}E_3$					5.9			14
$C_{12}E_4$					8.8			14
$C_{12}E_5$					9.9			14,34
		21.47[a]			13.5		CA	33

Table III.1.8. (continued)

Nonionic	Additive	Temperature (°C)								Method	Ref.
		5	10	15	20	25	30	35	40		
$C_{12}E_6$			24.30			14.8				CA	33
						16.3				ST	36
						28.8					21
$C_{12}E_6$	0% v/v Ethanol	35.4	30.4	25.6	20.9	16.4	12.1			GF	37
	0% v/v Ethanol	40.5[b]								GF	37
	5% v/v Ethanol	16.5	9.49	2.9	-3.6	-9.79				GF	37
	5% v/v Ethanol	23.2[b]								GF	37
	10% v/v Ethanol	16.2	9.79	3.6	-2.4	-8.24				GF	37
	10% v/v Ethanol	22.6[b]								GF	37
	15% v/v Ethanol	21.8	10.6	-0.3	-10.8	-21.0				GF	37
	15% v/v Ethanol	33.1[b]								GF	37
$C_{12}E_6$	0% w/w $HCONH_2$					16.3					46
	55% w/w $HCONH_2$					1.5					46
	100% w/w $HCONH_2$					-3.6					46
$C_{12}E_7$						12.5					14
						25.4					21
						21					47
$C_{12}E_8$			27.33			13.2				CA	14,34
						16.34			7.90		33
$C_{12}E_8$	0% w/w $HCONH_2$					18.6					46
	55% w/w $HCONH_2$					2.1					46
	100% w/w $HCONH_2$					-3.1					46
$C_{12}E_8$						15.3					13
$C_{12}E_{30}$						14.4					21

Table III.1.8. (continued)

Nonionic	Additive	T, °C 25	Method	Ref.
$C_{13}E_8$		13.7		13
$C_{14}E_8$		12.6		13
$C_{15}E_8$		11.4		13
$C_{16}E_{30}$		29.0		21

a at 11.75°C;

b at 0°C

Table III.1.9. Entropy of micellization (ΔS°_m, in J mol^{-1}K^{-1}) of alkylpolyoxyethylene glycol ethers (C$_n$E$_m$) at various temperatures and with various additives.

Nonionic	Additive	Temperature, °C							Method	Ref.
		5	10	15	20	25	30	35		
C$_8$E$_6$						121				21
C$_8$E$_9$						116				21
C$_{10}$E$_3$						128				21
C$_{10}$E$_6$						160				21
C$_{10}$E$_8$						152			ST	13
C$_{10}$E$_9$						119				21
C$_{11}$E$_8$						156			ST	13
C$_{12}$E$_2$						100				14
C$_{12}$E$_3$						102				14
C$_{12}$E$_4$						110				14
C$_{12}$E$_5$						113				14
						126			CA	33
C$_{12}$E$_6$						130			CA	33
						208				21
C$_{12}$E$_6$	0% v/v Ethanol	234	216	199	184	168	153		GF	37
	0% v/v Ethanol	254							GF	37
	5% v/v Ethanol	167	142	117	96	75			GF	37
	5% v/v Ethanol	188^b							GF	37

232

Table III.1.9. (continued)

Nonionic	Additive	5	10	15	20	25	30	Method	Ref.
$C_{12}E_6$	10% v/v Ethanol	167	142	121	184	79		GF	37
	10% v/v Ethanol	188[b]						GF	37
	15% v/v Ethanol	185	142	109	71	36		GF	37
	15% v/v Ethanol	226[b]						GF	37
$C_{12}E_6$	0% w/w $HCONH_2$					165			46
	55% w/w $HCONH_2$					94			46
	100% w/w $HCONH_2$					44			46
$C_{12}E_7$						120			14
						201			21
$C_{12}E_8$						120			14
						132		CA	33
$C_{12}E_8$	0% w/w $HCONH_2$					170			46
	55% w/w $HCONH_2$					88			46
	100% w/w $HCONH_2$					44			46
$C_{12}E_8$						164			13
$C_{12}E_{30}$						161			21
$C_{13}E_8$						167			13
$C_{14}E_8$						172			13
$C_{15}E_8$						176			13

[a] at 11.75°C; [b] at 0°C

Table III.1.10. Aggregation number of alkylpolyoxyethylene glycol ethers (C_nE_m) at various temperatures and concentrations and with various additives.

Nonionic	Concentration (mM)	Additive	Temperature (°C)											Method	Ref
			5	10	15	20	25	30	35	40	45	50	55		
C_8E_4							85[1]							PGNMR	20
C_8E_5							65 ±4[1,4]							QENS	31
C_8E_6						30[2]	32	41		51		82	210[3]	LS	24,77
$C_{10}E_6$							73		260		640	1330		LS	24
$C_{10}E_8$		+ 0 NaC[5]				69								FL	23
		+ 0.25 NaC				55								FL	23
		+ 0.75 NaC				42								FL	23
		+ 0.90 NaC				26								FL	23
						16								FL	23
		+ 2.3% decane					83								10
		+ 4.9% decane					90								10
		+ 3.4% decanol					105								10
		+ 8.5% decanol					89								10
$C_{10}E_{11}$					109		65						10		10
$C_{12}E_4$				165 ±62[6]										FPD	4
$C_{12}E_5$			111 ±35[6]				112 ±10							VPO	4
							3300							LS	12
$C_{12}E_6$	8					254								FL	1
	10.3					268								FL	1
	28					295								FL	1
	50					318								FL	1

Table III.1.10. (continued)

Nonionic	Concentration (mM)	Additive	Temperature (°C)											Method	Ref
			5	10	15	20	25	30	35	40	45	50	55		
$C_{12}E_6$	100					345								FL	1
	200					333								FL	1
					140	186²	400	710	1400	2220	4000			LS	7,77
					116		103			110				VPO	4
			99⁶											FPD	4
			109	104		255	555	910	1864					LS	5
			102	111	111									SE	6
				133		133								LS	8
							375	761	1466					MO	9
							330		1571					LS	9
				140^{1,8}										NS	31
				129										FL	11
											1330			LS	12
$C_{12}E_8$			38⁶				39			39		39	42⁹	VPO	4
							90					39	LS	12
						95^{1,4}								QENS	31
						55^{1,4}	123							PGNMR	86 77,80

Table III.1.10. (continued)

Nonionic	Concentration Additive (mM)	Temperature (°C) 20	25	30	35	Method	Ref
$C_{12}E_8$	0.30 N NaCl		127				77,80
	0.50 N NaCl		140				77,80
	0.30 N $CaCl_2$		120				77,80
	0.50 N $CaCl_2$		149				77,80
	0.30 N Na_3citrate		176				77,80
	0.50 N Na_3citrate		158				77,80
	0.30 N Na_2SO_4		310				77,80
	0.50 N Na_2SO_4		856				77,80
$C_{12}E_{12}$	–		81				77,80
	0.30 N NaCl		79				77,80
	0.50 N NaCl		69				77,80
	0.30 N $CaCl_2$		72				77,80
	0.50 N $CaCl_2$		63				77,80
	0.30 N Na_3citrate		64				77,80
	0.50 N Na_3citrate		66				77,80

Table III.1.10. (continued)

Nonionic	Concentration (mM)	Additive	Temperature (°C) 20	25	30	35	Method	Ref
$C_{12}E_{12}$		0.30 N Na_2SO_4		73				77,80
		0.50 N Na_2SO_4		73				77,80
$C_{12}E_{18}$		—		51				77,80
		0.30 N NaCl		53				77,80
		0.50 N NaCl		46				77,80
		0.30 N $CaCl_2$		45				77,80
		0.50 N $CaCl_2$		44				77,80
		0.30 N Na_3citrate		39				77,80
		0.50 N Na_3citrate		40				77,80
		0.30 N Na_2SO_4		46				77,80
		0.50 N Na_2SO_4		50				77,80
$C_{12}E_{23}$		—		40				77,80
		0.30 N NaCl		35				77,80
		0.50 N NaCl		33				77,80
		0.30 N $CaCl_2$		33				77,80

Table III.1.10. (continued)

Nonionic	Concentration (mM)	Additive	Temperature (°C)							Method	Ref
			10	15	20	25	30	35	40		
$C_{12}E_{23}$		0.50 N $CaCl_2$				32					77,80
		0.30 N Na_3citrate				28					77,80
		0.50 N Na_3citrate				28					77,80
		0.30 N Na_2SO_4				39					77,80
		0.50 N Na_2SO_4				37					77,80
$C_{13}E_8$						71				LS	77
$C_{13}E_{10}$						61				LS	77
						92					77,80
		0.30 N NaCl				110					77,80
		0.50 N NaCl				141					77,80
$C_{13}E_{10}$						55				LS	77
$C_{14}E_6$						3100	5400	7500		LS	24,38
								7500	11.70	LS	24,77
$C_{16}E_6$						2430				LS	77,41
						10500[15]	13.300[13]	16.00[14]		LS	77,24
$C_{16}E_7$				162	249	594				LS	24,77

Table III.1.10. (continued)

Nonionic	Concentration Additive (mM)	Temperature (°C)									Method	Ref
		15	20	25	30	35	40	45	50	55		
$C_{16}E_8$		220	150[1,4]	240		271	1010[16] 2800[17]				QENS LS	31 40,77
$C_{16}E_9$				219 279				243	549	1000[19]	LS	42,77
$C_{16}E_{12}$				152							LS	41,77
$C_{16}E_{21}$				70								

1) in D_2O; 2) at 18°C; 3) at 60°C; 4) temp. not specified; 5) NaC = sodium cholate; 6) at 0°C; 7) at 42°C; 8) at 11°C; 9) at 70°C; 10) at 5.4°C; 11) at 25.2°C; 12) at 49.1°C; 13) at 32°C; 15) at 28°C; 16) at 39°C; 17) 44.3°C; 18) at 7°C; 19) at 53.5°C.

Table III.1.11. Hydrodynamic radius of gyration (in nm) of alkyl-polyoxyethylene glycol ethers (C_nE_m) at various temperatures and concentrations and with various additives.

Nonionic	Concentration (wt%)	Additive	Temperature (°C)											Method	Ref
			5	10	15	20	25	30	35	40	45	50	55		
C_8E_4						2.5^4								PGNMR	20
							2.87^1							LS	73
C_8E_5	3.2						2.2^1			2.6^1				QENS	31
	7						2.4^1			3.0^1			3.8^1	QENS	31
	15						3.5^1			3.6^1				QENS	31
		0 M NaCl					2.3							QEIS	22
		1.3 M NaCl					2.55							QEIS	22
$C_{12}E_8$	2.5						2.9^1				3.1^1		$4.0^{1,9}$	QENS	31
	10		$3.03^{10,1}$				$3.10^{11,1}$	4.1^1			4.2^1			QENS	31
												$3.05^{12,1}$		PGNMR	86
						3.4^4								LS	73

1) in D_2O; 2) at 18°C; 3) at 60°C; 4) temp. not specified; 5) NaC = sodium cholate; 6) at 0°C; 7) at 42°C; 8) at 11°C; 9) at 70°C; 10) at 5.4°C; 11) at 25.2°C; 12) at 49.1°C; 13) at 32°C; 14) at 34°C; 15) at 28°C; 16) at 39°C; 17) 44.3°C; 18) at 7°C; 19) at 53.5°C.

Table III.1.12. Micellar weight of alkylpolyoxyethylene glycol ether (C_nE_m) micelles at various temperatures and with various additives.

Nonionic	Conc.	Additive	Temperature, °C				
			10	15	20	25	30
C_8E_4					$2.5*10^4$ [4]		
C_8E_6					$1.2*10^4$		$1.4*10^4$
$C_{10}E_6$						$3.1*10^4$	
$C_{12}E_6$					$5*10^4$ [4]		
				$6.3*10^4$	$8.4*10^4$ [2]	$1.8*10^5$	$3.2*10^5$
$C_{12}E_8$		—			$6.5*10^4$ [4]		
		0.30 N NaCl			$6.8*10^4$		
		0.50 N NaCl			$7.0*10^4$		
		0.30 N $CaCl_2$			$7.7*10^4$		
		0.50 N $CaCl_2$			$6.6*10^4$		
		0.30 N Na_3citrate			$8.2*10^4$		
		0.50 N Na_3citrate			$9.7*10^4$		
		0.30 N Na_2SO_4			$8.7*10^4$		
		0.30 N Na_2SO_4			$1.7*10^5$		
		0.50 N Na_2SO_4			$4.7*10^5$		
$C_{12}E_{12}$		—			$5.9*10^4$		
		0.30 N NaCl			$5.8*10^4$		
		0.50 N NaCl			$5.0*10^4$		
		0.30 N $CaCl_2$			$5.2*10^4$		
		0.50 N $CaCl_2$			$4.7*10^4$		
		0.30 N Na_3citrate			$4.6*10^4$		
		0.50 N Na_3citrate			$4.8*10^4$		
		0.30 N Na_2SO_4			$5.4*10^4$		
		0.50 N Na_2SO_4			$5.4*10^4$		

Table III.1.12. (continued)

	35	40	45	50	55	60	Method	Ref.
C_8E_4							LS	73
C_8E_6		$2.0*10^4$		$3.3*10^4$		$8.2*10^4$	LS	24
$C_{10}E_6$	$1.1*10^5$		$2.7*10^5$	$5.6*10^5$			LS	24
$C_{12}E_6$	$6.3*10^5$	$1*10^6$	$1.8*10^6$				LS	73
		7					LS	7,77
$C_{12}E_8$	–						LS	73
0.30 N NaCl								77,80
0.50 N NaCl								77,80
0.30 N CaCl$_2$								77,80
0.50 N CaCl$_2$								77,80
0.30 N Na$_3$Citrate								77,80
0.50 N Na$_3$Citrate								77,80
0.30 N Na$_2$SO$_4$								77,80
0.50 N Na$_2$SO$_4$								77,80
$C_{12}E_{12}$	–							
0.30 N NaCl								77,80
0.50 N NaCl								77,80
0.30 N CaCl$_2$								77,80
0.50 N CaCl$_2$								77,80
0.30 N Na$_3$Citrate								77,80
0.50 N Na$_3$Citrate								77,80
0.30 N Na$_2$SO$_4$								77,80
0.50 N Na$_2$SO$_4$								77,80

Table III.1.12. (continued)

Nonionic	Conc.	Additive	Temperature 25 °C	Method	Ref.
$C_{12}E_{18}$		—	$5.0*10^4$		77,80
		0.30 N NaCl	$5.3*10^4$		77,80
		0.50 N NaCl	$4.5*10^4$		77,80
		0.30 N $CaCl_2$	$4.5*10^4$		77,80
		0.50 N $CaCl_2$	$4.3*10^4$		77,80
		0.30 N Na_3citrate	$3.8*10^4$		77,80
		0.50 N Na_3citrate	$4.0*10^4$		77,80
		0.30 N Na_2SO_4	$4.6*10^4$		77,80
		0.50 N Na_2SO_4	$5.0*10^4$		77,80
$C_{12}E_{23}$		—	$4.9*10^4$		77,80
		0.30 N NaCl	$4.2*10^4$		77,80
		0.50 N NaCl	$4.0*10^4$		77,80
		0.30 N $CaCl_2$	$4.0*10^4$		77,80
		0.50 N $CaCl_2$	$3.9*10^4$		77,80
		0.30 N Na_3citrate	$3.5*10^4$		77,80
		0.50 N Na_3citrate	$3.4*10^4$		77,80
		0.30 N Na_2SO_4	$4.7*10^4$		77,80
		0.50 N Na_2SO_4	$4.5*10^4$		77,80
$C_{13}E_8$			$3.9*10^4$	LS	77
$C_{13}E_{10}$		—	$3.9*10^4$	LS	77
		0.30 N NaCl	$5.9*10^4$		77,80
		0.50 N NaCl	$7.0*10^4$		77,80
			$9.0*10^4$		77,80
$C_{13}E_{12}$			$4.0*10^4$	LS	77
$C_{13}E_{15}$		—	$3.3*10^4$		77,80
		0.30 N NaCl	$4.1*10^4$		77,80
		0.50 N NaCl	$4.0*10^4$		77,80

Table III.1.12. (continued)

Nonionic	Conc.	Additive	Temperature, °C			
			15	20	25	30
$C_{13}E_{22}$		-			$3.3*10^4$	
		0.30 N NaCl			$3.4*10^4$	
		0.50 N NaCl			$3.3*10^4$	
$C_{14}E_6$					$1.5*10^6$	$2.6*10^6$
$C_{16}E_6$					$1.2*10^6$ 15 $5.1*10^6$	$6.8*10^6$ 13
$C_{16}E_7$			$9.0*10^4$	$1.4*10^5$	$3.3*10^5$	
$C_{16}E_8$			$1.3*10^5$		$1.4*10^5$	
$C_{16}E_9$					$1.4*10^5$	
$C_{16}E_{12}$					$1.2*10^5$	
$C_{16}E_{21}$					$8*10^4$	

Table III.1.12. (continued)

	35	40	45	50	55	60	Method	Ref.
$C_{13}E_{22}$								77,80 77,80 77,80
$C_{14}E_6$	$3.6*10^6$	$5.6*10^6$					LS	24,77
$C_{16}E_6$	$8.4*10^6$ [14]						LS LS	77 77
$C_{16}E_7$		$6.0*10^5$ [16]	$1.7*10^6$ [17]				LS	24,77
$C_{16}E_8$	$1.6*10^5$						LS	24,77
$C_{16}E_9$			$1.8*10^5$	$3.5*10^5$	$6*10^5$ [19]		LS	42,77
$C_{16}E_{12}$								41,77
$C_{16}E_{21}$								41,77

1) in D_2O; 2) at 18°C; 3) at 60°C; 4) temp. not specified; 5) NaC = sodium cholate; 6) at 0°C; 7) at 42°C; 8) at 11°C; 9) at 70°C; 10) at 5.4°C; 11) at 25.2°C; 12) at 49.1°C; 13) at 32°C; 14) at 34°C; 15) at 28°C; 16) at 39°C; 17) 44.3°C; 18) at 7°C; 19) at 53.5°C.

Table III.1.13. Second virial coefficient (in cm^3/mol) of alkylpoly-oxyethylene glycol ether (C_nE_m)-water systems.

Nonionic	Temperature (°C)					Meth	Ref
	0	25	40	50	70		
$C_{12}EO_4$	$(-3.8 \pm 0.4)*10^5$					VPO	4
$C_{12}EO_5$	$(7.7 \pm 1.8)*10^3$	$(-3.6 \pm 0.6)*10^5$				VPO	4
$C_{12}EO_6$	$(7.4 \pm 5.2)*10^4$	$(-3.3 \pm 1.2)*10^5$	$(-4.2 \pm 0.7)*10^5$			VPO	4
$C_{12}E_8$	$(3.8 \pm 0.2)*10^4$	$(2.5 \pm 0.1)*10^4$	$(1.6 \pm 0.1)*10^4$	$(1.2 \pm 0.3)*10^4$	$(-8.1 \pm 0.4)*10^4$	VPO	4

Table III.1.14. Surface tension (γ, in mN/m) of alkylpolyoxy-ethylene glycol ethers (C_nE_m) at various temperatures and with various additives.

Nonionic	Additive	Temperature (°C)					Method	Ref.
		15	20	25	30	35		
C_9E_8				36.5				59
$C_{10}E_8$		37.3	36.5	35.5	34.4	33.5		13,59
$C_{11}E_8$		36.8	36.0	35.3	34.0	33.3		13,59
$C_{12}E_5$				30[b]				2
$C_{12}E_7$				34[b]				2
$C_{12}E_8$		36.0	35.0	34.3	33.4	32.7		13,59
$C_{12}E_8$	0.2 mol % SDS			33.8				83
	0.5 mol % SDS			32.9				83
	0.9 mol % SDS			31.8				83
$C_{12}E_9$				37				2
$C_{12}E_{12}$				40				2
$C_{13}E_8$		36.5	35.8	34.9	33.8	32.8		13,59
$C_{14}E_8$		35.7	35.0	34.0	33.2	32.3		13,59
$C_{15}E_8$		36.0	35.2	34.3	33.5	32.6		13,59

a   at 23.5°C;   b   temp. not specified.

Table III.1.15. Surface excess concentration (in 10^{-10} mol/cm^2) of alkylpolyoxyethylene glycol ethers (C_nE_m) at various temperatures

Nonionic	Temperature (°C)								Method	Ref.
	10	15	20	25	30	35	40	45		
C_6E_6				2.7						41
C_8E_4				3.3						19
C_9E_8				2.20						59
$C_{10}E_6$				3.0[a]						50
$C_{10}E_8$		2.26	2.38	2.42	2.48		2.53			59
$C_{11}E_8$		2.36	2.46	2.50	2.56		2.61			13,59
$C_{12}E_6$				3.7						50
$C_{12}E_8$		2.63	2.68	2.72	2.78		2.82			13,59
$C_{13}E_8$		2.83	2.87	2.87	2.90		2.92			13,59
$C_{14}E_8$		3.33	3.34	3.33	3.35		3.35			13,59
$C_{15}E_8$		3.69	3.68	3.67	3.68		3.67			13,59
$C_{16}E_6$				4.4						44

Table III.1.15. (continued)

Nonionic	Additive	25 $^\circ$C	Method	Ref.
$C_{16}E_7$		3.8		44
$C_{16}E_9$		3.1		44
$C_{16}E_{12}$		2.3		44
$C_{16}E_{15}$		2.0		44
$C_{16}E_{21}$		1.4		44

a   at 23.5°C;

b   Temp. not specified

Table III.1.16. Head group area (nm2) of alkyl-polyoxyethylene glycol ethers (C_nE_m) at various temperatures.

Nonionic	Temperature (°C)						Method	Ref.
	15	20	25	30	35	40		
C_6E_6			0.62					41
C_8E_4			0.49				RT	19
C_9E_8			0.755					59
$C_{10}E_6$			0.55[a]					50
$C_{10}E_8$	0.735	0.698	0.686	0.669		0.656		13,59
$C_{11}E_8$	0.704	0.675	0.664	0.649		0.636		13,59
$C_{12}E_6$			0.63 0.73 0.50[b] 0.45				DV WP	51 51 25 50
$C_{12}E_8$	0.631	0.620	0.611	0.597		0.589		13,59
$C_{13}E_8$	0.587	0.579	0.579	0.573		0.569		13,59
$C_{14}E_8$	0.499	0.497	0.499	0.496		0.496		13,59
$C_{15}E_8$	0.450	0.451	0.452	0.451		0.452		13,59
$C_{16}E_6$			0.38					44
$C_{16}E_7$			0.44					44
$C_{16}E_9$			0.53					44
$C_{16}E_{12}$			0.72					44
$C_{16}E_{15}$			0.81					44
$C_{16}E_{21}$			1.20					44

a   at 23.5°C

b   Tem. not specified

Table III.1.17. Hydration number of ethylene oxide units of alkyl-polyoxyethylene glycol ethers (C_nE_m) in the micellar state.

Nonionic	Number of H_2O molecules/EO	Method	Ref.
C_8E_4	5.5	PGNMR	20
C_8E_5	1 − 2.5	QENS	31
$C_{12}E_5$	ca 4	PGNMR	85
$C_{12}E_6$	1 − 2.5	QENS	31
$C_{12}E_8$	ca 3	PGNMR	85

Table III.1.18. Density of aqueous alkylpolyoxyethylene glycol ether (C_nE_m) solutions.

Nonionic	Temp (°C)	Concentration		Density g/cm^3	Ref
		wt %	molality ($*10^{-3}$)		
C_8E_4				1.017	20
$C_{12}E_5$	25.0	20	513	0.995	33
$C_{12}E_6$	25.0	0.87	20.02	0.99700	59
	25.0	20	461	0.994	33
$C_{12}E_8$	25.0	20	383	1.002	33
$C_{12}E_{9.0}$	25.0	1.13	20.08	0.99753	59
$C_{12}E_{14.8}$	25.0	1.64	20.00	0.99836	59
$C_{12}E_{19.3}$	25.0	2.03	19.56	0.99892	59
$C_{12}E_{23.9}$	25.0	2.44	20.11	0.99970	59

Table III.1.19. Partial molar volume of alkylpolyoxyethylene glycol ethers (C_nE_m) below (Φ^{MON}) and above (Φ^{MIC}) the critical micelle concentration.

Nonionic	Φ^{mon} [a] (cm^3/mole)	Φ^{mic} [a] (cm^3/mole)
$C_{10}E_3$	–	229.0
$C_{10}E_4$	–	335.0
$C_{10}E_9$	518.6	534.6
$C_{10}E_{11}$	593.0	616.9
$C_{10}E_{16}$	773.0	817.5

[a] At 25.0°C using density measurements, ref. 16

Table III.1.20. Group contributions for partial molar volume of alkylpolyoxyethylene glycol ethers (C_nE_m).

Group	Φ_V [a] (cm^3/mole)
CH_3	18.1
CH_2	16.0
OH	12.0
CH_2CH_2O	37.5 [b]
CH_2CH_2O	40.0 [c]

[a] at 25.0°C using density measurements ref. 16; [b] in solution; [c] in micelle.

Table III.1.21. Carbon-13 chemical shifts (δ, in ppm) of alkylpolyethylene glycol ethers (C_nE_m) as a function of structure[a] and solvent (Taken from references 57 and 58).

Compound	Solvent	a	b	c	d	e	j	i	h	k	g	l
$C_{11}E_8$	D_2O	13.8	22.6	31.9	29.4	29.8	71.0	69.8	69.6	71.9	60.5	
$C_{11}E_8$	$CDCl_3$	14.0	22.6	31.7	29.4	29.6	71.5	70.6	70.2	72.7	61.7	
$C_{11}E_8$	CD_3OD	14.3	23.5	32.9	30.3	30.5	72.2	71.4	71.1	73.5	62.1	
C_1E_1	$CDCl_3$									73.7	61.5	58.8
C_2E_1	$CDCl_3$	15.0					66.5			72.0	61.5	
C_3E_1	$CDCl_3$	13.9	19.3	31.8			71.1			72.2	61.6	
$C_1E_1C_1$	Dioxane						72.3					58.6

a Base structure: CH_3 CH_2 CH_2 CH_2 $(CH_2)_6$ CH_2 CH_2 $(OCH_2CH_2)_6OCH_2$ CH_2 OCH_2 CH_2OH $(-OCH_3)$

                      a   b   f   d   e   c   j   i   h   k   g   b

Table III.1.22. Carbon-13 chemical shift (δ , in ppm) of dodecylpentaoxyethylene glycol ether at various temperatures[a,b].

Carbon	Temperature	
Number	6°C	27°C
C 21	57.99	58.16
C 12	57.31	57.20
C 13	56.04	56.25
C 14–18	59.93	56.10
C 19	55.58	56.00
C 20	55.65	55.83
C 22	46.44	46.65
C 3	18.21	18.17
C 5, 7–10	16.21	16.11
C 5, 7–10		16.03
C 5, 7–10		15.95
C 4, 6	15.78	15.77
C 4, 6	15.68	15.68
C 11	12.33	12.36
C 2	8.84	8.81
C 1	0.00	0.00

[a] From ref. 84. [b] Terminal methyl (C_1) is reference.

Table III.1.23. Self-diffusion coefficient of octyltetraoxyethylene glycol ether (C_8E_4) (D^a, in 10^{-10} m^2/s), of water (D^w, in 10^{-10} m^2/s), and of solubilizate (D^s, in 10^{-10} m^2/s) in D_2O as a function of surfactant concentration [a,b,c].

M_t (mol/kg)	D^a (10^{-10} m^2/S)	D^w (10^{-10} m^2/S)	D^s (10^{-10} m^2/S)
$5.0 * 10^{-4}$		22.39 (1.0)	
$1.0 * 10^{-3}$	5.31 (3.1)	22.25 (1.5)	
$2.0 * 10^{-3}$	4.71 (2.6)	22.07 (1.8)	
$4.04 * 10^{-3}$	4.68 (1.5)	21.87 (1.0)	
$6.05 * 10^{-3}$	4.83 (2.0)	21.76 (2.7)	
$8.0 * 10^{-3}$	4.63 (2.8)	21.68 (2.3)	
$9.13 * 10^{-3}$	3.89 (3.1)		
$9.96 * 10^{-3}$	3.78 (2.4)	21.51 (0.8)	
$2.00 * 10^{-2}$	2.26 (3.0)	21.33 (4.6)	
$4.00 * 10^{-2}$	1.52 (2.3)	21.03 (0.66)	
$6.07 * 10^{-2}$	1.18 (3.4)		0.603
$7.8 * 10^{-2}$	1.01 (2.9)	20.69 (4.8)	0.509 (6.9)
0.100	0.923 (0.8)	20.16 (1.9)	0.471
0.1955	0.593 (1.1)	19.51 (1.7)	0.460 (1.7)
0.294	0.522 (1.3)	18.30 (1.1)	0.481 (1.8)
0.398	0.447 (2.7)	16.14 (1.3)	0.587 (2.5)
0.500	0.432 (0.8)	15.00 (2.8)	0.631 (2.4)
0.575	0.398 (1.0)	14.53 (1.0)	0.594 (3.7)
0.727	0.337 (1.5)	13.62 (1.6)	0.539 (3.7)
0.855	0.329 (1.0)	12.85 (2.1)	0.700 (0.6)
0.975	0.389 (0.9)	15.13 (1.4)	0.655 (0.8)

Table III.1.23. (continued)

M_t (mol/kg)	D^a (10^{-10} m^2/S)	D^w (10^{-10} m^2/S)	D^s (10^{-10} m^2/S)
1.184	0.370 (2.1)	12.96 (1.1)	
1.392	0.360 (1.4)	11.85 (1.3)	0.723 (0.8)
1.614	0.404 (1.2)	11.72 (2.0)	0.855 (2.9)
1.695	0.391 (1.9)	10.72 (2.0)	0.853 (3.1)
1.813	0.441 (0.8)	10.10 (0.8)	1.035
1.904	0.432 (0.9)	10.14 (1.0)	0.958 (2.9)
1.983	0.410 (1.5)	10.67 (1.4)	1.083 (1.9)
2.23	0.444 (2.5)	9.47 (1.4)	1.016 (0.9)
2.441	0.449 (1.4)	9.38 (0.2)	
2.498			1.178 (3.7)

a   From Ref. 20

b   At 25.0°C using pulsed field gradient NMR , solubilisate is tetramethylsilane or hexamethylsiloxane, random errors in percent in parentheses.

c   Infinite dilution values of the self diffusion coefficients are $D^{a,o}$ = $(4.83 \pm 0.12) * 10^{-10}$ m^2s^{-1}; $D^{w,o}$ = $(216 \pm 0.3) * 10^{-10}$ m^2s^{-1}; D^o_m = $(0.80 \pm 0.02) * 10^{-10}$ m^2s^{-1} (of the micelle).

Table III.1.24. Critical length (ν) and correlation length (ξ_0) for alkylpolyoxyethylene glycol ethers (C_nE_m) measured using quasi elastic light scattering.

Nonionic	Additive	ν	ξ_0 (nm)	Ref
C_6E_3		0.63 ± 0.03	0.34 ± 0.03	73
C_8E_4		0.57 ± 0.03	0.54 ± 0.05	73
C_8E_5	0 M NaCl	0.63	0.95	22
	0.5 M NaCl	0.63	0.97	22
	0.9 M NaCl	0.62	0.98	22
	1.3 M NaCl	0.63	0.90	22
$C_{12}E_6$		0.53 ± 0.05	2.0 ± 0.5	73
$C_{12}E_8$		0.44 ± 0.04	1.75 ± 0.5	73

Table III.1.25. Lifetime (τ) and intensity ratio of the first and third peak in the fluorescence spectrum (I_1/I_3) of pyrene solubilized in alkylpolyoxyethylene glycol ether (C_nE_m) micelles.

Nonionic	Additive	Temp (°C)	τ (nS)	I_1/I_3
$C_{10}E_8$ [a]	0 mol % NaC	25	340	1.24
	25 mol % NaC	25	340	1.14
	50 mol % NaC	25	360	1.08
	75 mol % NaC	25	368	1.03
	90 mol % NaC	25	386	0.90
	100 mol % NaC	25	393	0.79
$C_{12}E_6$ [b]		20	359	1.35

a   From ref. 23

b   Ref. 1

Table III.1.26. Activation energy (E_a, in kJ/mol) for solubilization of 2,6,10,15,19,23-hexamethyltetracosane (squalane) in alkylpoly-oxyethylene glycol ether (C_nE_m) solutions.

Nonionic	Activation energy E_a (kJ/mol)	Ref.
C_8E_4	120	18
$C_{12}E_5$	111	3
$C_{12}E_6$	125	3

1   Via Arrhenius type of plot.

2   Method: drop on fibre.

Table III.1.27. Dissociation constant (α) for mixtures of alkylpolyoxyethylene glycol ethers (C_nE_m) and sodium dodecylsulphate measured using either osmometry (OS) or an ion selective electrode (p_{Na}) (Taken from reference 53).

Nonionic	mole fraction SDS (%)	Temp (°C)	α	Method
$C_{12}E_5$	50	40	0.30	OS
		25	0.36	pNa
$C_{12}E_9$	20	40	0.65	OS
		25	0.66	pNa
	50	40	0.40	OS
		25	0.47	pNa
	80	40	0.25	OS
		25	0.30	pNa
	100	40	0.16	OS
		25	0.18	pNa
$C_{12}E_{15}$	50	40	0.51	OS
		25	0.58	pNa
$C_{12}E_{30}$	50	40	0.63	OS
		25	0.64	pNa

Table III.1.28. Surface interaction parameter (β_s) of the system octyltetraoxyethylene glycol ether/sodium dodecylsulphate as a function of the mole fraction sodium dodecylsulphate and added electrolyte[a,b].

[C_8E_4] (mol/l)	Mole fraction SDS (%)	Additive	β_s
$1 * 10^{-3}$	10	–	– 3.05
		0.1 mole/l NaCl	– 3.31
	30	–	– 2.28
		0.1 mole/l NaCl	– 3.12
	50	–	– 2.10
		0.1 mole/l NaCl	– 3.2
	67	–	– 2.11
	70	0.1 mole/l NaCl	– 3.35
	80	–	– 2.30
		0.1 mole/l NaCl	– 3.86
$3 * 10^{-3}$	10	–	– 2.84
	30	–	– 2.11
	50	–	– 1.90
	67	–	– 2.17
	80	–	– 2.52
$4 * 10^{-3}$	10	0.1 mole/l NaCl	– 3.24
	30	0.1 mole/l NaCl	– 3.02
	50	0.1 mole/l NaCl	– 3.07
	70	0.1 mole/l NaCl	– 3.38
	90	0.1 mole/l NaCl	– 3.95

[a] from ref 19   [b] at 25°C

Table III.1.29. Upper critical solution temperature (T_c, in °C) and critical concentration (W_c, in wt%) of hexylpentaoxyethylene glycol ether in alkanes.

Alkane	c	T_c	Ref
$C_{10}H_{20}$	12	18	72
$C_{12}H_{26}$	21	30	72
$C_{14}H_{30}$	27	42	72
$C_{16}H_{34}$	31	55	72

Table III.1.30. Phase inversion temperature (PIT, in °C) as a function of the alkylpolyoxyethylene glycol ether (C_nE_m) and hydrocarbon used.

Nonionic	Oil	PIT (°C)	Ref
$C_{12}E_4$	heptane	10 – 11	29
	hexadecane	30	35
	hexadecane/ squalane	37	35
$C_{12}E_5$	decane	38	28
	tetradecane	46	28
	hexadecane	51	28
	n–hexadecane	52	35
	n–hexadecane/ squalane	58	35

Table III.1.31. Aggregation number for dodecyltetraoxyethylene glycol ether ($C_{12}E_4$) in oil rich media measured using small-angle neutron scattering at 20 °C (Taken from reference 30).

Solvent	wt% $C_{12}E_4$	wt% water	N
Heptane	0.10	1	40
	0.10	2	120
	0.20	1	40
Decane	0.10	1	80
	0.10	2	200
	0.10	3	500
	0.15	1	50
	0.15	2	150
	0.15	3	300
Hexadecane	0.10	1	100
	0.10	2	250
	0.10	3	900
	0.15	1	100
	0.15	2	200
	0.15	3	600

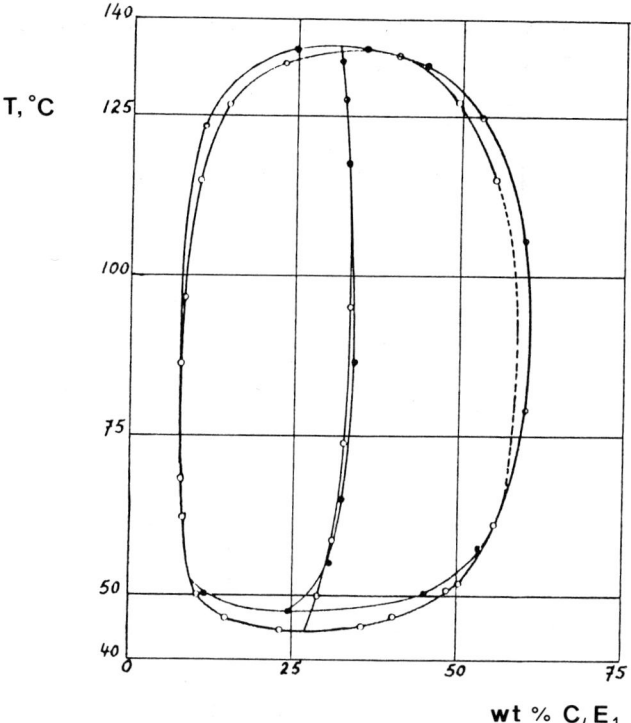

T, °C

wt % C_4E_1

Figure III.1.1. Phase diagram of butylmonooxyethylene glycol ether-water (C_4E_1-H_2O) system (© Comité van Beheer van het Bulletin, 1956. Reproduced with permission. Taken from reference 62).

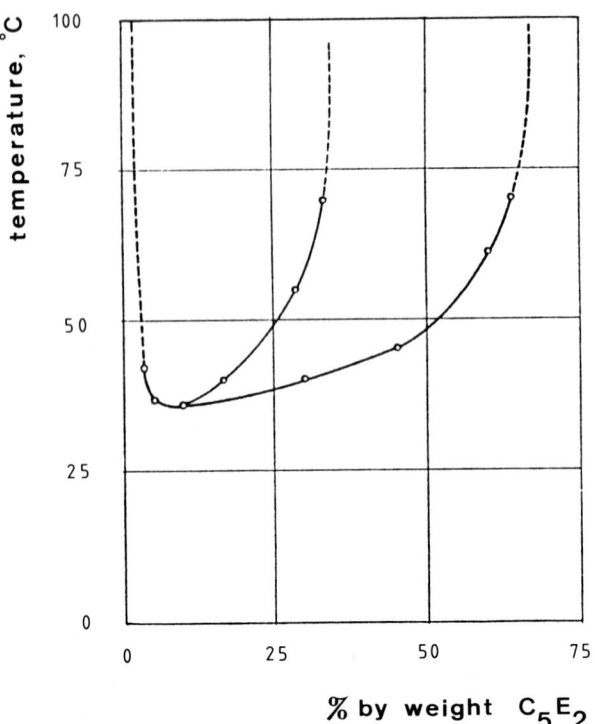

Figure III.1.2. Phase diagram of pentyldioxyethylene glycol ether-water (C_5E_2-H_2O) system (© Comité van Beheer van het Bulletin, 1956. Reproduced with permission. Taken from reference 62).

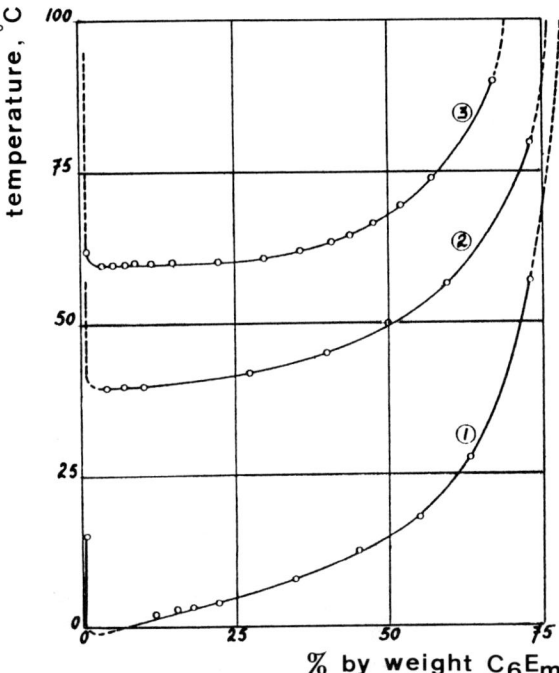

Figure III.1.3. Phase diagram of (1) : hexyldioxyethylene glycol ether-water (C_6E_2-H_2O) ; (2) : hexyltrioxyethylene glycol ether-water (C_6E_3-H_2O) ; (3) : hexyltetraoxyethylene glycol ether-water (C_6E_4-H_2O) systems (© Comité van Beheer van het Bulletin, 1956. Reproduced with permission. Taken from reference 62).

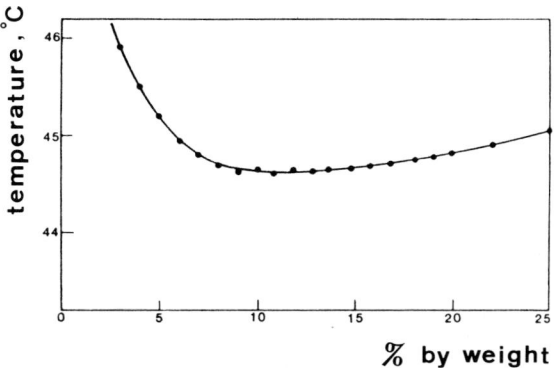

Figure III.1.4. Phase diagram of hexyltrioxyethylene glycol ether-water (C_6E_3-H_2O) system (© American Chemical Society, 1984. Reproduced with permission. Taken from reference 73).

266

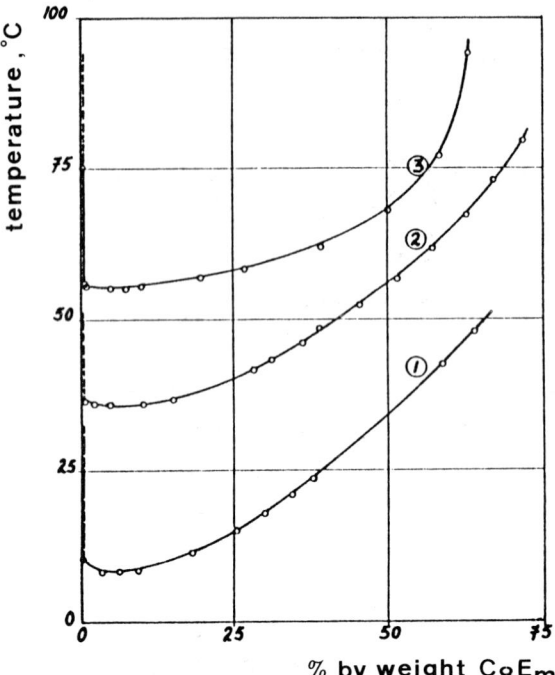

Figure III.1.5. Phase diagram of (1) : octyltrioxyethylene glycol ether-water (C_8E_3-H_2O) ; (2) : octyltetraoxyethylene glycol ether-water (C_8E_4-H_2O) ; (3) : octylpentaoxyethylene glycol ether-water (C_8E_5-H_2O) systems (© Comité van Beheer van het Bulletin, 1956. Reproduced with permission. Taken from reference 62).

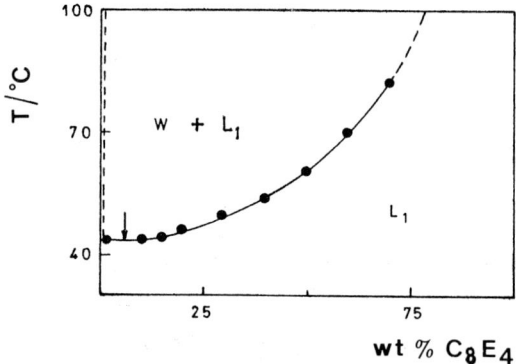

Figure III.1.6. Phase diagram of octyltetraoxyethylene glycol ether-water (C_8E_4-H_2O) system (© The Royal Society of Chemistry, 1983. Reproduced with permission. Taken from reference 17).

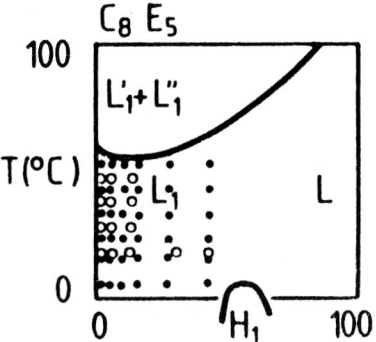

Figure III.1.7. Phase diagram of octylpentaoxyethylene glycol ether-water (C_8E_5-H_2O) system (© American Chemical Society, 1985. Reproduced with permission. Taken from reference 31).

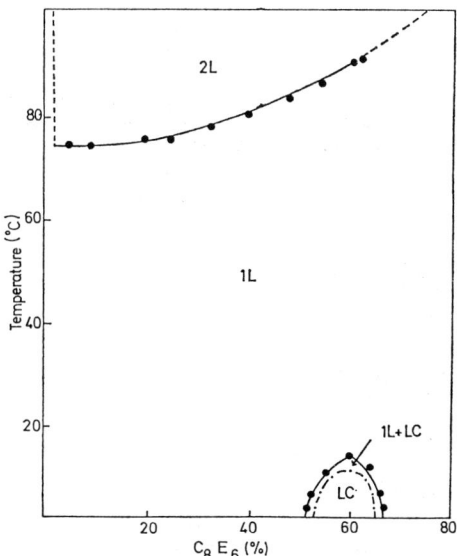

Figure III.1.8. Phase diagram of octylhexaoxyethylene glycol ether-water (C_8E_6-H_2O) system (© Pharmaceutical Society of Great Britain, 1971. Reproduced with permission. Taken from reference 67).

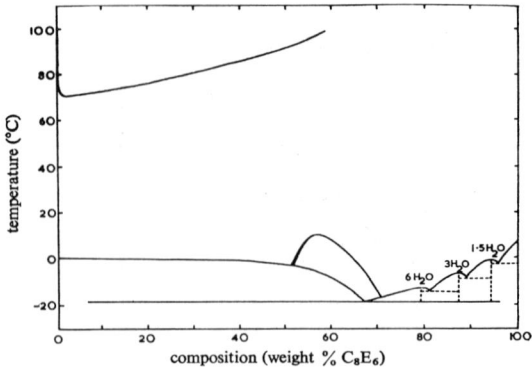

Figure III.1.9. Phase diagram of octylhexaoxyethylene glycol ether-water (C_8E_6-H_2O) system (© The Royal Society of Chemistry, 1967. Reproduced with permission. Taken from reference 15).

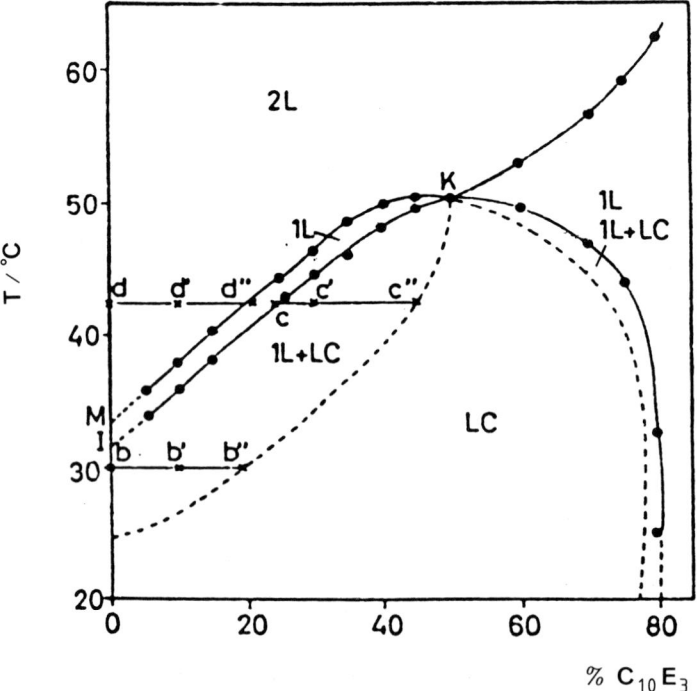

Figure III.1.10. Phase diagram of decyltrioxyethylene glycol ether-water ($C_{10}E_3$-H_2O) system (© Pharmaceutical Society of Great Britain, 1971. Reproduced with permission. Taken from reference 68).

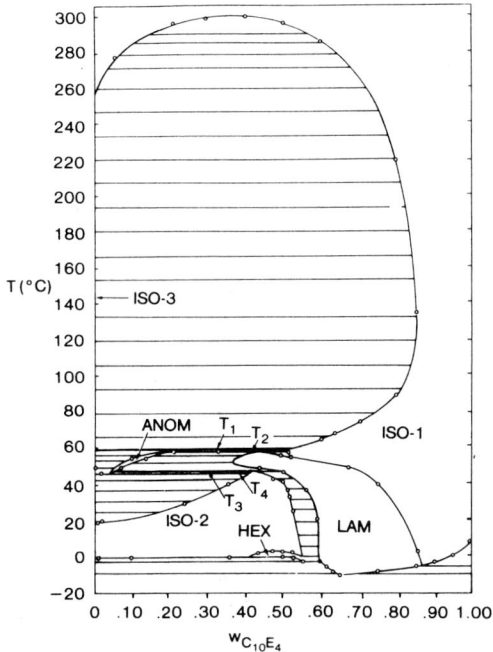

Figure III.1.11. Phase diagram of decyltetraoxyethylene glycol ether-water ($C_{10}E_4$-H_2O) system (© American Institute of Physics, 1980. Reproduced with permission. Taken from reference 63).

270

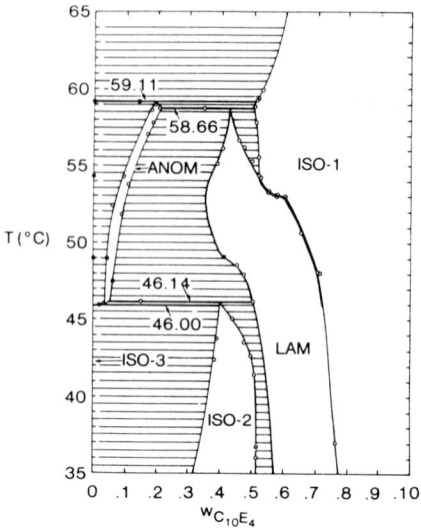

Figure III.1.12. Phase diagram of decyltetraoxyethylene glycol ether-water ($C_{10}E_4$-H_2O) system (© American Institute of Physics, 1980. Reproduced with permission. Taken from reference 63).

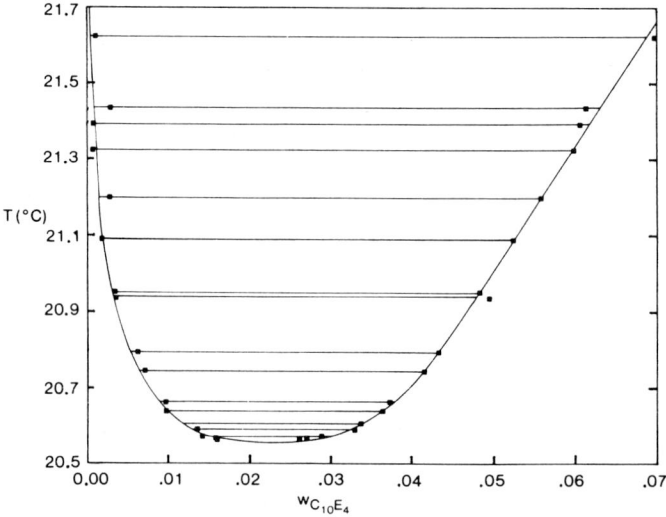

Figure III.1.13. Phase diagram of decyltetraoxyethylene glycol ether-water ($C_{10}E_4$-H_2O) system (© American Institute of Physics, 1980. Reproduced with permission. Taken from reference 63).

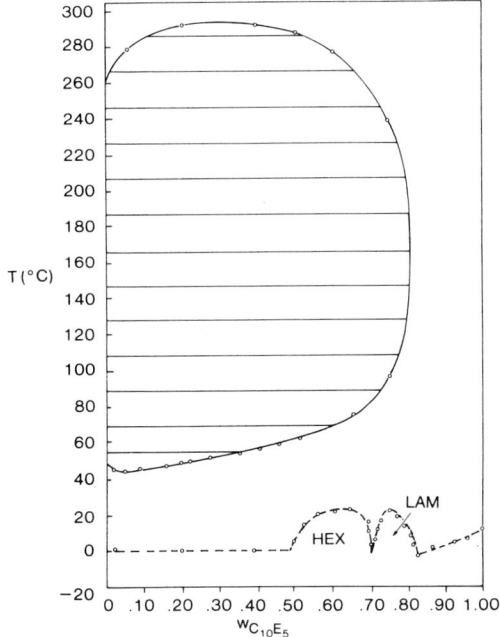

Figure III.1.14. Phase diagram of decylpentaoxyethylene glycol ether-water ($C_{10}E_5$-H_2O) system (© American Institute of Physics, 1980. Reproduced with permission. Taken from reference 63).

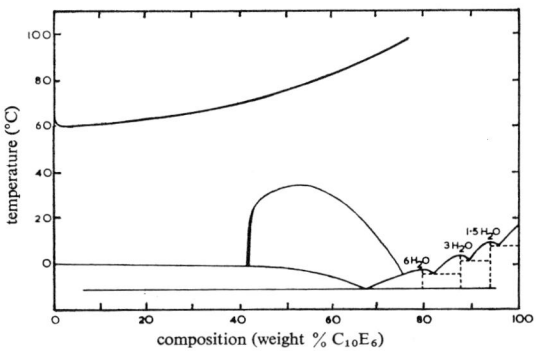

Figure III.1.15. Phase diagram of decylhexaoxyethylene glycol ether-water ($C_{10}E_6$-H_2O) system (© The Royal Society of Chemistry, 1967. Reproduced with permission. Taken from reference 15).

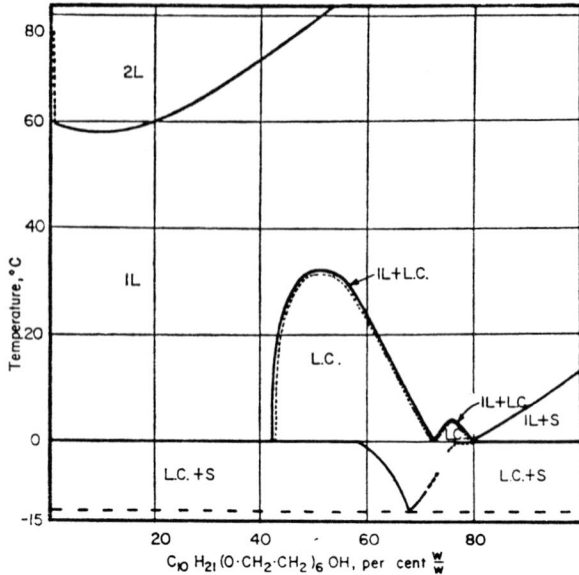

Figure III.1.16. Phase diagram of decylhexaoxyethylene glycol ether-water ($C_{10}E_6$-H_2O) system (© Academic Press, Inc., 1964. Reproduced with permission. Taken from reference 64).

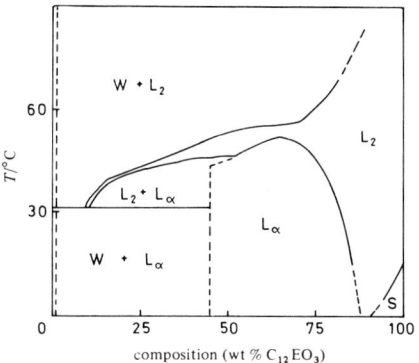

Figure III.1.17. Phase diagram of dodecyltrioxyethylene glycol ether-water ($C_{12}E_3$-H_2O) system (© The Royal Society of Chemistry, 1983. Reproduced with permission. Taken from reference 17).

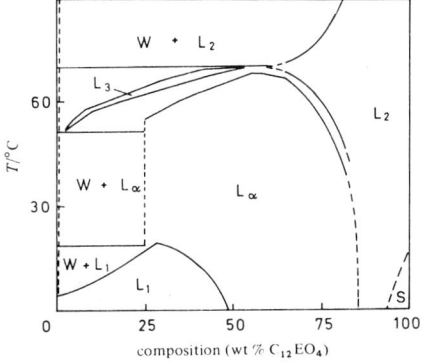

Figure III.1.18. Phase diagram of dodecyltetraoxyethylene glycol ether-water ($C_{12}E_4$-H_2O) system (© The Royal Society of Chemistry, 1983. Reproduced with permission. Taken from reference 17).

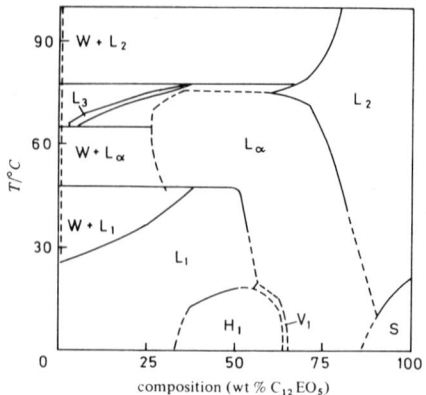

Figure III.1.19. Phase diagram of dodecylpentaoxyethylene glycol ether-water ($C_{12}E_5$-H_2O) system (© The Royal Society of Chemistry, 1983. Reproduced with permission. Taken from reference 17).

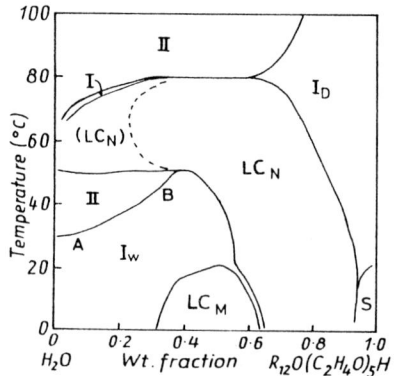

Figure III.1.20. Phase diagram of dodecylpentaoxyethylene glycol ether-water ($C_{12}E_5$-H_2O) system (© Dietrich Steinkopf Verlag GmbH & Co KG, 1974. Reproduced with permission. Taken from reference 69).

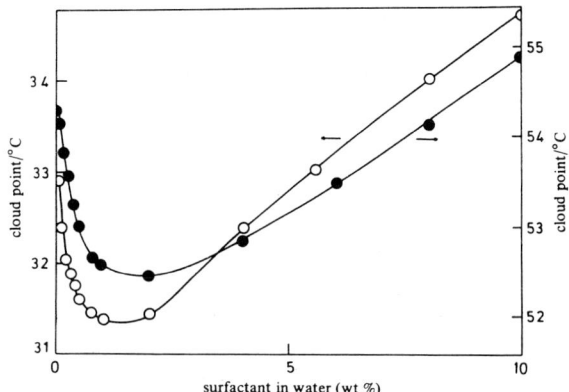

Figure III.1.21. Phase diagram of dodecylpentaoxyethylene glycol ether-water ($C_{12}E_5$-H_2O) and of dodecylhexaoxyethylene glycol ether-water ($C_{12}E_6$-H_2O) systems (© The Royal Society of Chemistry, 1986. Reproduced with permission. Taken from reference 25).

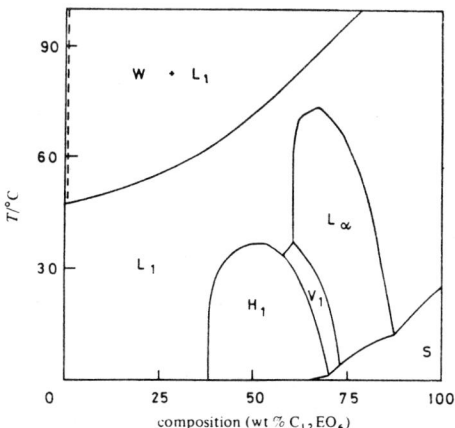

Figure III.1.22. Phase diagram of dodecylhexaoxyethylene glycol ether-water ($C_{12}E_6$-H_2O) system (© The Royal Society of Chemistry, 1983. Reproduced with permission. Taken from reference 17).

276

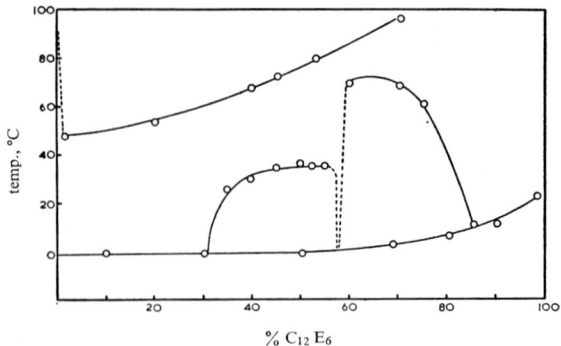

Figure III.1.23. Phase diagram of dodecylhexaoxyethylene glycol ether-water ($C_{12}E_6$-H_2O) system (© The Royal Society of Chemistry, 1962. Reproduced with permission. Taken from reference 7).

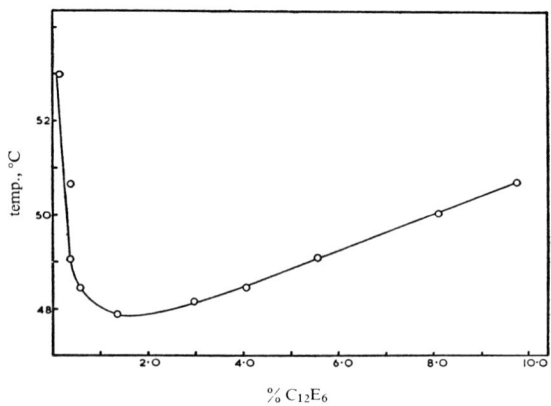

Figure III.1.24. Phase diagram of dodecylhexaoxyethylene glycol ether-water ($C_{12}E_6$-H_2O) system (© The Royal Society of Chemistry, 1962. Reproduced with permission. Taken from reference 7).

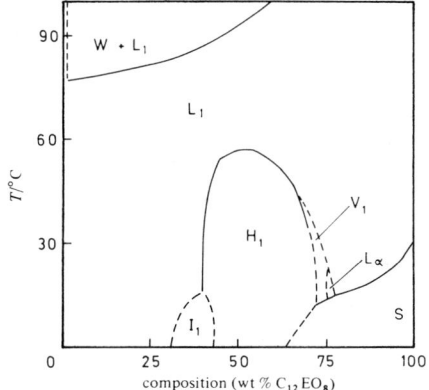

Figure III.1.25. Phase diagram of dodecyloctaoxyethylene glycol ether-water ($C_{12}E_8$-H_2O) system (© The Royal Society of Chemistry, 1983. Reproduced with permission. Taken from reference 17).

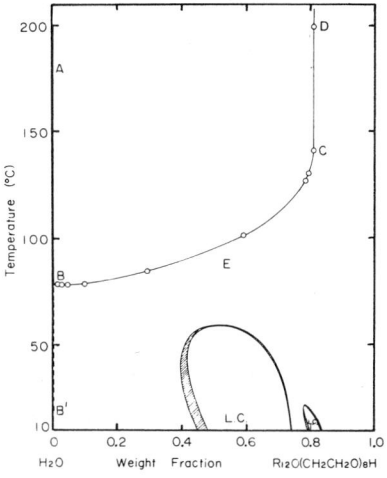

Figure III.1.26. Phase diagram of dodecyloctaoxyethylene glycol ether-water ($C_{12}E_8$-H_2O) system (© Academic Press Inc., 1970. Reproduced with permission. Taken from reference 65).

278

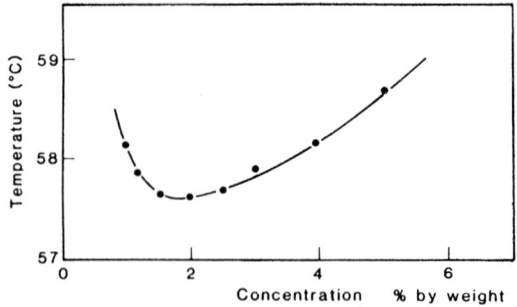

Figure III.1.27. Phase diagram of tetradecylheptaoxyethylene glycol ether-water ($C_{14}E_7$-H_2O) system (© American Chemical Society, 1984. Reproduced with permission. Taken from reference 73).

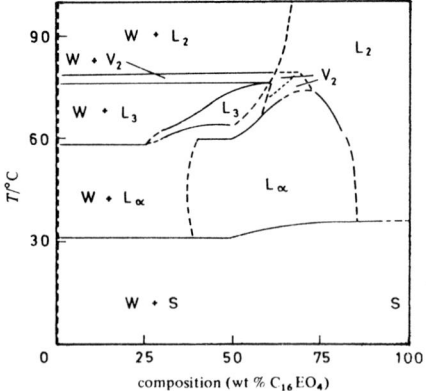

Figure III.1.28. Phase diagram of hexadecyltetraoxyethylene glycol ether-water ($C_{16}E_4$-H_2O) system (© The Royal Society of Chemistry, 1983. Reproduced with permission. Taken from reference 17).

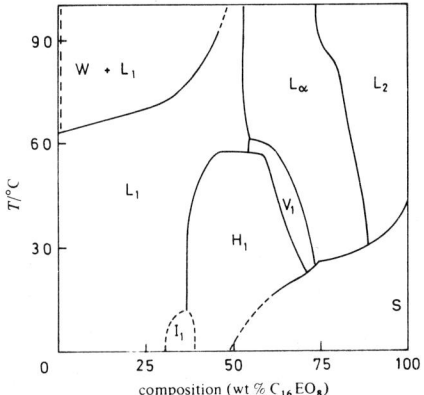

Figure III.1.29. Phase diagram of hexadecyloctaoxyethylene glycol ether-water ($C_{16}E_8$-H_2O) system (© The Royal Society of Chemistry, 1983. Reproduced with permission. Taken from reference 17).

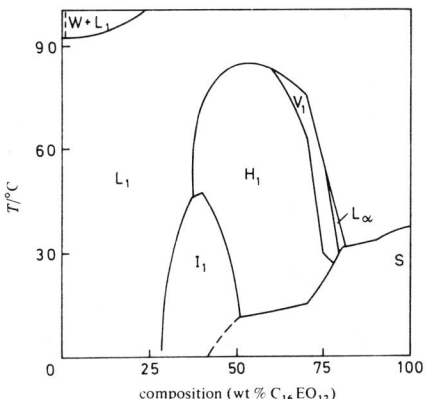

Figure III.1.30. Phase diagram of hexadecyldodecaoxyethylene glycol ether-water ($C_{16}E_{12}$-H_2O) system (© The Royal Society of Chemistry, 1983. Reproduced with permission. Taken from reference 17).

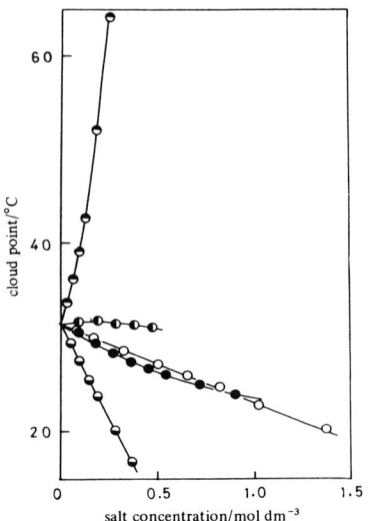

Figure III.1.31. Effect of electrolytes on the clouding temperature of 0.5 wt% solution of dodecylpentaoxyethylene glycol ether ($C_{12}E_5$). Points:◑,●, ○,◐,◒ refer, respectively, to Na_2CO_3, $CaCl_2$, NaCl, Et_4NBr, and $n\text{-}Bu_4NBr$ (© The Royal Society of Chemistry, 1986. Reproduced with permission. Taken from reference 25).

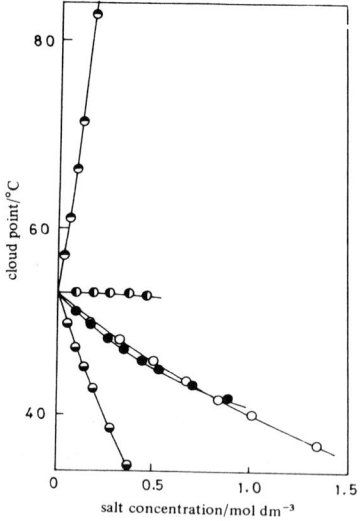

Figure III.1.32. Effect of electrolytes on the clouding temperature of 0.5 wt% solution of dodecylhexaoxyethylene glycol ether ($C_{12}E_6$). Points are as in Figure III.1.31. (© The Royal Society of Chemistry, 1986. Reproduced with permission. Taken from reference 25).

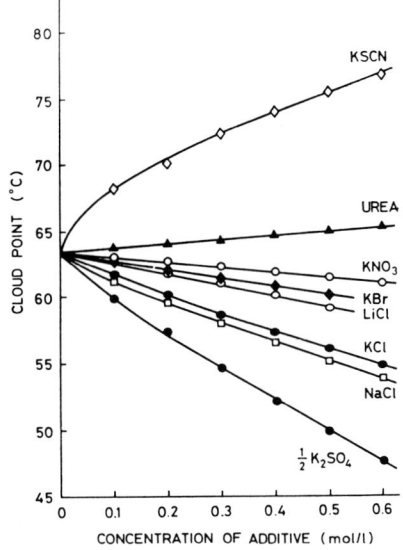

Figure III.1.33. Effect of electrolytes and of urea on the clouding temperature of 0.5 wt% solution of dodecylheptaoxyethylene glycol ether ($C_{12}E_7$). (© Marcel Dekker, Inc., 1987. Reproduced with permission. Taken from reference 59).

282

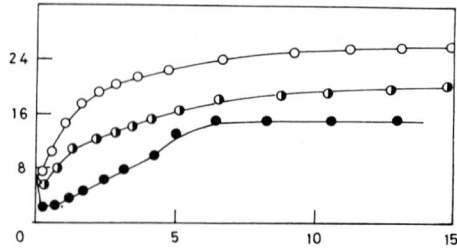

Figure III.1.34. Effect of alkanes on the clouding temperature of 0.5 wt% aqueous solution of dodecyltetraoxyethylene glycol ether ($C_{12}E_4$). Points are :◐, hexane;◑, heptane;●, nonane;◑, dodecane;

◐, tetradecane; and o, pentadecane (© The Royal Society of Chemistry, 1986. Reproduced with permission. Taken from reference 25).

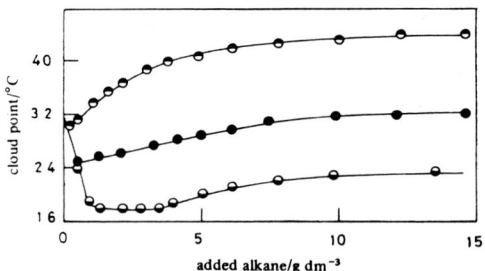

Figure III.1.35. Effect of alkanes on the clouding temperature of 0.5 wt% aqueous solution of dodecylpentaoxyethylene glycol ether ($C_{12}E_5$). Points are as in Figure III.1.34. (© The Royal Society of Chemistry, 1986. Reproduced with permission. Taken from reference 25).

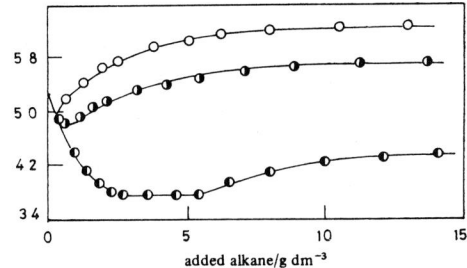

Figure III.1.36. Effect of alkanes on the clouding temperature of 0.5 wt% aqueous solution of dodecylhexaoxyethylene glycol ether ($C_{12}E_6$). Points are as in Figure III.1.34. (© The Royal Society of Chemistry, 1986. Reproduced with permission. Taken from reference 25).

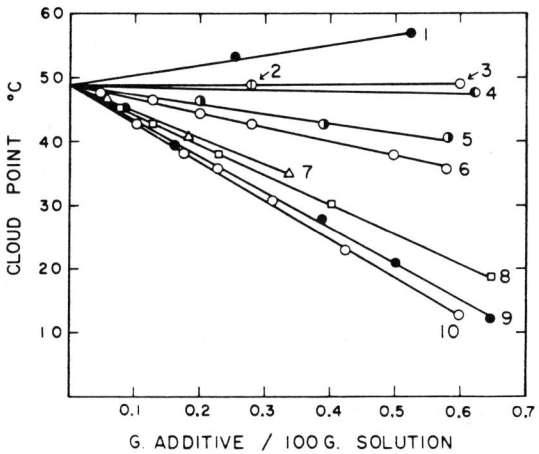

Figure III.1.37. Effect of added solubilizates on the clouding temperature of dodecylhexaoxyethylene glycol ether ($C_{12}E_6$)
(4.2 g in 100 mL). Solubilizates: (1) paraffin; (2)   butanol;
(3) hexaethylene glycol; (4) CARBOWAX 4000; (5)   hexane;
(6)   hexanol; (7)   octadecanol; (8)   octanol; (9)   dodecanol; (10)   decanol
(©   Marcel Dekker, Inc., 1966. Reproduced with permission. Taken from reference 78).

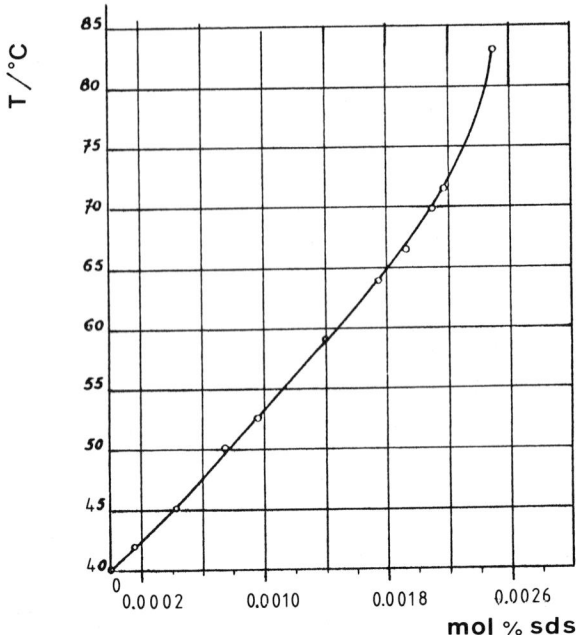

Figure III.1.38. Effect of sodium dodecylsulphate on the clouding temperature of butylmonooxyethylene glycol ether (© Comité van Beheer van het Bulletin, 1956. Reproduced with permission. Taken from reference 62).

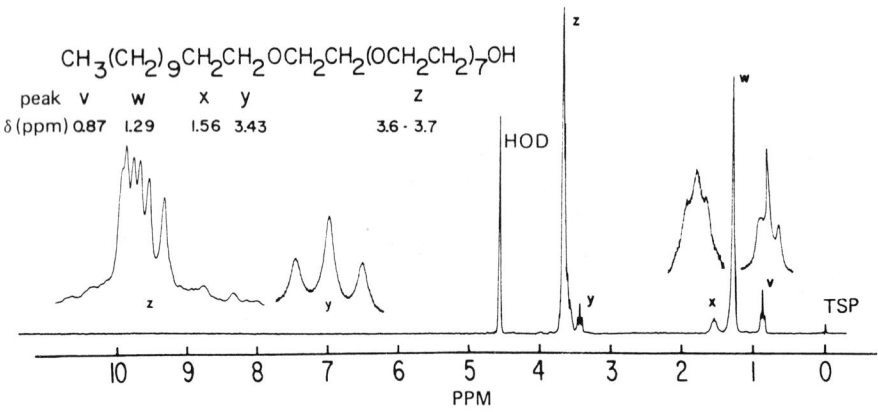

Figure III.1.39. Proton NMR spectrum (220 MHz) of dodecylocta-oxyethylene glycol ether in D$_2$O (© American Chemical Society, 1977. Reproduced with permission. Taken from reference 58).

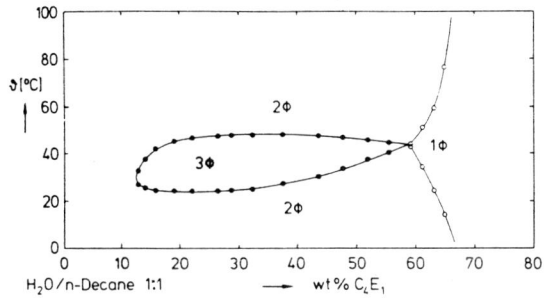

Figure III.1.40. Phase diagram of butylmonooxyethylene glycol ether-water-decane (C_4E_1-H_2O-$C_{10}H_{22}$) system at a water-decane ratio of 1:1 (© American Chemical Society, 1986. Reproduced with permission. Taken from reference 82).

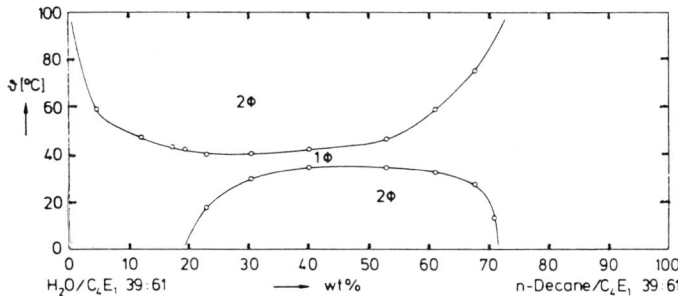

Figure III.1.41. Phase diagram of butylmonooxyethylene glycol ether-water-decane (C_4E_1-H_2O-$C_{10}H_{22}$) system at a water-butylmono-oxyethylene glycol ether ratio of 39:61 (© American Chemical Society, 1986. Reproduced with permission. Taken from reference 82).

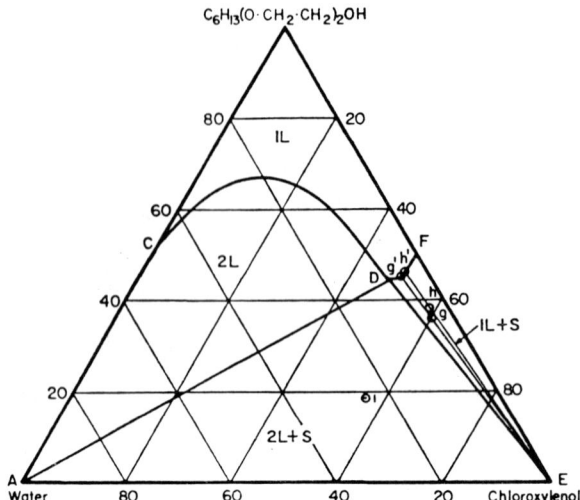

Figure III.1.42. Phase diagram of hexyldioxyethylene glycol ether-water-4-chloro-3,5-xylenol (C_6E_2-H_2O-C_8H_9ClO) system at 20 °C

(© Academic Press, Inc., 1964. Reproduced with permission. Taken from reference 64).

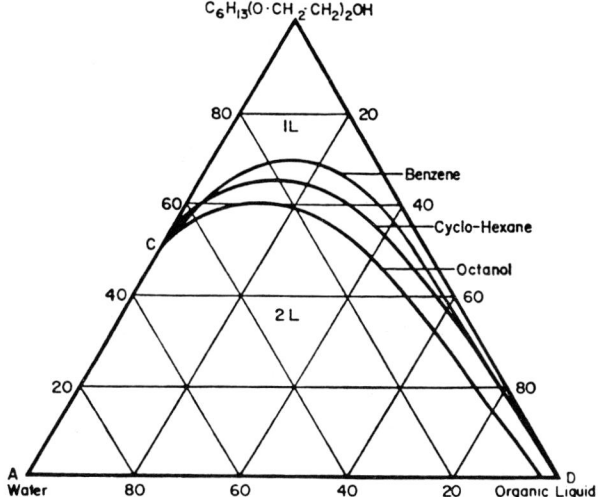

Figure III.1.43. Phase diagram of hexyldioxyethylene glycol ether-water-benzene (C_6E_2-H_2O-C_6H_6) , of hexyldioxyethylene glycol ether-water-cyclohexane (C_6E_2-H_2O-C_6H_{12}) , of hexyldioxyethylene glycol ether-water-octanol (C_6E_2-H_2O-$C_8H_{17}OH$) systems at 20 °C (© Academic Press, Inc., 1964. Reproduced with permission. Taken from reference 64).

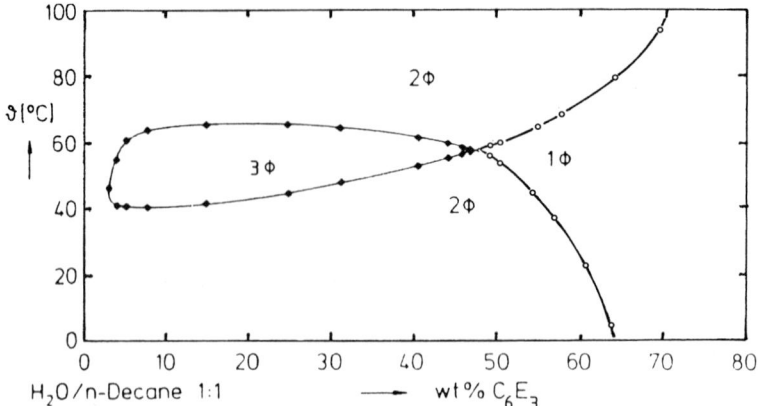

Figure III.1.44. Phase diagram of hexyltrioxyethylene glycol ether-water-decane (C_6E_3-H_2O-$C_{10}H_{22}$) system at a water-decane ratio of 1:1

(© American Chemical Society, 1986. Reproduced with permission. Taken from reference 82).

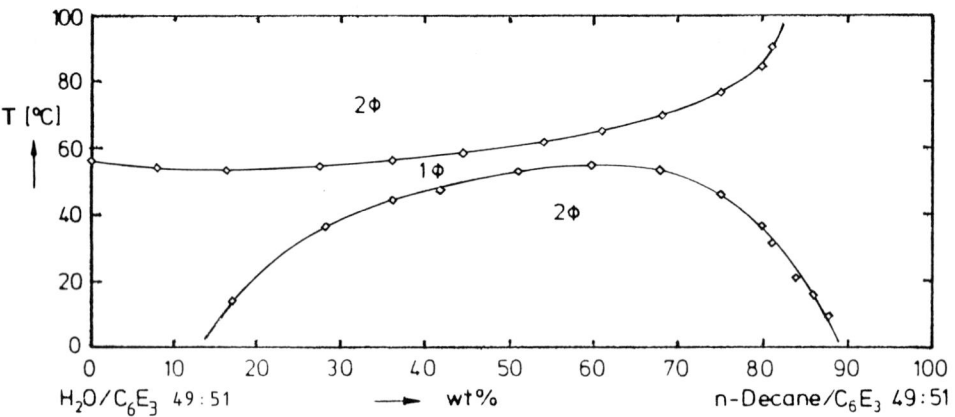

Figure III.1.45. Phase diagram of hexyltrioxyethylene glycol ether-water-decane (C_6E_3-H_2O-$C_{10}H_{22}$) system at a water-hexyltrioxyethylene glycol ether ratio of 49:51 (© American Chemical Society, 1986. Reproduced with permission. Taken from reference 82).

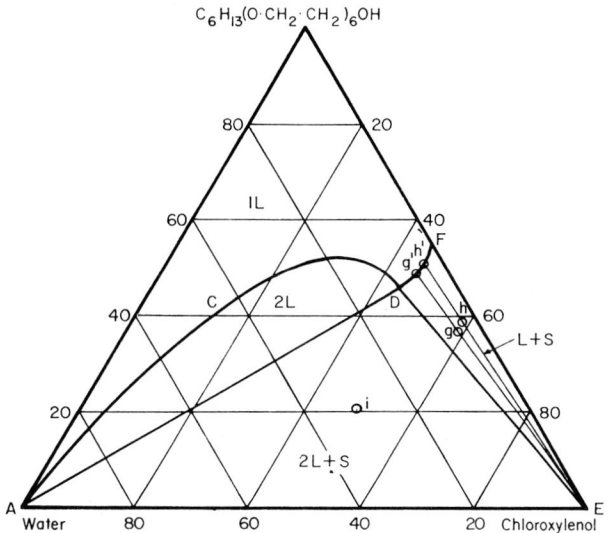

Figure III.1.46. Phase diagram of hexylhexaoxyethylene glycol ether-water-4-chloro-3,5-xylenol (C_6E_6-H_2O-C_8H_9ClO) system at 20 °C

(© Academic Press, Inc., 1964. Reproduced with permission. Taken from reference 64).

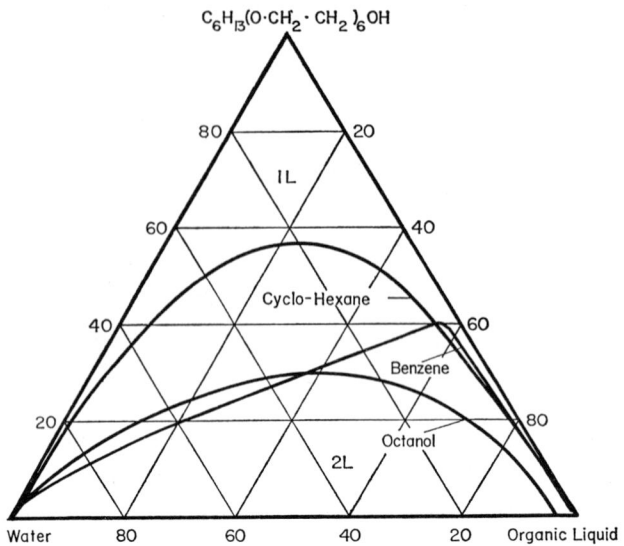

Figure III.1.47. Phase diagram of hexylhexaoxyethylene glycol ether-water-benzene (C_6E_6-H_2O-C_6H_6) , of hexylhexaoxyethylene glycol ether-water-cyclohexane (C_6E_6-H_2O-C_6H_{12}) , of hexylhexaoxyethylene glycol ether-water-octanol (C_6E_6-H_2O-$C_8H_{17}OH$) systems at 20 $\,^oC$

(© Academic Press, Inc., 1964. Reproduced with permission. Taken from reference 64).

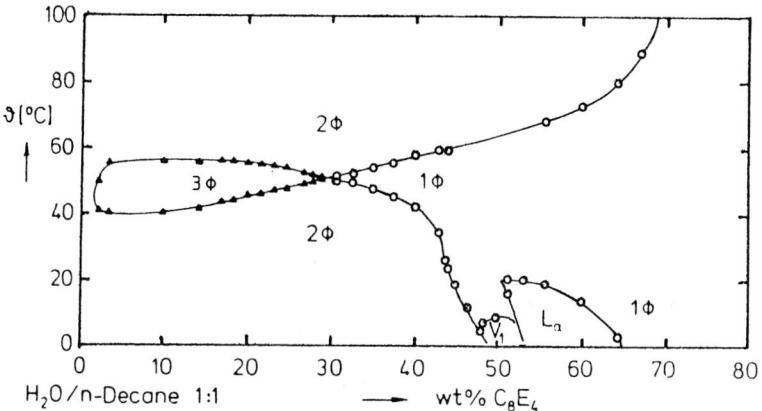

Figure III.1.48. Phase diagram of octyltetraoxyethylene glycol ether-water-decane (C_8E_4-H_2O-$C_{10}H_{22}$) system at a water-decane ratio of 1:1 (© American Chemical Society, 1986. Reproduced with permission. Taken from reference 82).

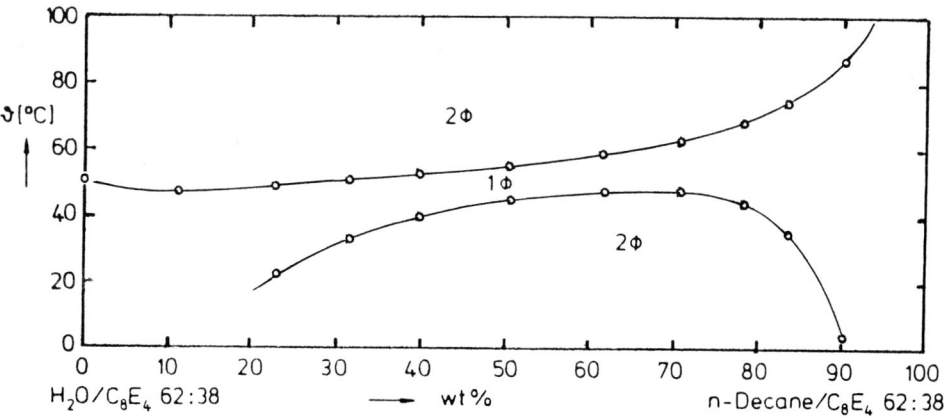

Figure III.1.49. Phase diagram of octyltetraoxyethylene glycol ether-water-decane (C_8E_4-H_2O-$C_{10}H_{22}$) system at a water-octyltetra-oxyethylene glycol ether ratio of 62:38 (© American Chemical Society, 1986. Reproduced with permission. Taken from reference 82).

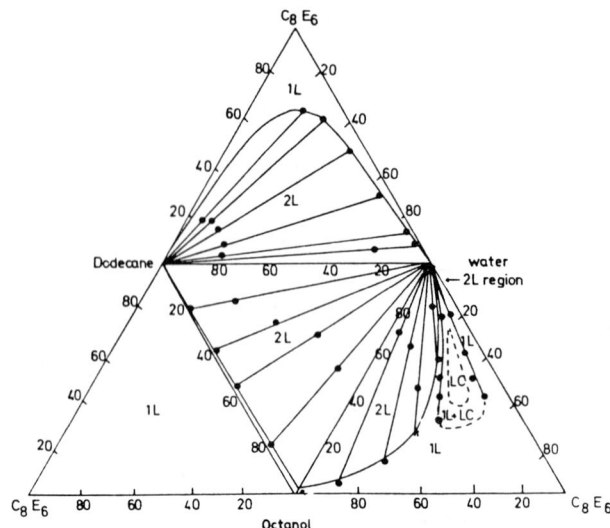

Figure III.1.50. Phase diagram of octylhexaoxyethylene glycol ether-water-dodecane-octanol (C_8E_6-H_2O-$C_{12}H_{26}$-$C_8H_{17}OH$) system at 25 ºC

(© Pharmaceutical Society of Great Britain, 1971. Reproduced with permission. Taken from reference 67).

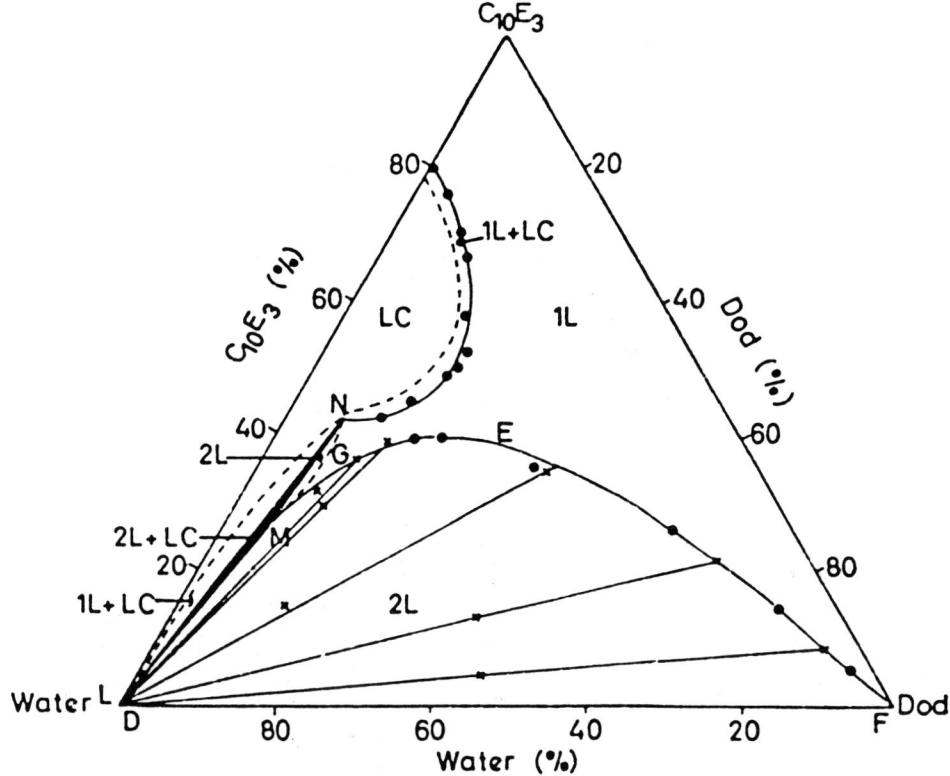

Figure III.1.51. Phase diagram of decyltrioxyethylene glycol ether-water-dodecane ($C_{10}E_3$-H_2O-$C_{12}H_{26}$) system at 25 °C (© Pharmaceutical Society of Great Britain, 1971. Reproduced with permission. Taken from reference 68).

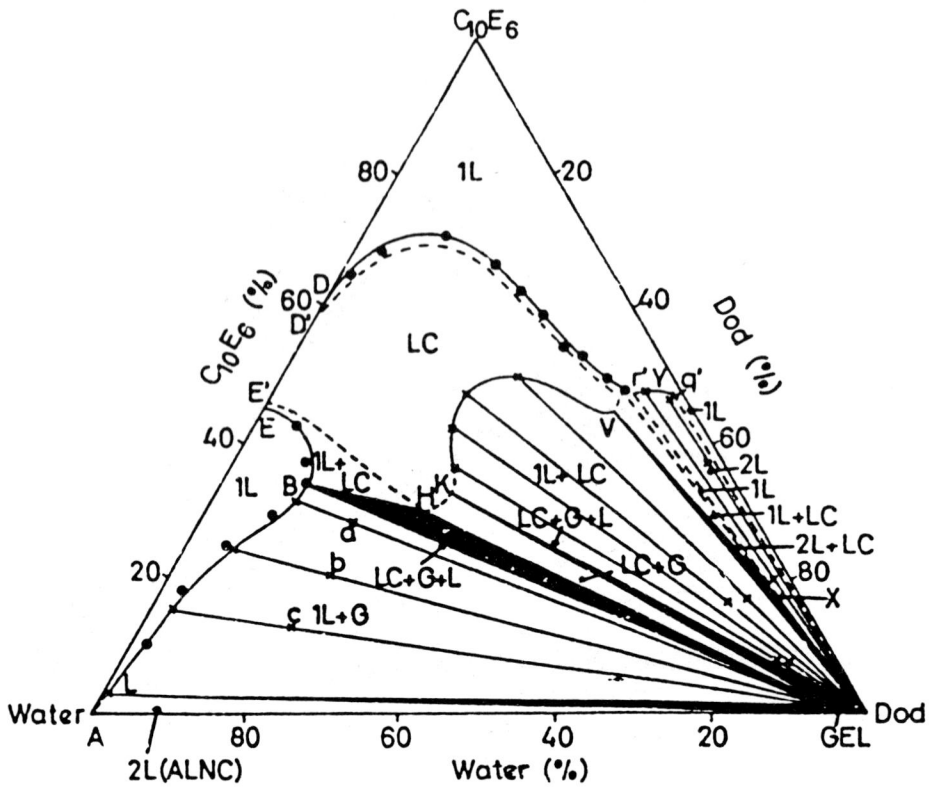

Figure III.1.52. Phase diagram of decylhexaoxyethylene glycol ether-water-dodecane ($C_{10}E_6$-H_2O-$C_{12}H_{26}$) system at 25 °C

(© Pharmaceutical Society of Great Britain, 1971. Reproduced with permission. Taken from reference 68).

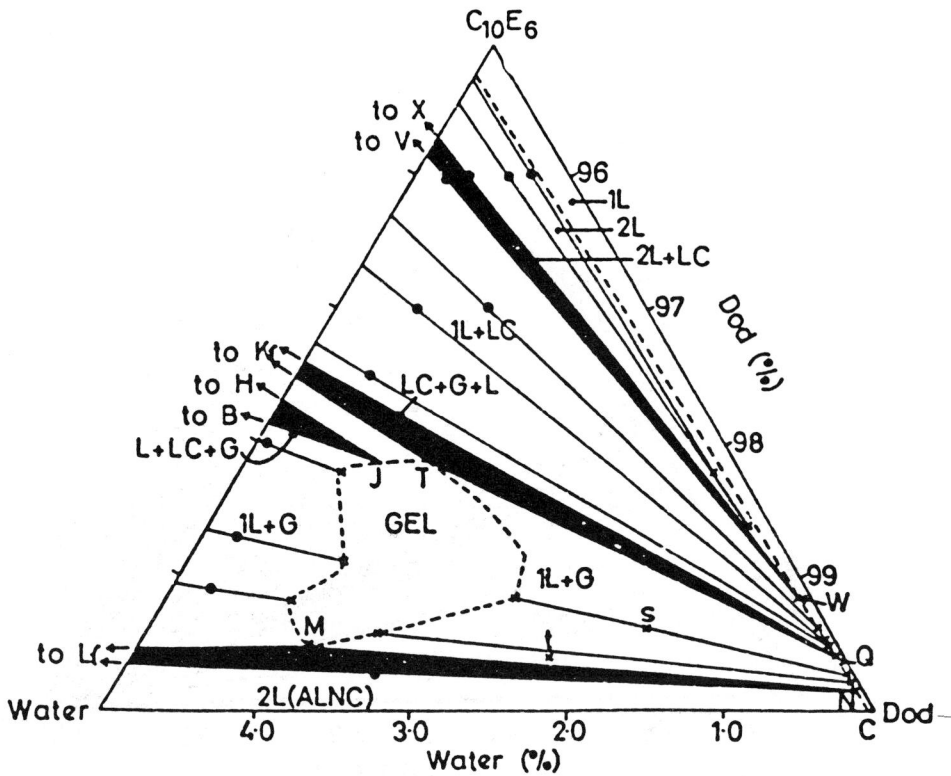

Figure III.1.53. Phase diagram (enlargement) of decylhexaoxy-ethylene glycol ether-water-dodecane ($C_{10}E_6$-H_2O-$C_{12}H_{26}$) system at

25 ºC (© Pharmaceutical Society of Great Britain, 1971. Reproduced with permission. Taken from reference 68).

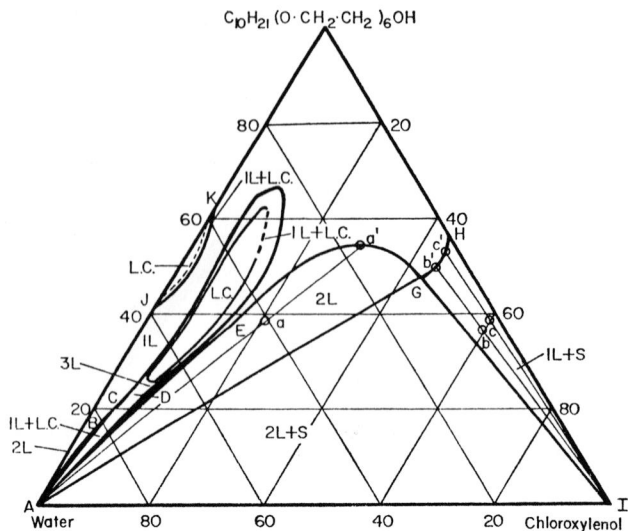

Figure III.1.54. Phase diagram of decylhexaoxyethylene glycol ether-water-4-chloro-3,5-xylenol ($C_{10}E_6$-H_2O-C_8H_9ClO) system at 25 °C

(© Academic Press, Inc., 1964. Reproduced with permission. Taken from reference 64).

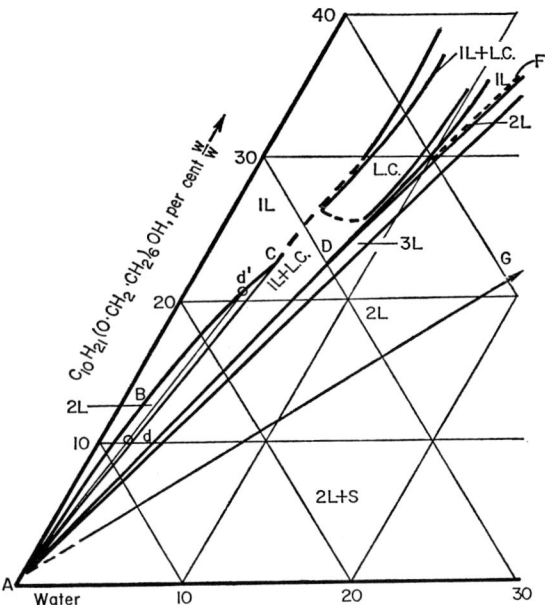

Figure III.1.55. Phase diagram (enlargement) of decylhexaoxy-ethylene glycol ether-water-4-chloro-3,5-xylenol ($C_{10}E_6$-H_2O-C_8H_9ClO) system at 25 ºC (© Academic Press, Inc., 1964. Reproduced with permission. Taken from reference 64).

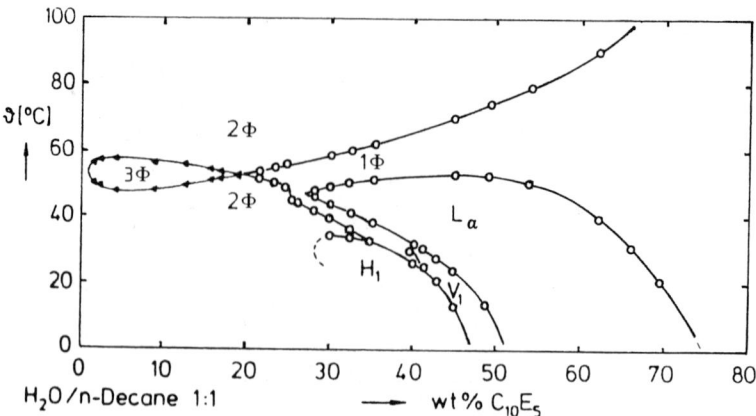

Figure III.1.56. Phase diagram of decylpentaoxyethylene glycol ether-water-decane ($C_{10}E_5$-H_2O-$C_{10}H_{22}$) system at a water-decane ratio of 1:1 (© American Chemical Society, 1986. Reproduced with permission. Taken from reference 82).

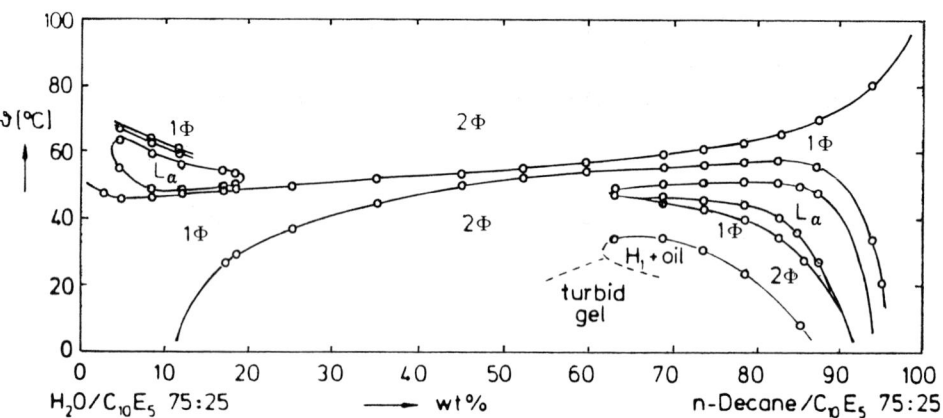

Figure III.1.57. Phase diagram of decylpentaoxyethylene glycol ether-water-decane ($C_{10}E_5$-H_2O-$C_{10}H_{22}$) system at a water-decylpenta-oxyethylene glycol ether ratio of 75:25 (© American Chemical Society, 1986. Reproduced with permission. Taken from reference 82).

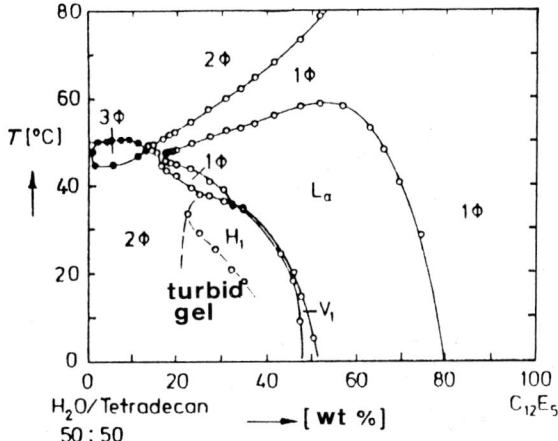

Figure III.1.58. Phase diagram of decylpentaoxyethylene glycol ether-water-tetradecane ($C_{10}E_5$-H_2O-$C_{14}H_{30}$) system at a water-tetradecane ratio of 1:1 (© VCH Verlagsgesellschaft, 1985. Reproduced with permission. Taken from reference 72).

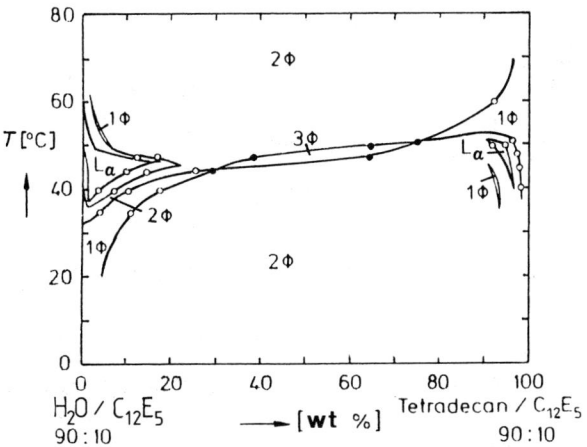

Figure III.1.59. Phase diagram of decylpentaoxyethylene glycol ether-water-tetradecane ($C_{10}E_5$-H_2O-$C_{14}H_{30}$) system at 10 wt% surfactant concentration (© VCH Verlagsgesellschaft, 1985. Reproduced with permission. Taken from reference 72).

300

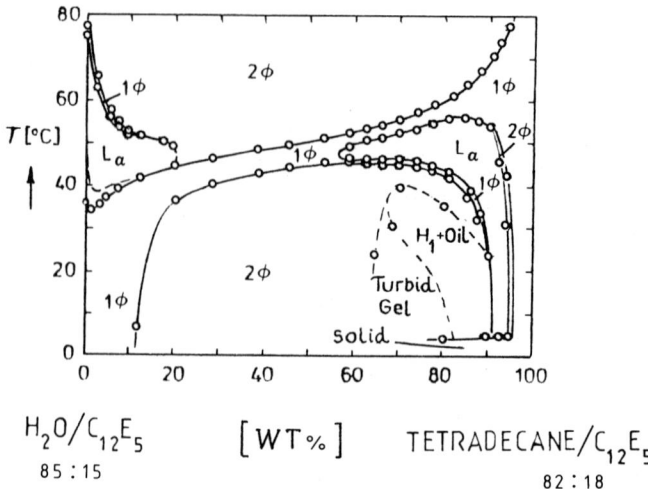

Figure III.1.60. Phase diagram: **do**decylpentaoxyethylene glycol ether-water-tetradecane ($C_{12}E_5$-H_2O-$C_{14}H_{30}$) system at 15 wt% surfactant concentration (© VCH Verlagsgesellschaft, 1985. Reproduced with permission. Taken from reference 72).

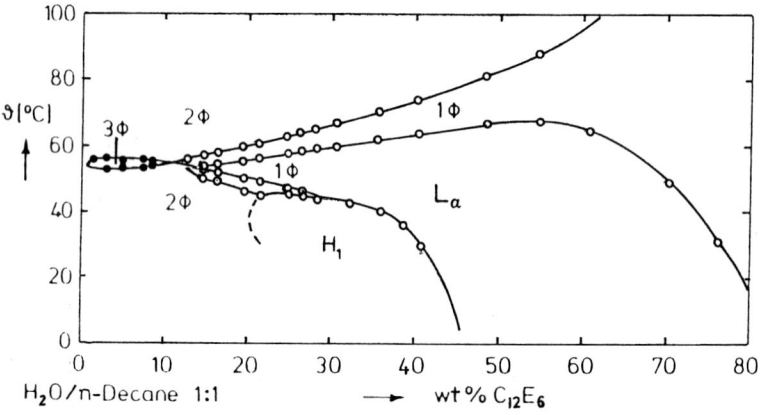

Figure III.1.61. Phase diagram of dodecylhexaoxyethylene glycol ether-water-decane ($C_{12}E_6$-H_2O-$C_{10}H_{22}$) system at a water-decane ratio of 1:1 (© American Chemical Society, 1986. Reproduced with permission. Taken from reference 82).

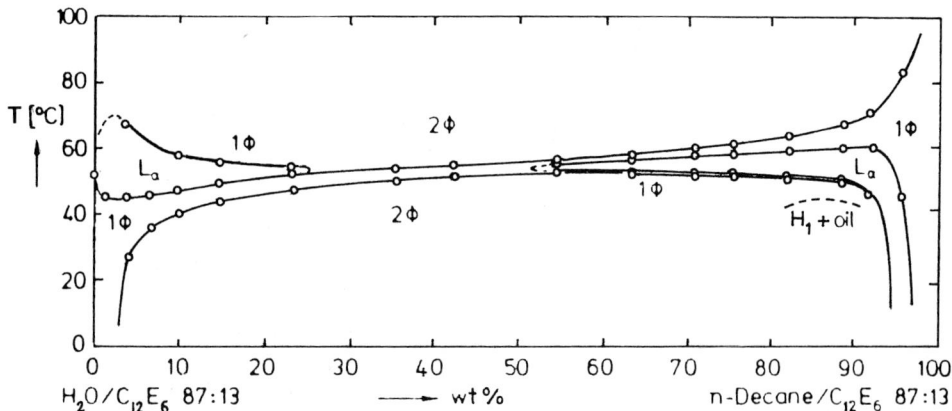

Figure III.1.62. Phase diagram of dodecylhexaoxyethylene glycol ether-water-decane ($C_{12}E_6$-H_2O-$C_{10}H_{22}$) system at a water-dodecylhexa-oxyethylene glycol ether ratio of 87:13 (© American Chemical Society, 1986. Reproduced with permission. Taken from reference 82).

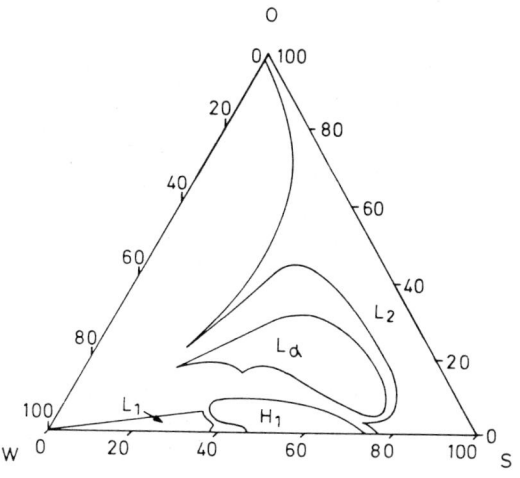

Figure III.1.63. Phase diagram of dodecyldecaoxyethylene glycol ether-water-oleic acid ($C_{12}E_{10}$-H_2O-$C_8H_{17}CH=CH(CH_2)_7COOH$) system at 20 ºC (© Springer Verlag, 1987. Reproduced with permission. Taken from reference 2).

REFERENCES Chapter III.1.

1) Lianos, P. and Zana, R., J. Colloid Interface Sci., **84**, 100 (1981).
2) Schambil, F. and Schwuger, M.J., Surfactants Consumer Products, 133 (1987).
3) O'Rourke, B.G.C., Ward, A.J.I., and Carroll, B.J., J. Pharm. Pharmacol., **39**, 865 (1987).
4) Herrington, T.M. and Sahi, S.S., J. Colloid Interface Sci., **121**, 107 (1988).
5) Corkill, J.M. and Walker, T., J. Colloid Interface Sci., **39**, 621 (1972).
6) Ottewill, R.H., Storer, C.C., and Walker, T., Trans. Faraday Soc., **63**, 2796 (1967).
7) Balmbra, R.R., Clunie, J.S., Corkill, J.M., and Goodman, J.F., Trans. Faraday Soc., **58**, 1661 (1962).
8) Brown, W., Johnsen, R., Stilbs, P., and Lindman, B., J. Phys. Chem., **87**, 4548 (1983).
9) Attwood, D., Elworthy, P.H., and Kayne, S.B., J. Phys. Chem., **74**, 3529 (1970).
10) Nakagawa, T., Kuriyama, K., and Inoue, H., J. Colloid Sci., **15**, 268 (1960).
11) Zana, R. and Weill, C., J. Phys. Lett., **46**, L-953 (1985).
12) Nishikido, N., J. Colloid Interface Sci., **120**, 495 (1987).
13) Meguro, K., Takasawa, Y., Kawahashi, N., Tabata, Y., and Uneo, M., J. Colloid Interface Sci., **83**, 50 (1981).
14) Kuwamura, T. in "Structure/Performance Relationships in Surfactants", Rosen, M.J. (Ed.), ACS Symposium Series 253, American Chemical Society, Washington DC (1984).
15) Clunie, J.S., Corkill, J.M., Goodman, J.F., Symons, P.C., and Tate, J.R., Trans. Faraday Soc., **63**, 2839 (1967).
16) Shil'nikov, G.V., Lobyshev, V.I., and Litvin, I.Ya., Kolloidnyi Zhurnal, **47**, 1002 (1985).
17) Mitchell, D.J., Tiddy, G.J.T., Waring, L., Bostock, T., and McDonald, M.P., J. Chem. Soc., Faraday Trans. I, **79**, 975 (1983).
18) Carroll, B.J., J. Colloid Interface Sci., **79**, 126 (1981).
19) Jayson, G.G. and Thompson, G., J. Colloid Interface Sci., **111**, 65 (1986).
20) Faucompre, B. and Lindman, B., J. Phys. Chem., **91**, 383 (1987).
21) Jolicoeur, C. and Philip, P.R., Can. J. Chem., **52**, 1834 (1974).
22) Weckström, K., Chem. Phys. Lett., **119**, 503 (1985).
23) Ueno, M., Kimoto, Y., Ikeda, Y., Momose, H., and Zana, R., J. Colloid Interface Sci., **117**, 179 (1987).
24) Balmbra, R.R., Clunie, J.S., Corkill, J.M., and Goodman, J.F., Trans. Faraday Soc., **60**, 979 (1964).
25) Aveyard, R. and Lawless, T.A., J. Chem. Soc., Faraday Trans. I, **82**, 2951 (1986).
26) Kunieda, H. and Shinoda, K., Bull. Chem. Soc. Japan, **55**, 1777 (1982).

27) Kunieda, H. and Shinoda, K., J. Dispersion Sci. Technol., **3**, 233 (1982).
28) Kunieda, H. and Shinoda, K., J. Colloid Interface Sci., **107**, 107 (1985).
29) Ravey, J.C., Progr. Colloid Polymer Sci., **73**, 107 (1987).
30) Ravey, J.C., Buzier, M., and Oberthur, R., Progr. Colloid Polymer Sci., **73**, 113 (1987).
31) Zulauf, M., Weckström, K., Hayter, J.B., Degiorgio, V., and Corti, M., J. Phys. Chem., **89**, 3411 (1985).
32) Anderson, B. and Olofsson, G., J. Colloid Polymer Sci., **265**, 318 (1982).
33) Olofsson, G., J. Phys. Chem., **89**, 1473 (1985).
34) Rosen, M.J., Cohen, A.W., Dahanayake, M., and Hua, X.-Y., J. Phys. Chem., **86**, 541 (1984).
35) Raney, K.H., Benton, W.J., and Miller, C.A., J. Colloid Interface Sci., **117**, 282 (1987).
36) Corkill, J.M., Goodman, J.F., and Harrold, S.P., Trans. Faraday Soc., **60**, 202 (1964).
37) Goto, A., Takemoto, M., and Endo, F., Bull. Chem. Soc. Japan, **58**, 247 (1985).
38) Rosen, M.J., "Surfactants and Interfacial Phenomena", John Wiley, New York (1978).
39) Becher, P., J. Colloid Sci., **16**, 49 (1961).
40) Elworthy, P.H. and Mac Farlane, C.B., J. Chem. Soc., 907 (1963).
41) Elworthy, P.H. and Florence, A.T., Kolloid-Z. Z. Polym., **195**, 23 (1964).
42) Elworthy, P.H. and McDonald, C., Kolloid-Z, **195**, 16 (1964).
43) Schick, M.J. and Gilbert, A.H., J. Colloid Sci., **20**, 464 (1965).
44) Elworthy, P.H. and Mac Pharlane, C.B., J. Pharm. Pharmacol. Suppl., **14**, 100 (1962).
45) Becher, P., J. Phys. Chem., **64**, 1221 (1960).
46) McDonald, C., J. Pharmacol., **22**, 774 (1970).
47) Schick, M.J., J. Phys. Chem., **67**, 1796 (1963).
48) Wrigley, A.M., Smith, F.D., and Stinton, A.J., J. Am. Oil Chem. Soc., **34**, 39 (1957).
49) Mulley, B.A. and Metcalf, A.D., J. Colloid Sci., **17**, 523 (1962).
50) Carless, J.E., Challis, R.A., and Mulley, B.A., J. Colloid Sci., **19**, 201 (1964).
51) Corkill, J.M., Goodman, J.F., and Ottewill, R.H., Trans. Faraday Soc., **57**, 1627 (1961).
52) Lange, H., Kolloid-Z, **201**, 131 (1965).
53) Tokiwa, F. and Moriyama, N., J. Colloid Interface Sci., **30**, 338 (1969).
54) Carroll, B.J., Doyle, P.J., Donegan, A.C., and Ward, A.J., J. Chem. Soc., Faraday Trans. I, **81**, 2975 (1985).
55) Clint, J.H., J. Chem. Soc., Faraday Trans. I, **71**, 1327 (1975).
56) Hua, X.-Y., and Rosen, M.J., J. Colloid Interface Sci., **90**, 212 (1982).
57) Ribeiro, A.A. and Dennis, E.A. in "Nonionic Surfactants", Schick, M.J. (Ed.), Marcel Dekker, New York (1987).
58) Ribeiro, A.A. and Dennis, E.A., J. Phys. Chem., **81**, 957 (1977).
59) Meguro, K., Ueno, M., and Esumi, K. in "Nonionic Surfactants", Schick, M.J. (Ed.), Marcel Dekker, New York (1987).

60) Ueno, M., Takasawa, Y., Miyashige, H., Tabata, Y., Meguro, K., Coll. Polym. Sci., **259**, 761 (1981).

61) Sjöblom, J., Stenius, P., and Danielson, J. in "Nonionic Surfactants", Schick, M.J. (Ed.), Marcel Dekker, New York (1987).

62) Chakhovskoy, N., Bull. Soc. Chem. Belg., **63**, 474 (1956).

63) Lang, J.C. and Margan, R.C., J. Chem. Phys., **73**, 5849 (1980).

64) Mulley, B.A. and Metcalf, A.D., J. Colloid Sci., **19**, 501 (1964).

65) Shinoda, K., J. Colloid Interface Sci., **34**, 278 (1970).

66) Schneider, G., Zeitschr. Phys. Chem. Neue Folge, **37**, 333 (1963).

67) Marland, J.S. and Mulley, B.A., J. Pharm. Pharmacol., **23**, 561 (1971).

68) Ali, A.A. and Mulley, B.A., J. Pharm. Pharmacol., **30**, 205 (1978).

69) Harusawa, F., Nakamura, S., and Mitsui, T., Colloid Polym. Sci., **252**, 613 (1974).

70) Leng, C.A. in "Physics of Amphiphiles: Micelles, Vesicles and Microemulsions", Degiorgio, V. and Corti, M. (Eds.), North Holland, Amsterdam (1985).

71) Corti, M. and Degiorgio, V., J. Phys. Chem., **85**, 1442 (1981).

72) Kahlweit, M. and Strey, R., Angew. Chem., **97**, 655 (1985).

73) Corti, M., Mineiro, C., and Degiorgio, V., J. Phys. Chem., **88**, 309 (1984).

74) Schott, H., J. Pharm. Sci., **58**, 1443 (1969).

75) Cox, K.R. and Benson, H.L., Fluid Phase Equilibria, **30**, 173 (1986).

76) Stache, H., "Tensid Taschenbuch", Carl Hansen Verlag, 2nd Ed., München (1981).

77) Becher, P. in "Nonionic Surfactants", Schick, M.J. (Ed.), Marcel Dekker, New York (1966).

78) Nakagawa, T. in "Nonionic Surfactants", Schick, M.J. (Ed.), Marcel Dekker, New York (1966).

79) Mulley, B.A. in "Nonionic Surfactants", Schick, M.J. (Ed.), Marcel Dekker, New York (1966).

80) Becher, P., J. Colloid Sci., **17**, 325 (1962).

81) Schick, M.J., J. Phys. Chem., **68**, 3585 (1964).

82) Kahlweit, M., Strey, R., and Firman, P., J. Phys. Chem., **90**, 671 (1986).

83) Hua, X.-Y. and Rosen, M.J., J. Colloid Interface Sci., **90**, 212 (1982).

84) Ahlnäs, T., Karlström, G., and Lindman, B., J. Phys. Chem., **91**, 4030 (1987).

85) Nilsson, P.G. and Lindman, B., J. Phys. Chem., **87**, 4756 (1983).

86) Nilsson, P.G., Wennerström, H., and Lindman, B., J. Phys. Chem., **87**, 1377 (1983).

III.2 Alkylphenol(ethylene oxide) ethers

Table III.2.1. Clouding temperature (T_c, in °C) of alkylphenol-(ethylene oxide) ethers ($C_n\phi EO_m$) with various additives.

Nonionic	Additive	Tc (°C)	ΔTc (°C)	Ref.
$C_8\phi EO_{7.3}$		21		5
$C_8\phi EO_{7.7}$		32		5
$C_8\phi EO_{8.0}$		40		5
$C_8\phi EO_{8.8}$		53		5
$C_8\phi EO_{9.5}$ [a]		64.5–67		1
$C_8\phi EO_{9.5}$ [a,b]	H^+		15	1
	Ag^+		11	1
	Li^+		4	1
	Na^+		−6	1
	K^+		−4.5	1
	NH_4^+		−0.5	1
	$(CH_3)_4N^+$		−9	1
	$(C_2H_5)_4N^+$		1.5	1
	$(HOC_2H_4)_3N^+H$		−2	1
	Mg^{2+}		4.5	1
	Ca^{2+}		2	1
	Ni^{2+}		4	1
	Zn^{2+}		4.5	1
	Cd^{2+}		5	1
	Pb^{2+}		9	1
	Al^{3+}		4	1
	OH^-		−24.5	1
	F^-		−23 [c]	1
	Cl^-		−10.5	1
	Br^-		−0.5	1
	I^-		17	1
	SCN^-		27	1
$C_8\phi EO_{9.5}$ [a,b]	$[Fe(CN)_5NO]^{2-}$		21.5	1
	SO_4^{2-}		−25.5	1
	PO_4^{3-}		−18	1
	$(H\ citrate)^{2-}$		−17.5	1
$C_8\phi EO_{9.5}$ [a]		67		2
	$10*10^{-5}M$ SLS [d]	72.5		2
	$20*10^{-5}M$ SLS	82		2
	$10*10^{-5}M$ SDeS [e]	71		2
	$10*10^{-5}M$ SDBS [f]	72.5		2

Table III.2.1. (continued)

Nonionic	Additive	Tc (°C)	ΔTc (°C)	Ref.
$C_8\phi EO_{9.5}$ [a]	0.50M Urea		4.4	3
	0.50M Methylurea		6.3	3
	0.50M Ethylurea		10.2	3
	0.50M 1,1 Dimethylurea		9.1	3
	0.50M 1,3 Dimethylurea		11.8	3
	0.50M 1,1 Diethylurea		23.5	3
	0.50M Tetramethylurea		22.4	3
	0.10M Phenylurea		−15.2	3
	0.50M Thiourea		6.8	3
	0.50M Methylthiourea		11.4	3
	0.50M 1,3 Dimethylthiourea		15.2	3
	0.50M Acetamide		3.9	3
	0.50M Guanidiniumchloride		4.3	3
$C_8\phi EO_{9.5}$ [a]		68.5		4
$C_8\phi EO_{9.6}$		64		5
$C_8\phi EO_{12.5}$		87		5
$n-C_9\phi EO_8$ [g]		68		6
$C_9\phi EO_9$		51		7
$C_9\phi EO_{30}$		>100		7
$C_{10}\phi EO_9$		79		6

a : Triton X-100
b : Total molal strength salt 2.0 M
c : Extrapolated
d : Sodium laurylsulphate
e : Sodium decylsulphate
f : Sodium dodecylbenzenesulphonate
g : Pure straight chain nonionic

Table III.2.2. Critical micelle concentration (CMC, in μmol/L) of alkylphenol(ethylene oxide) ethers ($C_n\phi EO_m$) at various temperatures and with various additives.

Nonionic	Additive	Temperature (°C)									Method	Ref.
		15	20	25	30	35	40	45	50	55		
$C_8\phi EO_1$ [a]		43.0		49.5		56.0		63.0		76.5	ST	8,9
$C_8\phi EO_1$				48.5							ST	8,9
$C_8\phi EO_2$ [a]		73.0		76.5		79.0		86.5		96.5	ST	8,9
$C_8\phi EO_2$				68							ST	8,9
$C_8\phi EO_3$ [b]				90							ST	10
$C_8\phi EO_3$ [a]		102		103		107		113		123	ST	8,9
$C_8\phi EO_3$				114							ST	8,9
				190							ST	27
$C_8\phi EO_4$ [a]		134		129		130		141		159	ST	8,9
$C_8\phi EO_4$				105							ST	8,9
$C_8\phi EO_5$ [a]		181		172		164		164		172	ST	8,9
$C_8\phi EO_5$				117							ST	8,9
				200							ST	27
$C_8\phi EO_6$ [b]				130							ST	10
$C_8\phi EO_6$ [a]		270		250		237		228		230	ST	8,9
$C_8\phi EO_6$				180							ST	8,9

Table III.2.2. (continued)

Nonionic	Additive	15	20	25	30	35	40	45	50	55	Method	Ref.
						Temperature (°C)						
$C_8\varnothing EO_7$ [a]		291		268		244		243		241	ST	8,9
$C_8\varnothing EO_7$				184							ST	8,9
$C_8\varnothing EO_8$ [b]				220							ST	10
$C_8\varnothing EO_8$ [a]		294		283		258		250		246	ST	8,9
$C_8\varnothing EO_8$				247							ST	8,9
$C_8\varnothing EO_9$ [a]		322		304		275		262		270	ST	8,9
$C_8\varnothing EO_9$				290							ST	8,9
$C_8\varnothing EO_{9.5}$ [c]				270							MC	11
				230								12
				280								13
				240							CV	4
$C_8\varnothing EO_{9.5}$ [c]				300 [i]							MC	26
	0.001M NaCl			270 [i]							MC	26
	0.010M NaCl			330 [i]							MC	26
	0.100M NaCl			340 [i]							MC	26
$C_8\varnothing EO_{9.5}$ [c]			175								ST	44
	5.3*10^{-3} M CD^s		360								ST	44
	17.1*10^{-3} M CD		580								ST	44
$C_8\varnothing EO_{9.5}$ [c]			300								ST	46
	0.36M KCl		200								ST	46
	0.36M KBr		250								ST	46
	0.36M KI		280								ST	46

Table III.2.2. (continued)

Nonionic	Additive	Temperature (°C)									Method	Ref.
		15	20	25	30	35	40	45	50	55		
$C_8\emptyset EO_{9.5}$ c	0.01% PEG 400 t					234	225	216			IS	47
	0.02% PEG 400					222	214	210			IS	47
	0.05% PEG 400					218	214	203			IS	47
	0.10% PEG 400					214	207	202			IS	47
	0.50% PEG 400					210	206	204			IS	47
$C_8\emptyset EO_{9.5}$ c	0.75% PEG 400					212	208	202			IS	47
	1.00% PEG 400 u					214	210	200			IS	47
	0.01% ACA					211	202	213			IS	47
	0.05% ACA					211	214	227			IS	47
	0.10% ACA					212	222	235			IS	47
	1.00% ACA					218	228	243			IS	47
$C_8\emptyset EO_{10}$ b				250							ST	10
$C_8\emptyset EO_{10}$ a		330		323		303		280		290	ST	8,9
$C_8\emptyset EO_{10}$				320							ST	8,9
$C_8\emptyset EO_{13}$				280							ST	27
$C_8\emptyset EO_{16}$				430							ST	8,9
$C_8\emptyset EO_{17}$				340							ST	27
$C_8\emptyset EO_{40}$				810							ST	8,9
$C_9\emptyset EO_5$				32 d							ST	14

Table III.2.2. (continued)

Nonionic	Additive	20	25	30	35	40	Method	Ref.
				Temperature (°C)				
$C_9\emptyset EO_5$			55				ST	15
	0.10 SDS [e]		50.1				ST	15
	0.25 SDS		41.7				ST	15
	0.40 SDS		58.2				ST	15
	0.50 SDS		66.8				ST	15
	0.75 SDS		158				ST	15
	0.90 SDS		295				ST	15
	0.95 SDS		407				ST	15
	0.98 SDS		794				ST	15
$C_9\emptyset EO_{8.5}$ [j]			50–70 [k]				ST	43
$C_9\emptyset EO_9$ [f]			75					16
$C_9\emptyset EO_{9.5}$			85				ST	8,17
$C_9\emptyset EO_{10}$			75				ST	8,18
$C_9\emptyset EO_{10}$			97				IS,LS	8,19
$C_9\emptyset EO_{10}$ [g]	0.03M NaCl			53				20
$C_9\emptyset EO_{10}$			75.8				ST	15
	0.10 SDS [e]		54.9				ST	15
	0.25 SDS		73.2				ST	15
	0.40 SDS		97.7				ST	15
	0.50 SDS		104				ST	15
	0.75 SDS		165				ST	15
	0.90 SDS		280				ST	15
	0.95 SDS		849				ST	15
	0.98 SDS		1174				ST	15

Table III.2.2. (continued)

Nonionic	Additive	Temperature (°C)			Method	Ref.
		20	25	30		
C9ØEO10 [e]	0.889 SDS [e]		390		FL	21
	0.909		450		FL	21
	0.995		2400		FL	21
	0.998		3400		FL	21
	0.999		4300		FL	21
C9ØEO10 [h]			75		ST	22
	3M Urea		100		ST	22
	6M Urea		240		ST	22
	3M Guanidinium Cl		140		ST	22
	1.5M Dioxane		100		ST	22
	3M Dioxane		180		ST	22
C9ØEO10.5 [l]			60–80 [k]		ST	43
C9ØEO10.5			82		ST	8,17
C9ØEO15			87		IS/LS	8,19
			110		ST	8,18
			120		ST	8,17
C9ØEO15			110		ST	8,18
	0.43M NaCl		65		ST	8,18
	0.86M NaCl		55		ST	8,18
	1.29M NaCl		45		ST	8,18
C9ØEO15 [m]			110–150 [k]		ST	43

Table III.2.2. (continued)

Nonionic	Additive	Temperature (°C) 20	25	30	Method	Ref.
$C_9 \varnothing EO_{15}$	0.10 SDS e		107		ST	15
	0.25 SDS		100		ST	15
	0.40 SDS		97		ST	15
	0.50 SDS		117		ST	15
	0.75 SDS		129		ST	15
	0.90 SDS		218		ST	15
	0.95 SDS		295		ST	15
	0.98 SDS		442		ST	15
			1010		ST	15
$C_9 \varnothing EO_{20}$	0.10 SDS e		132		ST	15
	0.25 SDS		125		ST	15
	0.40 SDS		115		ST	15
	0.50 SDS		128		ST	15
	0.60 SDS		147		ST	15
	0.70 SDS		200		ST	15
	0.90 SDS		239		ST	15
	0.98 SDS		292		ST	15
			955		ST	15
$C_9 \varnothing EO_{20}$			140		ST	8,18
$C_9 \varnothing EO_{20}$			150		ST	8,17
$C_9 \varnothing EO_{20}$ n			130–160 k		ST	43

Table III.2.2. (continued)

Nonionic	Additive	Temperature (°C) 20	25	30	Method	Ref.
$C_9\emptyset EO_{30}$	0.10 SDS [e]		173		ST	15
	0.25 SDS		125		ST	15
	0.40 SDS		123		ST	15
	0.50 SDS		138		ST	15
	0.60 SDS		190.5		ST	15
	0.75 SDS		240		ST	15
	0.90 SDS		263		ST	15
	0.95 SDS		384.5		ST	15
	0.98 SDS		851		ST	15
			1589		ST	15
$C_9\emptyset EO_{30}$			153		IS/LS	8,19
			185		ST	8,18
			275		ST	8,17
$C_9\emptyset EO_{30}$ [p]			150–200 [k]		ST	43
$C_9\emptyset EO_{31}$ [h]	3M Urea		180		ST	22
	6M Urea		350		ST	22
	3M Guanidinium Cl		740		ST	22
	3M Dioxane		430		ST	22
			570		ST	22
$C_9\emptyset EO_{40}$ [q]			170–210 [k]		ST	43
$C_9\emptyset EO_{50}$	0.43N NaCl		280		ST	8,18
	0.86N NaCNS		200		ST	8,18
	0.86N LiCl		225		ST	8,18
	0.86N NaCl		200		ST	8,18
	0.86N NaCl		150		ST	8,18
	0.86N $(CH_3)_4NCl$		100		ST	8,18
	0.86N Na_2SO_4		7		ST	8,18
	1.29N NaCl		100		ST	8,18

316

Table III.2.2. (continued)

Nonionic	Additive	Temperature (°C)			Method	Ref.
		20	25	30		
$C_9 \emptyset EO_{50}$ r			$190\text{--}230^k$		ST	43
$C_9 \emptyset EO_{100}$			1000		ST	8,17

a : Homogeneous by distillation
b : Pure compound
c : Triton X-100
d : At 23°C
e : In mole fraction SDS (Sodium Dodecyl Sulphate)
f : Igepal CO-630
g : Igepal CO-660
h : Not homogeneous, distribution reduced by molecular distillation
i : In mol/kg
j : Igepal CO-610
k : At 27.5°C
l : Igepal CO-710
m : Igepal CO-730
n : Igepal CO-850
p : Igepal CO-880
q : Igepal CO-890
r : Igepal CO-970
s : α-cyclodextrin
t : Percentage of polyethyleneglycol (MW 400) in the solvent
u : Percentage of acetamide

Table III.2.3. Relation between the critical micelle concentration (CMC, in mol/L) and the number of ethylene oxide (m) units in the hydrophile of alkylphenol(ethylene oxide) ethers ($C_n\phi EO_m$) at 25 °C.

Nonionic	Relation	Ref.
p–t– $C_8H_{17}C_6H_4(OCH_2CH_2)_mOH$	Log CMC = −3.8 + 0.029 m	23
$C_9H_{19}C_6H_4(OCH_2CH_2)_mOH$	Log CMC = −4.3 + 0.020 m	18

Table III.2.4. Relation between the critical micelle concentration (CMC, in mol/L) and various electrolyte concentrations (in mol/L) for Triton X-100[a] at 20 °C.

Salt	Relation	Ref.
KBr	CMC = $-1.3*10^{-5}$ log $\left[KBr\right]$ + $2.4*10^{-4}$	46
KI	CMC = $-4.6*10^{-6}$ log $\left[KI\right]$ + $2.7*10^{-4}$	46

a : p–t– $C_8H_{17}C_6H_4(OCH_2CH_2)_{9.5}$–OH

Table III.2.5. Critical micelle concentration (CMC, in mole fraction) of alkylphenol(ethylene oxide) ethers ($C_n\phi EO_m$) at various temperatures in different solvents.

Nonionic	Solvent	Temp. (°C)	CMC [b]	Ref.
$C_8\phi EO_{9.5}$ [a]	Ethylammonium nitrate	21	$5.93*10^{-3}$	24
		50	$6.11*10^{-3}$	24
$C_9\phi EO_9$ [c]	H_2O	27.5	$5.6*10^{-5}$	25
	Glycerol	27.5	$8.7*10^{-5}$	25
	Ethyleneglycol	27.5	$1.25*10^{-2}$	25
	2-Aminoethanol	27.5	$1.53*10^{-2}$	25
	Formamide	27.5	$1.57*10^{-2}$	25
	1,3-Propanediol	27.5	$6.3*10^{-2}$	25
	1,4-Butanediol	27.5	$1.6*10^{-1}$	25
	Ethylenediamine	27.5	$3.39*10^{-1}$	25
	1,3-Butanediol	27.5	$4.0*10^{-1}$	25
	2-Mercaptoethanol	27.5	$4.0*10^{-1}$	25
	Formic Acid	27.5	$4.5*10^{-1}$	25
	1,2-Propanediol	27.5	$5.0*10^{-1}$	25
	1-Amino-2-propanol	27.5	$5.0*10^{-1}$	25

a : Triton X-100
b : In mole fraction
c : Igepal CO-630

Table III.2.6. Gibbs free energy of micellization (ΔG°_m, in kJ/mol) of alkylphenol(ethylene oxide) ethers ($C_n\phi EO_m$).

Nonionic	Additive	20	25	30	35	40	45	50	55	Method	Ref.
$C_8\phi EO_{9.5}$ [a]			-20.4							MC	11
			-30.6		-31.8	-32.2	-32.9			UV	28
	0.01% PEG 400 [b]				-31.6	-32.4	-32.9			IS	47
	0.02% PEG 400				-31.8	-32.4	-33.0			IS	47
	0.05% PEG 400				-31.8	-32.5	-33.0			IS	47
	0.10% PEG 400				-31.9	-32.5	-33.0			IS	47
	0.50% PEG 400				-31.8	-32.4	-33.0			IS	47
	0.75% PEG 400				-31.9	-32.4	-33.0			IS	47
	1.00% PEG 400				-31.9	-32.5	-33.0			IS	47
	0.01% ACA [c]				-31.9	-32.4	-32.9			IS	47
	0.05% ACA				-31.9	-32.3	-32.7			IS	47
	0.10% ACA				-31.8	-32.3	-32.6			IS	47
	1.00% ACA				-31.7	-32.1	-32.5			IS	47
$C_8\phi EO_{9.5}$ [a]	10 vol % EG [d]				-30.7					UV	28
	15 vol % EG				-30.1					UV	28
	20 vol % EG				-29.6					UV	28
	40 vol % EG				-26.5					UV	28
	50 vol % EG				-24.8					UV	28
$C_8\phi EO_{12.5}$ [e]	10 vol % EG		-30.4		-31.6		-32.5			UV	28
	15 vol % EG				-30.5					UV	28
	20 vol % EG				-30.0					UV	28
	30 vol % EG				-29.3					UV	28
	40 vol % EG				-26.4					UV	28
	50 vol % EG				-24.6					UV	28

Table III.2.6. (continued)

Nonionic	Additive	Temperature (°C)								Method	Ref.
		20	25	30	35	40	45	50	55		
$C_8\phi EO_{30}$ f	50 vol % EG		-27.7		-29.1		-30.1			UV	28
$C_9\phi EO_{30}$ g	10.2 vol % EG		-31.3		-32.7		-33.9			UV	28
	15.9 vol % EG				-31.5					UV	28
	19.8 vol % EG				-30.8					UV	28
					-30.3					UV	28
	47.2 vol % EG				-25.5					UV	28

a : Triton X-100
b : Percentage poly(ethylene glycol) (MW 400) in solution
c : Percentage acetamide in solution
d : EG: Ethylene Glycol
e : Triton X-102
f : Triton X-305
g : Igepal CO-880

Table III.2.7. Gibbs free energy of micellization (ΔG°_m, in kJ/mol) of alkylphenol(ethylene oxide) ethers ($C_n \phi EO_m$) at various temperatures in different solvents.

Nonionic	Solvent	Temp. (°C)	ΔG°_m (kJ/mol)	Ref.
$C_8 \phi EO_{9.5}$ [a]	Ethylammonium nitrate	25.0	−12.5	24
$C_9 \phi EO_9$ [b]	H_2O	27.5	−24.4	25
	Glycerol	27.5	−22.6	25
	Ethylene glycol	27.5	−10.9	25
	2-Aminoethanol	27.5	−10.4	25
	Formamide	27.5	−10.4	25
	1,3-Propanediol	27.5	− 6.90	25
	1,4-Butanediol	27.5	− 4.56	25
	Ethylenediamine	27.5	− 2.71	25
	1,3-Butanediol	27.5	− 2.25	25
	2-Mercaptoethanol	27.5	− 2.25	25
	Formicacid	27.5	− 2.01	25
	1,2-Propanediol	27.5	− 1.71	25
	1-Amino-2-propanol	27.5	− 1.71	25

a : Triton X-100
b : Igepal CO-630

Table III.2.8. Enthalpy of micellization (ΔH°_m, in kJ/mol) of alkylphenol(ethylene oxide) ethers ($C_n\phi EO_m$).

Nonionic	Additive	15	20	25	30	35	40	45	50	55	Method	Ref.
$C_8\phi EO_{9.5}$ [a]				5.55							MC	26
	0.001M NaCl			5.68							MC	26
	0.010M NaCl			6.60							MC	26
	0.100M NaCl			6.43							MC	26
$C_8\phi EO_{9.5}$ [a]		13.1		9.0		4.9					MC	11
				8.8		3.8		-0.84			UV	28
$C_8\phi EO_{9.5}$ [a]	10%wt NaCl		13.5			4.2					MC	29
			3.9			0					MC	29
$C_8\phi EO_{9.5}$ [a]	0.02% PEG 400 [i]					6.00	6.20	6.40			IS	47
	0.05% PEG 400					3.11	3.21	3.32			IS	47
	0.10% PEG 400					4.10	4.24	4.37			IS	47
	0.50% PEG 400					2.89	2.99	3.08			IS	47
	0.75% PEG 400					2.73	2.81	2.91			IS	47
	1.00% PEG 400					4.20	4.34	4.48			IS	47
						4.15	4.28	4.41			IS	47
$C_8\phi EO_{9.5}$ [a]	0.05% ACA [j]					-5.96	-6.15	-6.35			IS	47
	0.10% ACA					-6.43	-6.64	-6.85			IS	47
	1.00% ACA					-6.64	-6.85	-7.07			IS	47
	10 vol % EG [k]			-4.6		-4.2		-3.8			UV	28
	15 vol % EG			-9.6		-10.4		-11.2			UV	28
	20 vol % EG			-12.5		-11.3		-9.6			UV	28
	40 vol % EG			-16.7		-13.8		-11.2			UV	28
	50 vol % EG			-12.1		-18.8		-25.1			UV	28
	60 vol % EG			-11.7		-18.8		-25.1			UV	28

323

Table III.2.8. (continued)

Nonionic	Additive	Temperature (°C)								Method	Ref.
		20	25	30	35	40	45	50	55		
$C_8ØEO_{12.5}$ [b]	10 vol % EG		8.8		0		-8.4			UV	28
	15 vol % EG		-1.3		-4.2		-7.1			UV	28
	20 vol % EG		-3.8		-8.4		-19.6			UV	28
	40 vol % EG		-2.5		-9.6		-16.3			UV	28
	50 vol % EG		-14.2		-13.7		-13.3			UV	28
	60 vol % EG		-14.2		-16.7		-19.6			UV	28
			-4.2		-10.4		-16.7			UV	28
$C_8ØEO_{12.5}$ [b]	10%wt NaCl	15.3			6.5					MC	29
		14.0			3.3					MC	29
$C_8ØEO_{16}$ [c]	10%wt NaCl	18.0			10.5					MC	29
		19.7			8.6					MC	29
$C_8ØEO_{30}$ [d]	10%wt NaCl	17.1			13.2					MC	29
			18.0		8.8		0			UV	28
		20.5			13.5					MC	29
$C_8ØEO_{40}$ [e]	10%wt NaCl	15.0			14.3					MC	29
		17.4			16.0					MC	29
$C_9ØEO_{9.5}$ [f]		7.4								MC	29
$C_9ØEO_{11}$ [g]		9.0								MC	29
$C_9ØEO_{15}$ [h]		12.0								MC	29

Table III.2.8. (continued)

Nonionic	Additive	Temperature (°C)								Method	Ref.
		20	25	30	35	40	45	50	55		
$C_9 \emptyset EO_{30}$ [l]	10.2 vol % EG		16.7		8.8					UV	28
	15.9 vol % EG		5.8		3.3					UV	28
	19.8 vol % EG		0.4		0.4					UV	28
	47.2 vol % EG		-0.4		-2.9		-5.4			UV	28
			-13.3		-9.6		-8.4			UV	28

a : Triton X-100
b : Triton X-102
c : Triton X-165
d : Triton X-305
e : Triton X-405
f : Triton N-01
g : Triton N-111
h : Triton N-150
i : Percentage poly(ethylene glycol) (MW 400) in solution
j : Percentage acetamide in solution
k : EG: ethylene glycol
l : Igepal CO-880

Table III.2.9. Enthalpy of micellization (ΔH°_m, in kJ/mol) of Triton X-100 at 25 °C in water and in ethylammonium nitrate.

Nonionic	Solvent	ΔH°_m (kJ/mol)	Ref.
$C_8\phi EO_{9.5}$	H_2O	9.0	11
	Ethylammonium nitrate	0	24

Table III.2.10. Entropy of micellization (ΔS°_m, in J mol^{-1}K^{-1}) of alkylphenol(ethylene oxide) ethers ($C_n\phi EO_m$).

Nonionic	Additive	Temperature (°C)					Method	Ref.
		25	30	35	40	45		
$C_8\phi EO_{9.5}$ a		99		117		100	MC	11
		134		122	123	123	UV	28
	0.02% PEG 400 b			113	114	114	IS	47
	0.05% PEG 400			117	117	117	IS	47
	0.10% PEG 400			113	113	113	IS	47
	0.50% PEG 400			112	113	113	IS	47
	0.75% PEG 400			117	117	118	IS	47
	1.00% PEG 400			117	117	118	IS	47
	0.05% ACA c			84	83	82	IS	47
	0.10% ACA			82	82	81	IS	47
	1.00% ACA			81	81	80	IS	47
	100% Ethylammoniumnitrate d	42					IS	24
$C_8\phi EO_{9.5}$ a	10 vol % EG e			88			UV	28
	15 vol % EG			63			UV	28
	20 vol % EG			59			UV	28
	40 vol % EG			42			UV	28
	50 vol % EG			21			UV	28
$C_8\phi EO_{12.5}$ f		134		104		75	UV	28
	10 vol % EG			84			UV	28
	15 vol % EG			71			UV	28
	20 vol % EG			63			UV	28
	40 vol % EG			42			UV	28
	50 vol % EG			25			UV	28

Table III.2.10. (continued)

Nonionic	Additive	Temperature (°C) 20	25	30	35	40	45	Method	Ref.
$C_8\emptyset EO_{30}$ g			155		121		96	UV	28
$C_9\emptyset EO_{30}$ h			159		133		109	UV	28
	10.2 vol % EG				113			UV	28
	15.9 vol % EG				100			UV	28
	19.8 vol % EG				88			UV	28
	47.2 vol % EG				50			UV	28

a : Triton X-100
b : Percentage poly(ethylene glycol) (MW 400) in solution
c : Percentage acetamide in solution
d : solvent is ethylammonium nitrate
e : EG: Ethylene Glycol
f : Triton X-102
g : Triton X-305
h : Igepal CO-880

Table III.2.11. Aggregation number of alkylphenol(ethylene oxide) ethers ($C_n\phi EO_m$) at various temperatures and with various additives.

Nonionic	Additive	5	10	15	20	25	30	35	40	45	50	Method	Ref.
$C_8\phi EO_{9.0}$						111						LS	8,30
$C_8\phi EO_{9.5}$[a]		25[b]	35[c]		75[d]		95[e]		105[f]		130[g]	FL	4
			70[h]			105		127[i]		156[j]		LS	
		10[k]	18[l]		35[m]			42[n]				FL	12
$C_8\phi EO_{9.5}$[a]						135						LS	32
	40mM NaCl				140[p]							LS	32
	120mM NaCl				140[p]							LS	32
	100mM NaOAc				140[p]							LS	32
					134							UC	32
$C_8\phi EO_{9.5}$[a]						100						UC	32
$C_8\phi EO_{9.5}$[a]						150	170	209	260	322		MO	33
$C_8\phi EO_{9.5}$[a]			25		43		66		95		133	FP	34
$C_8\phi EO_{9.5}$[a]			80[q]	100[r]	142	290	400					LS	46
	0.36M KCl				87							LS	46
	0.36M KBr				102							LS	46
	0.36M KI				122							LS	46
$C_8\phi EO_{9.7}$						100						UC	35
$C_8\phi EO_{10}$						139						LS	36

Table III.2.11. (continued)

Nonionic	Additive	20	25	30	35	40	45	50	55	Method	Ref.
					Temperature (°C)						
$C_8\emptyset EO_{12.5}$			11							VP	37
$C_8\emptyset EO_{16}$			8							VP	37
$C_8\emptyset EO_{20}$			6							VP	37
$C_8\emptyset EO_{30}$			4							VP	37
$C_9\emptyset EO_9$		280^s	350							LS	7
$C_9\emptyset EO_{9.5}$			265							UC	35
$C_9\emptyset EO_{10}$			276							LS	8,18
			100							LS	8,19
			142							MO	33
				125							39
		95	152			178		257		UC	38
$C_9\emptyset EO_{10}$	0.30M NaCl		100								8,40
	0.50M NaCl		113								8,40
			116								8,40
$C_9\emptyset EO_{10.3}$			125							UC	8,35
$C_9\emptyset EO_{12.5}$			100		7^t					VP	8,37
											39
$C_9\emptyset EO_{13}$			83							MO	33
$C_9\emptyset EO_{15}$			80							LS	8,18
	0.43M NaCl		83							LS	8,18

Table III.2.11. (continued)

Nonionic	Additive	Temperature (°C) 20	25	30	40	50	Method	Ref.
$C_9\emptyset EO_{15}$			52				LS	8,19
	0.30M NaCl		58				LS	8,40
	0.50M NaCl		53				LS	8,40
$C_9\emptyset EO_{18}$			50				MO	33
$C_9\emptyset EO_{20}$		36		40	44	49	UC	38
			56				LS	39
			62				LS	8,18
$C_9\emptyset EO_{30}$			60				LS	7
			55				LS	8,18
$C_9\emptyset EO_{30}$			19				LS	8,19
	0.30M NaCl		20				LS	8,40
	0.50M NaCl		16				LS	8,40
$C_9\emptyset EO_{30}$		20		22	24	28	UC	38
$C_9\emptyset EO_{40}$		24		26	30	34	UC	38
$C_9\emptyset EO_{50}$	0.43N NaCl		20				LS	8,18,40
	0.86N NaCl		19				LS	8,18,40
	0.86N LiCl		26				LS	8,18,40
	0.86N NaCNS		19				LS	8,18,40
	0.86N TMACl[u]		24				LS	8,18,40
	0.86N Na$_2$SO$_4$		32				LS	8,18,40
	1.29N NaCl		26				LS	8,18,40

a : Triton X-100
b : 2°C
c : 12°C
d : 22°C
e : 32°C
f : 41°C
g : 51°C
h : 10.3°C
i : 34.2°C
j : 44.1°C
k : 2.6°C
l : 12.6°C
m : 21.6°C
n : 32.8°C
p : r.t.
q : 11°C
r : 16°C
s : 21°C
t : 37°C
u : Tetramethyl ammonium chloride

Table III.2.12. Micellar radius (R, in nm) of alkylphenol(ethylene oxide) ethers ($C_n\phi EO_m$) at various temperatures and with various additives.

Nonionic	Additive	Temperature (°C)								Method	Ref.
		20	25	30	35	40	45	50	55		
$C_8\phi EO_9$ [a]	0.113 DMPC [d]			3.2[b]						SEC	42
	0.176 DMPC			3.8[b]						SEC	42
	0.258 DMPC			4.4[b]		4.4				SEC	42
	0.293 DMPC			5.4[b]						SEC	42
	0.411 DMPC			5.8[b]						SEC	42
	0.500 DMPC			7.0[b]						SEC	42
	0.117 PS [e]			8.6[b]						SEC	42
	0.104 PE [f]			3.7[b]						SEC	42
	0.043 LPC [g]			3.9[b]						SEC	42
	0.145 LPC			3.8[b]						SEC	42
	0.111 PA [h]			4.5[b]						SEC	42
				3.7[b]						SEC	42
	0.117 DPPC [i]					5.4				SEC	42
	0.191 DPPC					6.1				SEC	42
	0.247 DPPC					7.6				SEC	42
	0.336 DPPC					9.0				SEC	42
	0.09 SM [j]					4.7				SEC	42
$C_8\phi EO_{9.5}$ [k]				4.1[b]						SEC	42

Table III.2.12. (continued)

Nonionic	Additive		10	15	20	25	30	35	40	45	50	Method	Ref.
								Temperature (°C)					
$C_8\emptyset EO_{9.5}$[k]			4.0[q]	5.6[r]		8.5	8.9					LS	46
$C_8\emptyset EO_{9.5}$[k]			4.5[l]			5.2		7.2[m]		9.2[n]		QELS	4
			3.1[l]			3.2		3.2[m]		3.3[n]		PGNMR	4
			3.0		2.9		2.6		2.2		2.0	FP	34
$C_8\emptyset EO_{9.5}$[k]	SDS[p]	Ion strength											
	0	0.20M				6.0						QELS	41
		0.20M				4.7						SEC	41
		0.40M				6.2						QELS	41
		0.40M				4.7						SEC	41
		0.60M				7.0						QELS	41
		1.00M				9.5						QELS	41
		1.00M				6.8						SEC	41
	0.33	0.20M				6.0						QELS	41
		0.20M				4.6						SEC	41
		0.40M				10.0						QELS	41
		0.40M				7.2						SEC	41

Table III.2.12. (continued)

Nonionic	Additive			Temperature (°C)								Method	Ref.
				5	10	15	20	25	30	35	40		
$C_8\varnothing EO_{9.5}$ [a]	SDS^e Ion strength												
	0.33	0.60M						13.0				QELS	41
		1.00M						19.5				QELS	41
	0.50	0.20M						4.5				QELS	41
		0.40M						10.0				QELS	41
		0.60M						13.5				QELS	41
		1.00M						34.5				QELS	41
		1.00M						24.3				SEC	41
	0.60	0.60M						14.0				QELS	41
		0.60M						9.8				SEC	41
	0.90	0.20M						2.5				QELS	41
		0.20M						2.3				SEC	41
		0.40M						7.0				QELS	41
		0.40M						5.3				SEC	41
		1.00M						39.0				QELS	41
$C_9\varnothing EO_{10}$					3.43		3.81		4.18		4.60	UC	38
$C_9\varnothing EO_{20}$					3.25		3.25		3.21		3.21	UC	38
$C_9\varnothing EO_{30}$					2.86		2.90		2.94		3.01	UC	38
$C_9\varnothing EO_{40}$					2.52		2.55		2.62		2.67	UC	38

a : Pure p-(1,1,3,3-tetramethylbutyl) phenoxynonaoxyethylene glycol
b : 28°C
c : mole fraction phospholipid
d : DMPC: Dimyristoylphosphatidylcholine
e : PS: Phosphatidylserine
f : PE: Phosphatidylethanolamine
g : lAC: Lysophosphatidylcholine
h : PA: Palmitic Acid
i : DPPC: Dipalmitoylphosphatidylcholine
j : SM: Sphingomyelin
k : Triton X-100
l : 10.3°C
m : 34.2°C
n : 44.1°C
p : mole fraction SDS (Sodium Dodecylsulphate)
q : 11°C
r : 16°C

Table III.2.13. Second virial coefficient (in cm³ mol g^{-2}) for alkylphenol(ethylene oxide) ether-water ($C_n\phi EO_m-H_2O$) systems at various temperatures.

Nonionic	Temperature (°C)								Method	Ref.
	10	15	20	25	30	35	40	50		
$C_8\phi EO_{9.5}$[c]	$1.17*10^{-4}$[d,f]	$1.32*10^{-4}$[e,f]		$1.36*10^{-4}$[f]	$1.44*10^{-4}$[f]				LS	46
$C_9\phi EO_9$			$-1.03*10^7$[a,b]	$-1.51*10^7$[b]					LS	7
$C_9\phi EO_{10}$			$1.19*10^4$		$0.85*10^4$		$0.56*10^4$	$0.36*10^4$	UC	38
$C_9\phi EO_{20}$			$2.27*10^4$		$1.82*10^4$		$1.50*10^4$	$1.16*10^4$	UC	38
$C_9\phi EO_{30}$			$2.60*10^4$		$2.22*10^4$		$1.88*10^4$	$1.54*10^4$	UC	38
				$4.37*10^6$[b]					UC	38
$C_9\phi EO_{40}$			$2.90*10^4$		$2.42*10^4$		$2.01*10^4$	$1.64*10^4$	UC	38

a : 21°C
b : cm³ mol^{-1}
c : Triton X-100
d : 11°C
e : 16°C
f : in cm³/gr

Table III.2.14. Surface tension (mN/m) of alkylphenol-
(ethylene oxide) ethers at various temperatures and
with various additives

Nonionic	Additive	Temperature (°C)			Method	Ref.
		20	25	30		
$C_8\emptyset EO_1$[a]			35		ST	9
$C_8\emptyset EO_2$[a]			31		ST	9
$C_8\emptyset EO_3$[a]			28		ST	9
$C_8\emptyset EO_3$			31.8		ST	27
$C_8\emptyset EO_4$[a]			27		ST	9
$C_8\emptyset EO_5$[a]			28		ST	9
$C_8\emptyset EO_5$			32.6		ST	27
$C_8\emptyset EO_6$[a]			29		ST	9
$C_8\emptyset EO_7$[a]			31		ST	9
$C_8\emptyset EO_7$			33.4		ST	27
$C_8\emptyset EO_8$[a]			33		ST	9
$C_8\emptyset EO_9$[a]			34		ST	9
$C_8\emptyset EO_{9.5}$[b]			30.9		ST	24
$C_8\emptyset EO_{10}$[a]			36		ST	9
$C_8\emptyset EO_{10}$			34.7		ST	27

336

Table III.2.14. (continued)

Nonionic	Additive	Temperature (°C)			Method	Ref.
		20	25	30		
$C_8 \emptyset EO_{13}$			34.8		ST	27
$C_8 \emptyset EO_{16}$			34.8		ST	27
$C_8 \emptyset EO_{17}$			34.8		ST	27
$C_9 \emptyset EO_5$			27.7		ST	15
	0.10 SDS [c]		29.8		ST	15
	0.25 SDS		30.2		ST	15
	0.40 SDS		30.6		ST	15
	0.50 SDS		31.2		ST	15
	0.75 SDS		31.8		ST	15
	0.90 SDS		30.6		ST	15
	0.95 SDS		31.4		ST	15
	0.98 SDS		30.4		ST	15
$C_9 \emptyset EO_{10}$			32		ST	15
	0.10 SDS		32.2		ST	15
	0.25 SDS		32		ST	15
	0.40 SDS		33.6		ST	15
	0.50 SDS		33.2		ST	15
	0.75 SDS		33		ST	15
	0.90 SDS		33.6		ST	15
	0.95 SDS		34.3		ST	15
	0.98 SDS		34.5		ST	15

Table III.2.14. (continued)

Nonionic	Additive	Temperature (°C)			Method	Ref.
		20	25	30		
$C_9 \emptyset EO_{15}$			37.6		ST	15
	0.10 SDS		37.8		ST	15
	0.25 SDS		37.8		ST	15
	0.40 SDS		38.3		ST	15
	0.50 SDS		37.8		ST	15
	0.75 SDS		38.2		ST	15
	0.90 SDS		37.2		ST	15
	0.95 SDS		37.4		ST	15
	0.98 SDS		37.6		ST	15
$C_9 \emptyset EO_{20}$			38.2		ST	15
	0.10 SDS		40.2		ST	15
	0.25 SDS		40		ST	15
	0.40 SDS		39.8		ST	15
	0.50 SDS		40.4		ST	15
	0.60 SDS		40.2		ST	15
	0.75 SDS		40.2		ST	15
	0.90 SDS		40.7		ST	15
	0.95 SDS		39.4		ST	15
	0.98 SDS		40		ST	15
$C_9 \emptyset EO_{30}$			41		ST	15
	0.10 SDS		41.3		ST	15
	0.25 SDS		41.2		ST	15
	0.40 SDS		41.6		ST	15
	0.50 SDS		41.4		ST	15
	0.60 SDS		42.6		ST	15
	0.75 SDS		41.8		ST	15
	0.90 SDS		42.5		ST	15
	0.95 SDS		41.4		ST	15

a : Single Compounds
b : Triton X-100 in ethylammonium nitrate
c : Mole fraction SDS (Sodium Dodecyl Sulphate)

Table III.2.15. Surface excess concentration (in 10^{-10} mol/cm^2) of alkylphenol(ethylene oxide) ethers ($C_n\phi EO_m$) at various temperatures

Nonionic [a]	Temperature (°C)					Method	Ref.
	25	55	65	75	85		
$C_8\phi EO_3$	3.7				3.2	ST	23,45
$C_8\phi EO_4$	3.35					ST	23,45
$C_8\phi EO_5$	3.1					ST	23,45
$C_8\phi EO_6$	3.0	2.9			2.7	ST	23,45
$C_8\phi EO_7$	2.9					ST	9,23,45
$C_8\phi EO_8$	2.6					ST	9,23,45
$C_8\phi EO_9$	2.5					ST	9,23,45
$C_8\phi EO_{10}$	2.1	2.2			2.1	ST	23,45

a : Homogeneous single compounds

Table III.2.16. Head group area (nm^2) of alkylphenol-(ethylene oxide) ethers at various temperatures

Nonionic	Temperature (°C)			Method	Ref.
	25	55	85		
$C_8 \emptyset EO_3$[a]	4.5		5.2	ST	23,45
$C_8 \emptyset EO_4$[a]	5.0			ST	23,45
$C_8 \emptyset EO_5$[a]	5.3			ST	23,45
$C_8 \emptyset EO_6$[a]	5.6	5.8	6.1	ST	23,45
$C_8 \emptyset EO_7$[a]	5.8			ST	9,23,45
$C_8 \emptyset EO_8$[a]	6.4			ST	9,23,45
$C_8 \emptyset EO_9$[a]	6.6			ST	9,23,45
$C_8 \emptyset EO_{10}$[a]	8.0	7.9	7.45	ST	23,45
$C_9 \emptyset EO_{9.5}$	5.5			ST	17
$C_9 \emptyset EO_{10.5}$	6.0			ST	17
$C_9 \emptyset EO_{15}$	7.2			ST	17
$C_9 \emptyset EO_{20}$	8.2			ST	17
$C_9 \emptyset EO_{30}$	10.1			ST	17
$C_9 \emptyset EO_{100}$	17.2			ST	17

a : Homogeneous compounds

Table III.2.17. Hydration number of ethylene oxide units in alkylphenol(ethylene oxide) ether micelles

Nonionic	Temperature (°C)							Method	Ref.
	20	25	30	35	40	45	50		
$C_8\phi EO_{9.5}$ [a]		2.5	3.5	5	5	5.5		Vis	33
		4	4	3.5	3	3		Vol	33
			4.5					IR	48
			0.8					Co	48
			4					Compr.	48
	3.5							PFGNMR	49
	3.0							Vis	49
	1.5							PFGNMR	49
								2H–NMR	
$C_9\phi EO_{10}$		1.5^b						DE/NMR	50
		1.0^c						DE/NMR	50
	3.0		2.8		2.3		1.9	UC	38
		2						Vis	33
		4.5						Vol	33
$C_9\phi EO_{13}$		3.5						Vis	33
		7.5						Vol	33
$C_9\phi EO_{18}$		4.5						Vis	33
		18						Vol	33
$C_9\phi EO_{20}$	3.9		3.2		2.6		2.0	UC	38
$C_9\phi EO_{30}$	3.0		2.7		2.4		2.0	UC	38
$C_9\phi EO_{40}$	2.3		1.9		1.7		1.4	UC	38

a : Triton X-100
b : Hexagonal phase
c : Lamellar phase

Table III.2.18. Molar volume (V, in cm^3/mol) of alkylphenol (ethylene oxide) ethers ($C_n\phi EO_m$) at 25 °C

Nonionic	V (cm^3/mol)	Method	Ref.
$C_8\phi_3$	393	ST	27
$C_8\phi_5$	495	ST	27
$C_8\phi_7$	598	ST	27
$C_8\phi_{10}$	743	ST	27
$C_8\phi_{13}$	905	ST	27
$C_8\phi_{16}$	1006	ST	27
$C_8\phi_{17}$	1129	ST	27
$C_9\phi_{10}$	602	De	51
$C_9\phi_{12.5}$	719	De	51
$C_9\phi_{15}$	792	De	51
$C_9\phi_{20}$	971	De	51
$C_9\phi_{30}$	1361	De	51
$C_9\phi_{50}$	2125	De	51

Table III.2.19. Proton chemical shifts[b] (δ , in ppm) and proton spin-lattice relaxation times (T_1, in ms) for Igepal CO-630[a] (4 mmol/L) in micelles in D_2O (Taken from reference 52).

C–H	δ (ppm)	T_1 (ms)	C–H	δ (ppm)	T_1 (ms)
1	0.58	230	5	3.65	240
2	0.64	200	6	3.80	130
3	1.00	95	7	6.64	200
4	3.40–3.60	240, 280, 350, 505, 515, 545	8	6.92	165

a : $CH_3\,CH_2-(CH_2)_7-$⬡$-\,OCH_2\,CH_2-(OCH_2\,CH_2)_8\,OH$

     1 2   3   8 7   6   5   4

b : Reference is TMS

Table III.2.20. Carbon-13 chemical shifts[b] (δ ,in ppm) and carbon-13 spin-lattice relaxation times (T_1, in ms) for Triton X-100[a] (4 mmol/L) in micelles in D_2O (Taken from reference 52).

^{13}C	δ (ppm)	T_1 (ms)	^{13}C	δ (ppm)	T_1 (ms)
1	32.4	380	7	128.0	480
2	57.6	320	8	158.0	2000
3	32.8	1800	9	68.0	350
4	38.0	1850	10	70.8	380
5	143.0	1900	11	61.2	1100
6	115.0	460	12	72.8	1000

a :

b : Reference is TMS

Table III.2.21. Diffusion coefficient (D_{mic}, in m^2s^{-1}) of Triton X-100 micelles[a,b].

Temperature (°C)	D_{mic} ($m^2 s^{-1}$)
15	$4.4 * 10^{-9}$
25	$6.2 * 10^{-9}$
35	$8.2 * 10^{-9}$
45	$9.7 * 10^{-9}$

a : Triton X-100 : $i-C_8H_{17}$—⬡—$(EO)_{9.5}$—OH

b : From ref. 53

Table III.2.22. Critical parameters for the system Triton X-100/water as obtained by light scattering measurements.

Equation [a]	γ	C ($m^2 s^2$)	Ref.
$\dfrac{\delta \pi}{\delta c} = \left[\dfrac{T_c - T}{T}\right]^{-\gamma}$	0.93	$6.3 * 10^{-4}$	4
$M_{mic} = \left[\dfrac{T_c - T}{T}\right]^{-\gamma}$	2.03		46

a : $\dfrac{\delta \pi}{\delta c}$: Osmotic compressibility

M_{mic}: micellar molecular weight
T_c : Clouding or critical temperature
T : Experimental temperature
C and γ : Critical parameters

Table III.2.23. Various physical properties of alkylphenol(ethylene oxide) ether ($C_n\phi EO_m$) micelles from probe studies[a].

Nonionic	Temp. (°C)	P	η (cP)	ε	λ_{EM} (nm)	I_1/I_3	τ (ns)	$k_e *10^{-6}$ (dm^3/mol.s)	Probe[b]	Ref.
$C_8\phi EO_{9.5}$ [c]	2.6						272	4.2[d]	PY	12
	10	0.040	15				2.9	645	2MA	34
	10						434		PY	34
	12.6						211	4.6[d]	PY	12
	rt			27					2MA	54,55
	rt		35	30					ET(30)	54,56
	e			15					ET(30)	54,56
	e			35					PYCHO	54,57
	e			32					ANS	54,57
	e			32					HHC	54,59
	rt		12						HHC	54,59
	rt		12						Bz-9-Ant	54,60
	rt		12						Bz-1-Pn	54,60
									Bz-1-Np	54,60

Table III.2.23. (continued)

Nonionic	Temp. (°C)	p	η (cP)	ε	λ_EM (nm)	I₁/I₃	τ (ns)	$k_e *10^{-6}$ (dm³/mol.s)	Probe[b]	Ref.
C₈ØEO₉.₅ [c]	20	0.042	18				3.25	545	2MA	34
	20		156						9MA	54,61
	20		90						9MA	54,61
	20		879						A	54,61
	20		60						A	54,61
	20						400		PY	34
	20		191						2AP	54,63
	20		222						2AP	54,63
	20		206						12AS	54,63
	20		210						12AS	54,63
	20		227						9AS	54,63
	20		228						9AS	54,63
	20		235						6AS	54,63
	20		198						6AS	54,63
	20		11						DMA	54,61
	21.6						175	4.3[d]	PY	12
	22		158						PER	54,62
	24		28	28					DQEB	54,58
	25		145						PY	54,64
	25		29.7						PY(3)PY	54,64
	25		250						PY(3)PY	54,64
	25		54.1						PY(10)PY	54,64
	25		300						PY(10)PY	54,64
	30		105						PY(10)PY	54,65
	30	0.055	24				3.10	415	2MA	34
	30						384		PY	34
	32.8						120	6.1[d]	PY	12
	40	0.097	48				2.8	215	2MA	34
	40						370		PY	34
	50	0.128	66				2.4	165	2MA	34
	50						345		PY	34

Table III.2.23. (continued)

Nonionic	Temp. (°C)	P	η (cP)	ε	λ_{EM} (nm)	I_1/I_3	τ (ns)	k_e*10^{-6} (dm^3/mol.s)	Probe[b]	Ref.
$C_9\varnothing EO_{7.5}$	rt				451				PYCHO	66
	rt				331				In	66
	rt					1.41			PY	66
$C_9\varnothing EO_{10}$	rt				452				PYCHO	66
	rt				330				In	66
	rt					1.41			PY	66
	e			33.8					PY	21
$C_9\varnothing EO_{15}$	rt				452				PYCHO	66
	rt				330				In	66
	rt					1.43			PY	66
$C_9\varnothing EO_{18}$	rt				453				PYCHO	66
	rt				331				In	66
	rt					1.46			PY	66
$C_9\varnothing EO_{20}$	rt				452				PYCHO	66
	rt				332				In	66
	rt					1.47			PY	66

Table III.2.23. (continued)

a : P   : Fluorescence polarization
     η   : Microscopic viscosity
     ε   : Effective dielectric constant
     λ_{EM} : Wavelength of maximum intensity of the emission signal
     I_1/I_3 : ratio of the first and third vibronic peak in the emission spectrum of pyrene
     τ   : Lifetime of the fluorescence probe in the excited state
     k_e*10^{-6} : rate constant for intramicellar eximer formation

b : PY   : Pyrene
     2MA   : 2-methylanthracene
     ET(30)   : 2,6-diphenyl-4-(2,4,6-triphenyl-1-pyridinio)phenoxide
     PYCHO   : 1-pyrene-3-carboxaldehyde
     ANS   : 1anilinonaphthalene-8-sulphonate
     HHC   : 4-heptadecyl-7-hydroxycoumarin
     Bz-9-Ant: benzyl-9-anthroate
     Bz-1-Pm : benzyl-1-pyrenoate
     Bz-1-Np : Benzyl-2-naphthoate
     9MA   : 9-methylanthracene
     A   : Anthracene
     2AP   : 2-(9-anthroyloxy)palmitate
     12AS   : 12-(9-anthroyloxy)stearate
     9AS   : 9-(9-anthroyloxy)stearate
     6AS   : 6-(9-anthroyloxy)stearate
     DMA   : N,N-dimethylaniline
     PER   : perylene
     DCQEB   : 2[6-(2,2-dicyanovinyl)-3,4-dihydro-2,2,4-trimethyl-1(2H)-quinoyl]ethyl benzoate
     PY(3)PY : 1,3-di(1-pyrenyl)propane
     PY(10)PY: 1,10-di(1-pyrenyl)decane
     In   : Indole surfactant

c : Triton X-100

d : Unit s^{-1}

e : Temperature not specified

Table III.2.24. Partition coefficient (K_D) of single-species (S) and normal-distributed (N) octylphenol(ethylene oxide) ethers ($C_8\phi EO_m$) between isooctane and water at 25 °C (Taken from reference 67).

Nonionic	Type	K_D
$C_8\phi EO_1$	S	$1.84 * 10^{-4}$
$C_8\phi EO_2$	S	$7.17 * 10^{-4}$
$C_8\phi EO_{2.00}$	N	$4.14 * 10^{-3}$
$C_8\phi EO_{2.98}$	N	$1.43 * 10^{-2}$
$C_8\phi EO_3$	S	$3.13 * 10^{-3}$
$C_8\phi EO_4$	S	$9.83 * 10^{-3}$
$C_8\phi EO_{4.07}$	N	$4.75 * 10^{-2}$
$C_8\phi EO_5$	S	$2.46 * 10^{-2}$
$C_8\phi EO_{5.01}$	N	$1.29 * 10^{-1}$
$C_8\phi EO_6$	S	$5.92 * 10^{-2}$
$C_8\phi EO_{6.03}$	N	$2.54 * 10^{-1}$
$C_8\phi EO_7$	S	$1.82 * 10^{-1}$
$C_8\phi EO_{7.05}$	N	$4.49 * 10^{-1}$
$C_8\phi EO_8$	S	$5.04 * 10^{-1}$
$C_8\phi EO_{8.03}$	N	$8.22 * 10^{-1}$
$C_8\phi EO_9$	S	1.42
$C_8\phi EO_{9.06}$	N	1.55
$C_8\phi EO_{9.93}$	N	1.89
$C_8\phi EO_{10}$	S	3.85
$C_8\phi EO_{16}$	N	31.3
$C_8\phi EO_{40}$	N	47.2

Table III.2.25. Phase inversion temperature (PIT, in °C) of alkyl-phenol(ethylene oxide) ethers ($C_n\phi EO_m$) in surfactant-water-cyclohexane systems[a].

Nonionic	Oil	Composition (Non/wa/oil) (wt%)	PIT (°C)
$C_6\phi EO_{7.5}$	Cyclohexane	3/48.5/48.5	65
$C_6\phi EO_{7.5}$	Cyclohexane	10/45/45	52
$C_8\phi EO_{8.4}$	Cyclohexane	3/48.5/48.5	60.6
$C_8\phi EO_{8.5}$	Cyclohexane	5/47.5/47.5	57
$C_9\phi EO_{8.1}$	Cyclohexane	5/47.5/47.5	50
$C_9\phi EO_{8.6}$	Cyclohexane	3/48.5/48.5	60.5
$C_9\phi EO_{8.6}$ + 5wt% $C_{12}\phi SO_3\frac{1}{2}Ca$	Cyclohexane	3/48.5/48.5	> 100
$C_{12}\phi EO_{5.3}$[b]	Cyclohexane	4/48/48	54
$C_{12}\phi EO_8$[c]	Cyclohexane	4/48/48	51.5
$C_{12}\phi EO_{8.0}$[d]	Cyclohexane	4/48/48	64.7
$C_{12}\phi EO_{8.2}$[d]	Cyclohexane	4/48/48	53.5
$C_{12}\phi EO_{9.4}$	Cyclohexane	5/47.5/47.5	51
$C_{12}\phi EO_{9.7}$	Cyclohexane	3/48.5/48.5	61
$C_{16}\phi EO_{12.4}$	Cyclohexane	3/48.5/48.5	49.5
$C_{16}\phi EO_{12.4}$	Cyclohexane	5/47.5/47.5	48

a : From ref. 68
b : Commercial sample
c : Homogeneous sample
d : Molecularly distilled sample

Table III.2.26. Interfacial tension (γ, in dynes/cm) of various octylphenol(ethylene oxide) ethers ($C_8\phi EO_m$) between water and isooctane at 25°C (Taken from reference 9).

Nonionic	Interfacial Tension (Dynes/cm)
$C_8\phi EO_3$	0.5
$C_8\phi EO_4$	0.5
$C_8\phi EO_5$	0.5
$C_8\phi EO_6$	0.5
$C_8\phi EO_7$	1
$C_8\phi EO_8$	2.5
$C_8\phi EO_9$	3
$C_8\phi EO_{10}$	4

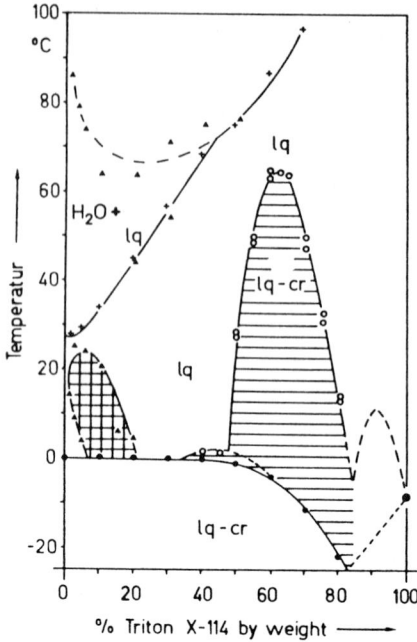

Figure III.2.1. Phase diagram of Triton X-114 ($C_8 \phi EO_{7.5}$). Phase boundary determined using turbidity measurements (+), polarization microscopy (o), thermography (•), and viscometry (▲) (© VCH Verlagsgesellschaft, 1987. Reproduced with permission. Taken from reference 86).

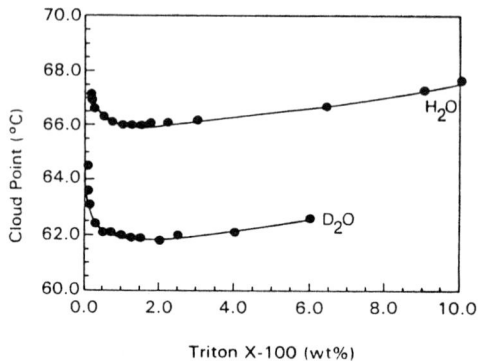

Figure III.2.2. Phase diagram of Triton X-100 ($C_8 \phi EO_{9.5}$).(© Academic Press, Inc., 1988. Reproduced with permission. Taken from reference 69).

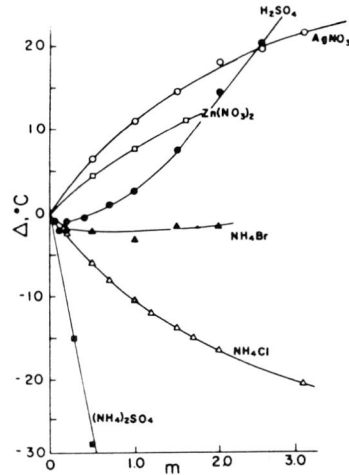

Figure III.2.3. Effect of electrolytes on the clouding temperature of a 2 wt% Triton X-100 ($C_8 \phi EO_{9.5}$) solution as a function of electrolyte concentration (in mol/kg) (© Academic Press, Inc., 1984. Reproduced with permission. Taken from reference 1).

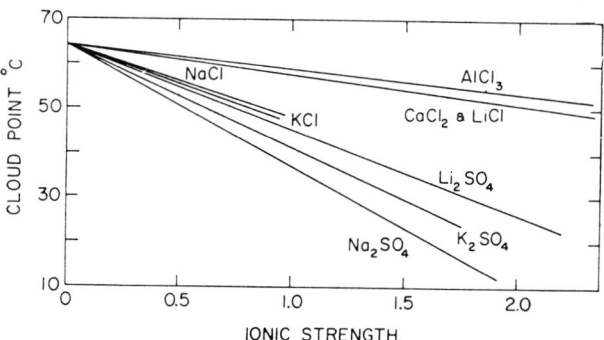

Figure III.2.4. Effect of electrolytes on the clouding temperature (Cloud point, in ºC) of 2 wt% Triton X-100 ($C_8 \phi EO_{9.5}$) solution as a function of ionic strength (© Academic Press, Inc., 1956. Reproduced with permission. Taken from reference 71).ϕ

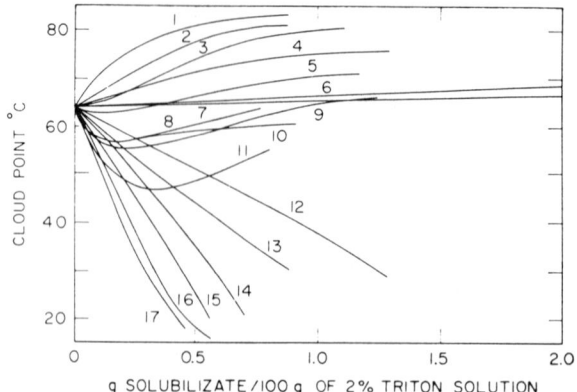

Figure III.2.5. Effect of solubilizates on the clouding temperature (cloud point, in ₀C) of 2 wt% Triton X-100 ($C_8 \phi EO_{9.5}$) solution.
(1) hexadecane; (2) dodecane; (3) decane; (4) 1-tetradecene;
(5) tetradecanethiol; (6) acetone; (7) citric acid; (8) 1-octene; (9) hexane; (10) 2-ethylhexene; (11) cyclohexane; (12) aniline; (13) butylacetaat; (14) ethylene dichloride; (15) phenol and oleic acid; (16) dodecanol and nitrobenzene; (17) benzene (© Academic Press, Inc., 1956. Reproduced with permission. Taken from reference 71).

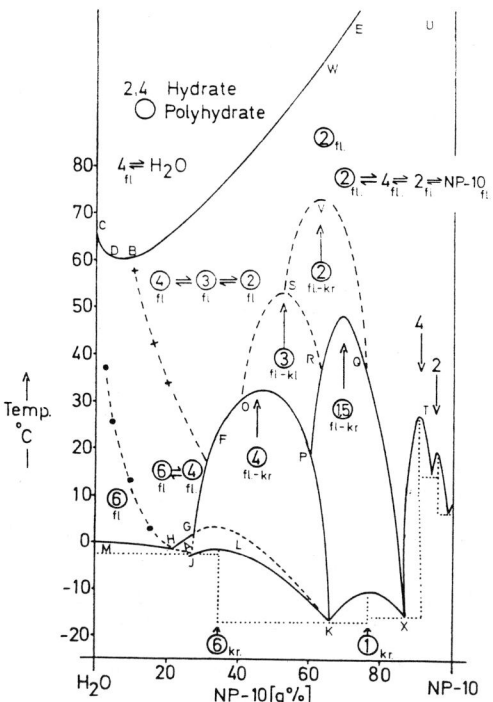

Figure III.2.6. Phase diagram for nonylphenol(ethylene oxide)$_{10}$ ether ($C_9 \phi EO_{10}$) (© Carl Hanser Verlag, 1983. Reproduced with permission. Taken from reference 73).

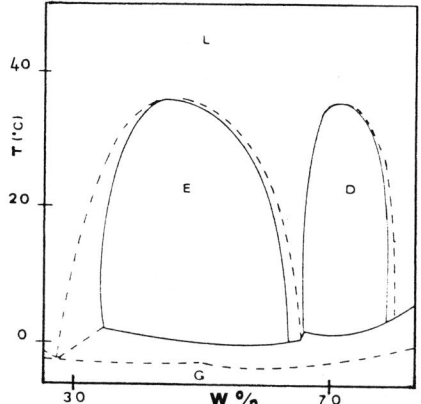

Figure III.2.7. Phase diagram for nonylphenol(ethylene oxide)$_{10}$ ether ($C_9 \phi EO_{10}$) (© Academic Press, Inc., 1988. Reproduced with permission. Taken from reference 50).

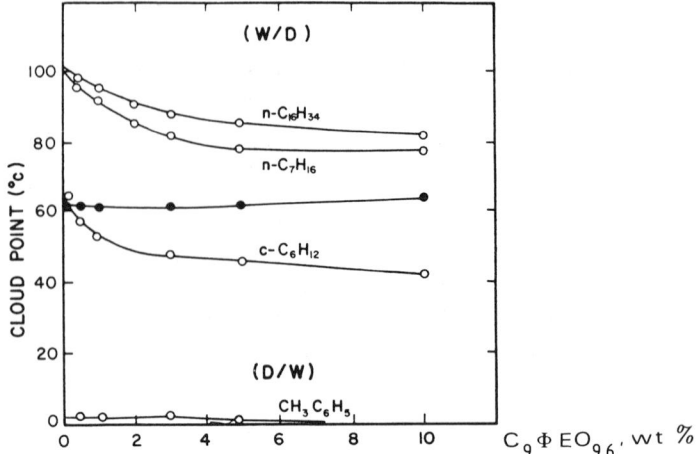

Figure III.2.8. Effect of solubilizates on the clouding temperature (cloud point, in °C) of nonylphenol(ethylene oxide)$_{9.6}$ ether ($C_9 \phi EO_{9.6}$). The solid circles represent the clouding temperature in the absence of hydrocarbon. (© American Chemical Society, 1964. Reproduced with permission. Taken from reference 74).

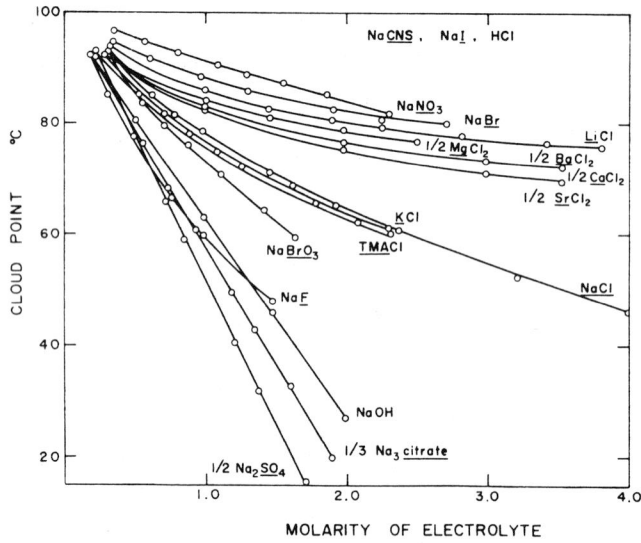

Figure III.2.9. Effect of electrolytes on the clouding temperature (cloud point, in °C) of nonylphenol(ethylene oxide)$_{15}$ ether ($C_9 \phi EO_{15}$). (© Academic Press, Inc., 1962. Reproduced with permission. Taken from reference 75).

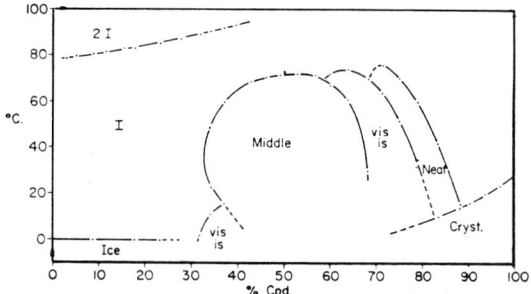

Figure III.2.10. Phase diagram of decylphenol(ethylene oxide)$_9$ ether (C$_{10}$ φ EO$_9$) (© Academic Press, Inc., 1976. Reproduced with permission. Taken from reference 76).

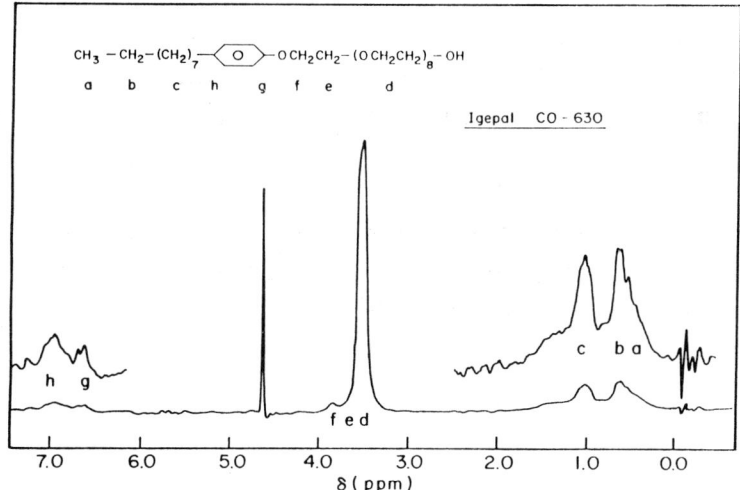

Figure III.2.11. Proton-NMR spectrum of 5 mmol/L Igepal CO-630 (C$_9$ φ EO$_8$) in D$_2$O (© Plenum Press, 1987. Reproduced with permission. Taken from reference 52).

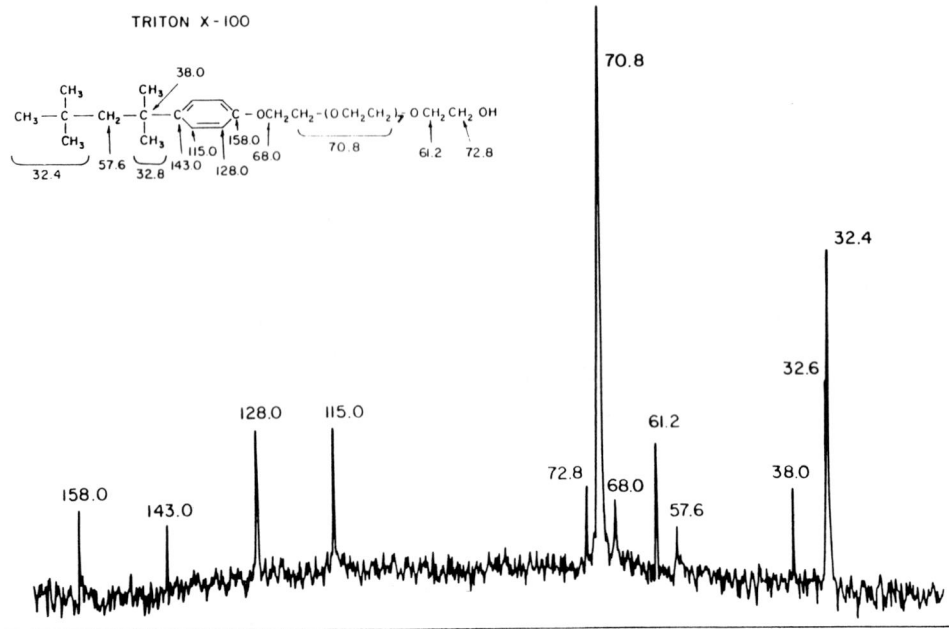

Figure III.2.12. Carbon-13 NMR spectrum of Triton X-100 ($C_8 \phi EO_{9.5}$) above the critical micelle concentration in D_2O (© Plenum Press, 1987. Reproduced with permission. Taken from reference 52).

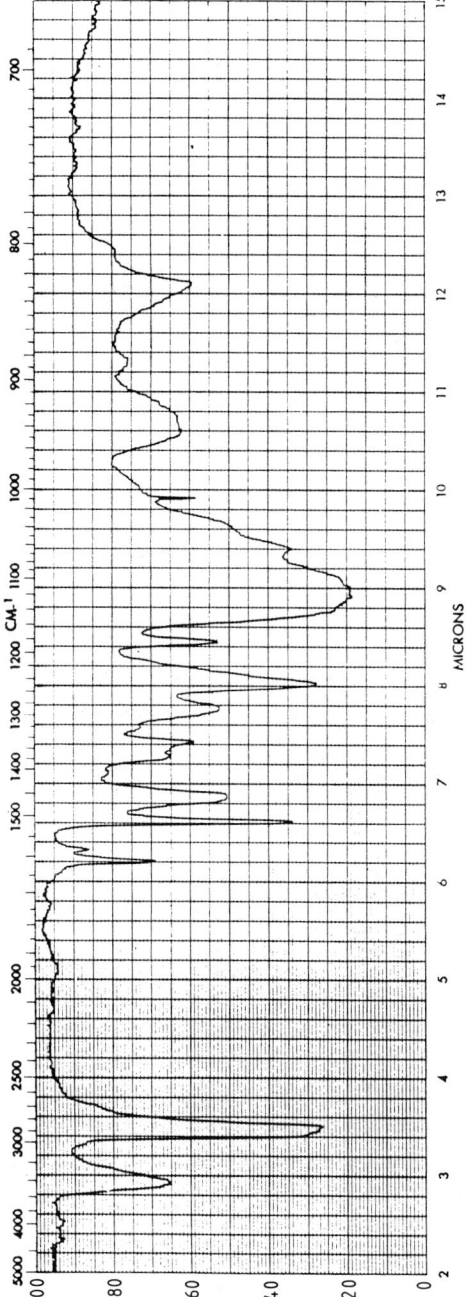

Figure III.2.13. Infrared spectrum of Polytergent G-200 (C$_8$ φ EO$_7$) (© Marcel Dekker, 1966. Reproduced with permission. Taken from reference 77).

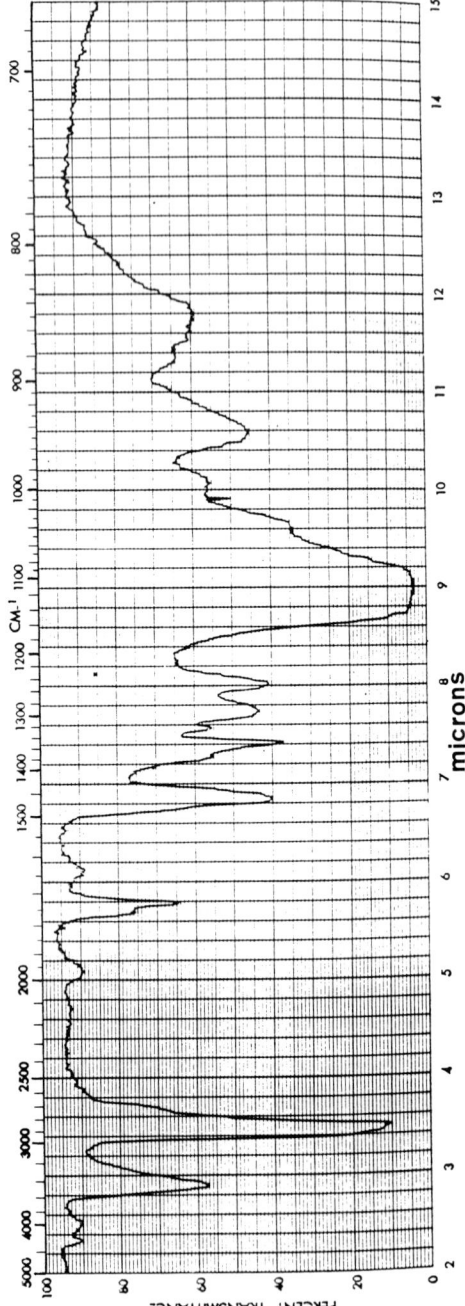

Figure III.2.14. Infrared spectrum of Polytergent B-200 ($C_9 \phi EO_7$) (© Marcel Dekker, 1966. Reproduced with permission. Taken from reference 77).

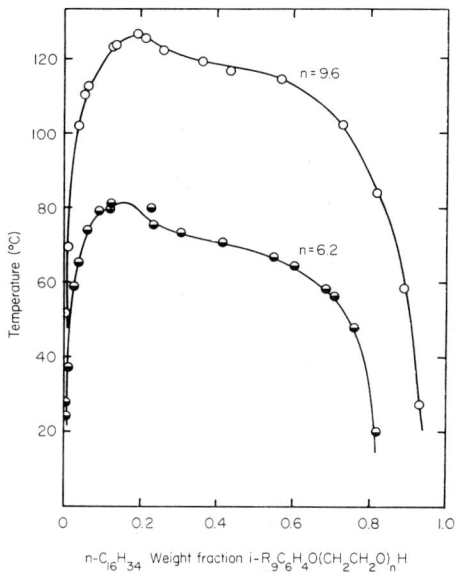

Figure III.2.15. Liquid-liquid solubility curve of nonylphenol (ethylene oxide) ethers ($C_9 \phi EO_m$) in hexadecane (© Academic Press, Inc., 1965. Reproduced with permission. Taken from reference 78).

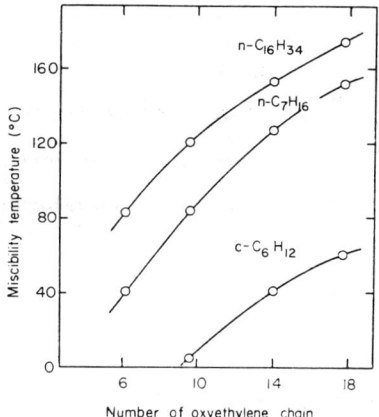

Figure III.2.16. Phase separation temperature of nonylphenol (ethylene oxide) ethers ($C_9 \phi EO_m$) in various 10 wt% hydrocarbon solutions as a function of the EO number (© Academic Press, Inc., 1965. Reproduced with permission. Taken from reference 78).

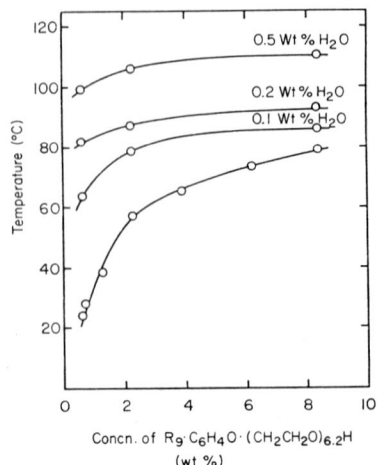

Figure III.2.17. Phase separation temperature of nonylphenol (ethylene oxide)$_{6.2}$ ether (C$_9$ ϕ EO$_{6.2}$) in hexadecane at various amounts of added water (© Academic Press, Inc., 1965. Reproduced with permission. Taken from reference 78).

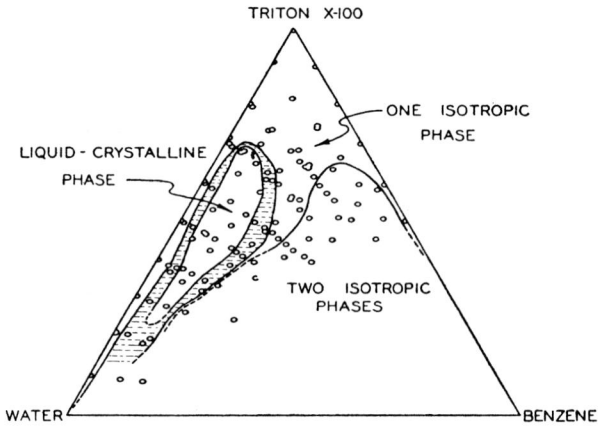

Figure III.2.18. Phase diagram of Triton X-100-water-benzene (C$_8$ ϕ EO$_{9.5}$-H$_2$O-C$_6$H$_6$). Temperature not specified. (© Marcel Dekker, 1966. Reproduced with permission. Taken from reference 72).

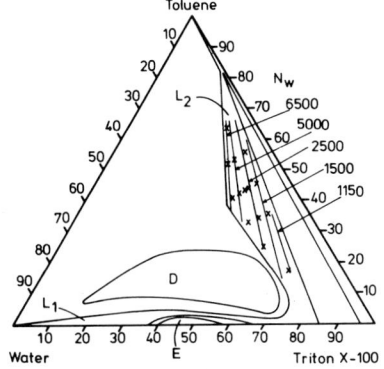

Figure III.2.19. Phase diagram of Triton X-100-water-toluene ($C_8 \phi EO_{9.5}$-H_2O-C_6H_6) at 24 oC. D represents a lamellar liquid crystalline phase. (© American Chemical Society, 1986. Reproduced with permission. Taken from reference 80).

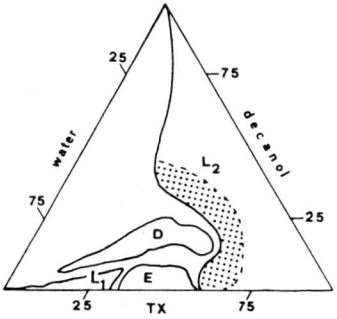

Figure III.2.20. Phase diagram of Triton X-100-water-decanol ($C_8 \phi EO_{9.5}$-H_2O-$C_{10}H_{21}OH$) at 20 oC. D represents a lamellar liquid crystalline phase and E a hexagonal phase. (© Elsevier Science Publishers BV, 1989. Reproduced with permission. Taken from reference 81).

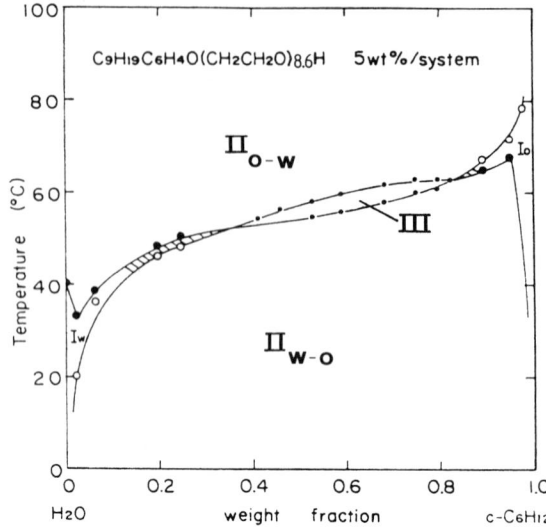

Figure III.2.21. Phase diagram of a 5 wt% nonylphenol(ethylene oxide)$_{8.6}$ ether-water-cyclohexane (C$_9$ ϕ EO$_{8.6}$-H$_2$O-C$_6$H$_{12}$). IIw-o represents a two phase oil-in-water (O/W) microemulsion and IIo-w represents a two phase water-in-oil (W/O) microemulsion, respectively.

(© Academic Press, Inc., 1973. Reproduced with permission. Taken from 82).

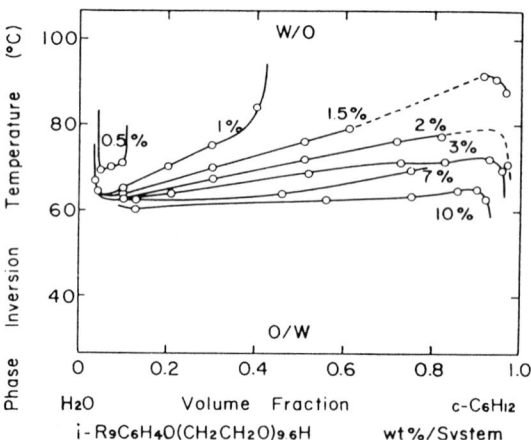

Figure III.2.22. Phase inversion temperature (PIT, in ºC) for nonylphenol(ethylene oxide)$_{9.6}$ ether-water-cyclohexane (C$_9$ ϕ EO$_{9.6}$-H$_2$O-C$_6$H$_{12}$) system at various surfactant concentrations as a function of the cyclohexane to water ratio (© Academic Press, Inc., 1967. Reproduced with permission. Taken from reference 83).

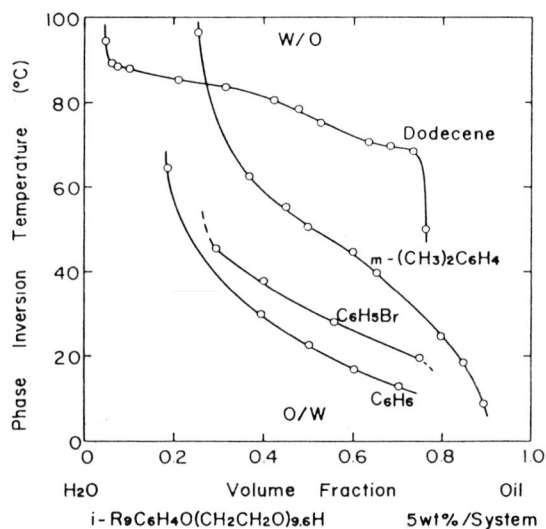

Figure III.2.23. Phase inversion temperature (PIT, in ∘C) for nonylphenol(ethylene oxide)$_{9.6}$ ether-water-oil ($C_9 \phi EO_{9.6}$-H_2O-oil) systems at a 5 wt% surfactant concentration as a function of the oil to water ratio (© Academic Press, Inc., 1967. Reproduced with permission. Taken from 83).

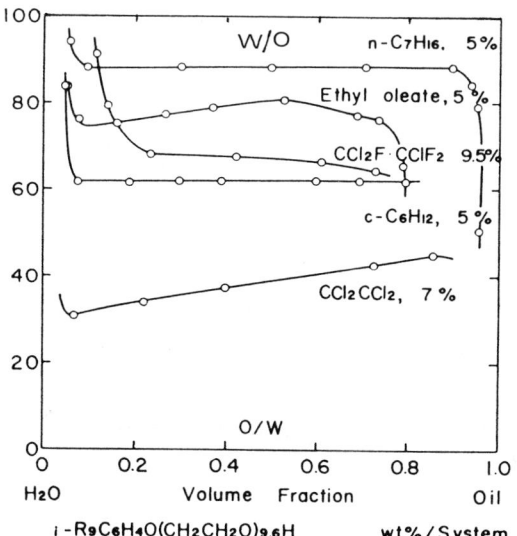

Figure III.2.24. Phase inversion temperature (PIT, in ∘C) for nonylphenol(ethylene oxide)$_{9.6}$ ether-water-oil ($C_9 \phi EO_{9.6}$-H_2O-oil) systems at various surfactant concentrations as a function of the oil to water ratio (© Academic Press, Inc., 1967. Reproduced with permission. Taken from reference 83).

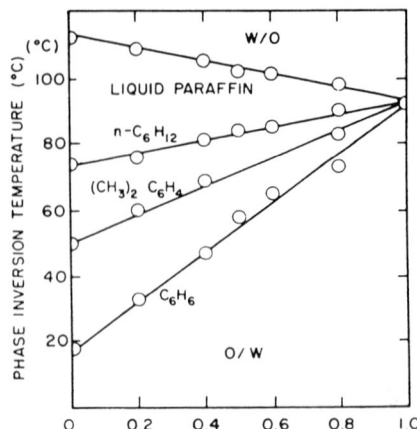

Figure III.2.25. Phase inversion temperature (PIT, in ᵒC) for nonylphenol(ethylene oxide)$_{9.6}$ ether-water-oil (C$_9$ ϕ EO$_{9.6}$-H$_2$O-oil) systems at a 3 wt% surfactant concentration as a function of the volume fraction heptane. The oil consists of n-heptane and the indicated hydrocarbon. (© Chemical Society of Japan, 1965. Reproduced with permission. Taken from reference 84).

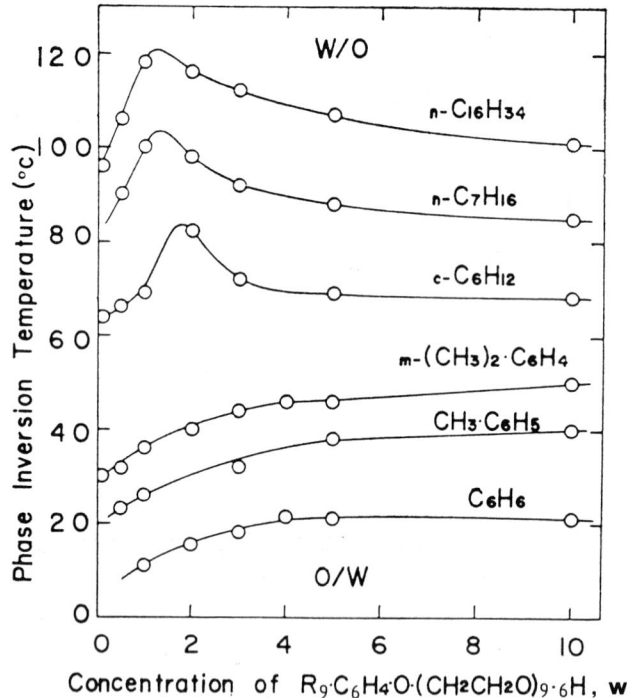

Figure III.2.26. Phase inversion temperature (PIT, in ᵒC) for nonylphenol(ethylene oxide)$_{9.6}$ ether-water-hydrocarbon (C$_9$ ϕ EO$_{9.6}$-H$_2$O-C$_n$H$_m$) systems as a function of surfactant concentration with various hydrocarbon structures at an oil-water ratio of 1:1. (© Marcel Dekker, 1966. Reproduced with permission. Taken from reference 8).

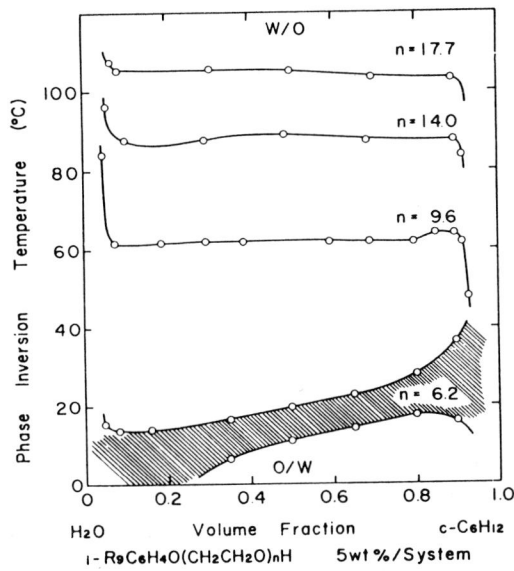

Figure III.2.27. Phase inversion temperature (PIT, in oC) for nonylphenol(ethylene oxide)$_n$ ethers-water-cyclohexane ($C_9 \phi EO_n$-H_2O-C_6H_{12}) systems at a 5 wt% surfactant concentration as a function of the oil to water ratio at various ethylene oxide numbers (n)

(© Academic Press, Inc., 1967. Reproduced with permission. Taken from reference 83).

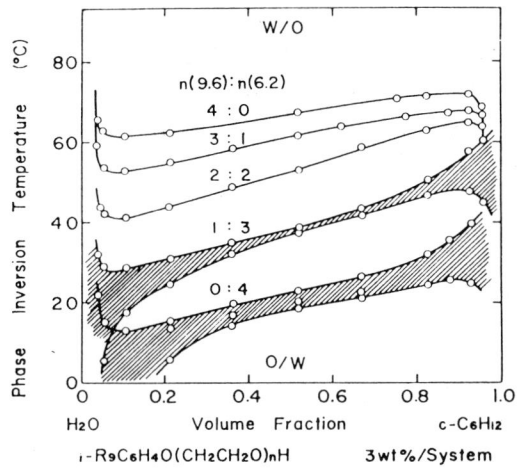

Figure III.2.28. Phase inversion temperature (PIT, in oC) for nonylphenol(ethylene oxide)$_n$ ethers-water-cyclohexane ($C_9 \phi EO_n$-H_2O-C_6H_{12}) systems at a 3 wt% surfactant concentration as a function of the oil to water ratio at various ratios of surfactant (© Academic Press, Inc., 1967. Reproduced with permission. Taken from reference 83).

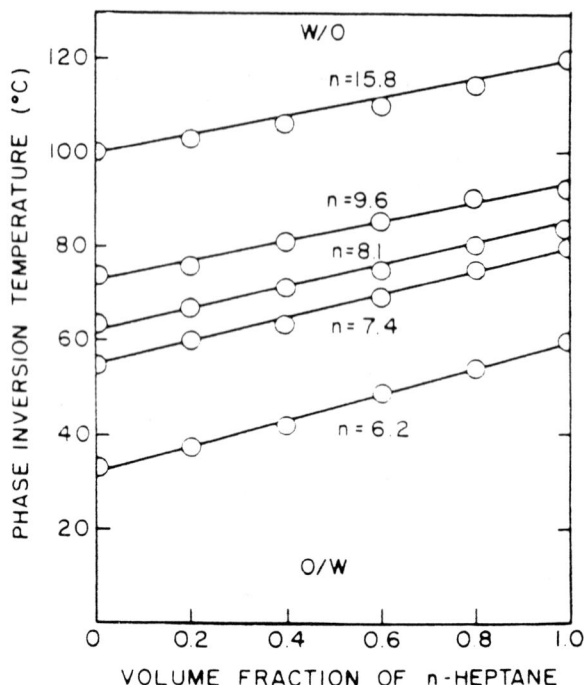

Figure III.2.29. Phase inversion temperature (PIT, in °C) for nonylphenol(ethylene oxide)$_n$ ethers-water-cyclohexane/n-heptane ($C_9 \phi EO_n$-H_2O-C_6H_{12}/C_7H_{16}) systems at a 3 wt% surfactant concentration as a function of the volume fraction heptane at various ethylene oxide numbers (n) (© Chemical Society of Japan, 1965. Reproduced with permission. Taken from reference 84).

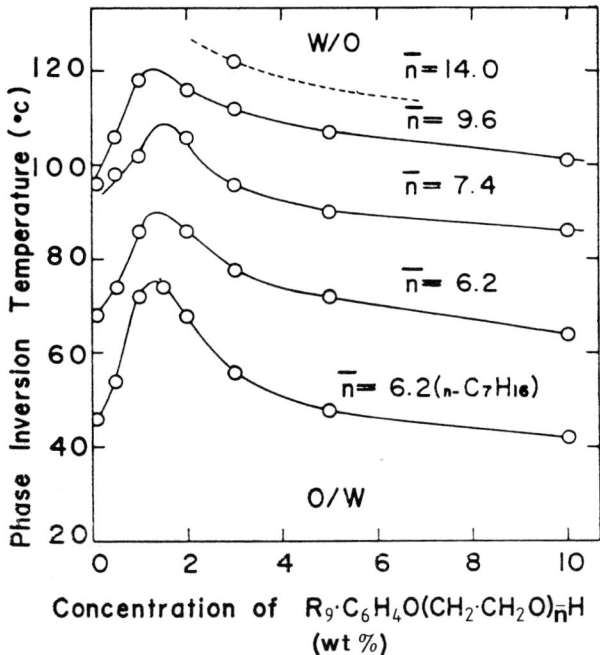

Figure III.2.30. Phase inversion temperature (PIT, in ºC) for nonylphenol(ethylene oxide)$_n$ ethers-water-hexadecane ($C_9 \phi EO_n$-H_2O-$C_{16}H_{34}$) systems at various ethylene oxide numbers (n) as a function of surfactant concentration. The hydrocarbon-to-water ratio is 1:1 (v/v). (© American Chemical Society, 1964. Reproduced with permission. Taken from reference 74).

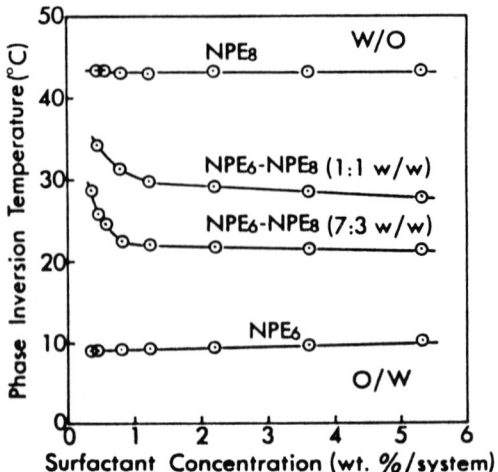

Figure III.2.31. Phase inversion temperature (PIT, in oC) for nonylphenol(ethylene oxide)$_n$ ethers-water-cyclohexane ($C_9 \phi EO_n$-H_2O-C_6H_{12}) systems at various ratios of two different surfactants as a function of total surfactant concentration. The cyclohexane-to-water ratio is 1:1 (v/v)

(© Academic Press, Inc., 1980. Reproduced with permission. Taken from 85).

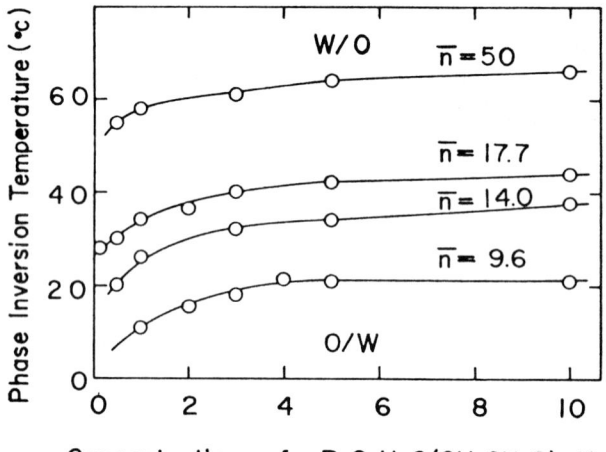

Figure III.2.32. Phase inversion temperature (PIT, in oC) for nonylphenol(ethylene oxide)$_n$ ethers-water-cyclohexane ($C_9 \phi EO_n$-H_2O-C_6H_{12}) systems as a function of surfactant concentration. The cyclohexane-to-water ratio is 1:1 (v/v) (© American Chemical Society, 1980. Reproduced with permission. Taken from reference 74).

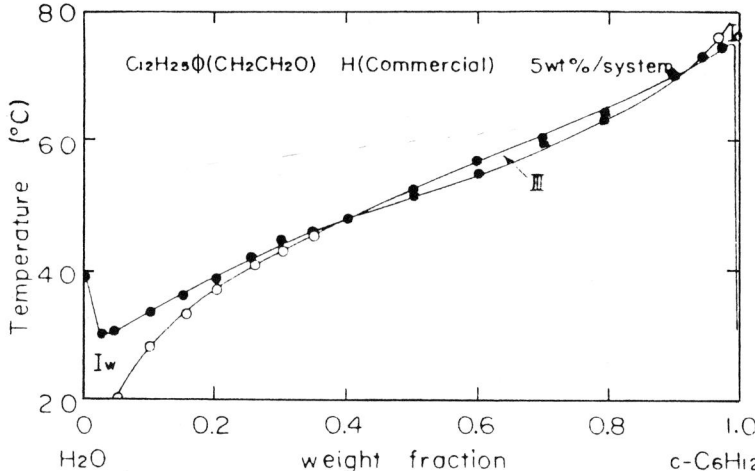

Figure III.2.33. Phase diagram of dodecylphenol(ethylene oxide)$_{5.3}$ ether-water-cyclohexane ($C_{12} \phi EO_{5.3}$-H_2O-C_6H_{12}). Concentration of surfactant (technical grade) is 5 wt%. (© Academic Press, Inc., 1973. Reproduced with permission. Taken from reference 82).

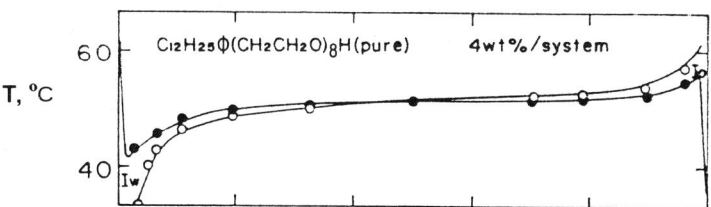

Figure III.2.34. Phase diagram of dodecylphenol(ethylene oxide)$_8$ ether-water-cyclohexane ($C_{12} \phi EO_8$-H_2O-C_6H_{12}). Concentration of surfactant (pure) is 4 wt%. (© Academic Press, Inc., 1973. Reproduced with permission. Taken from reference 82).

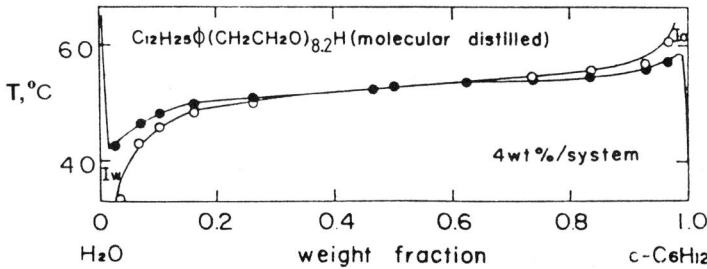

Figure III.2.35. Phase diagram of dodecylphenol(ethylene oxide)$_{8.2}$ ether-water-cyclohexane ($C_{12} \phi EO_{8.2}$-H_2O-C_6H_{12}). Concentration of surfactant (molecular distilled) is 4 wt%. (© Academic Press, Inc., 1973. Reproduced with permission. Taken from reference 82).

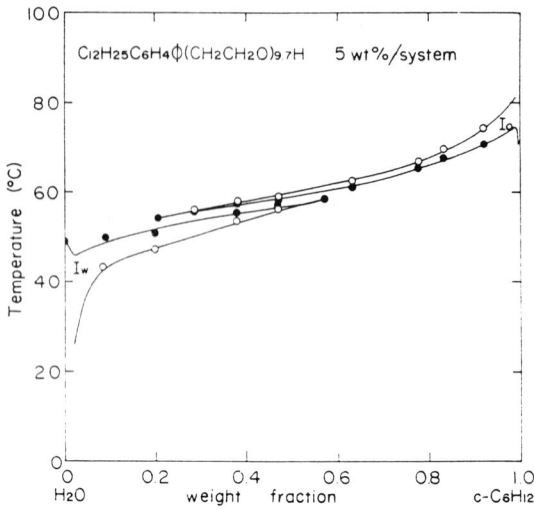

Figure III.2.36. Phase diagram of dodecylphenol(ethylene oxide)$_{9.7}$ ether-water-cyclohexane ($C_{12} \phi EO_{9.7}$-H_2O-C_6H_{12}). Concentration of surfactant is 5 wt%. (© Academic Press, Inc., 1973. Reproduced with permission. Taken from reference 82).

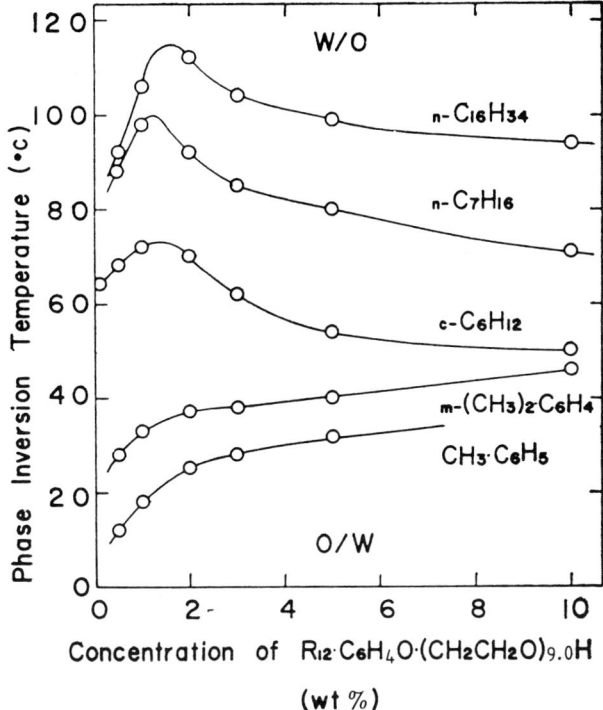

Figure III.2.37. Phase inversion temperature (PIT, in ºC) for dodecylphenol(ethylene oxide)$_{9.0}$ ether-water-hydrocarbon ($C_9 \phi EO_n$-H_2O-C_nH_m) systems as a function of surfactant concentration. The hydrocarbon-to-water ratio is 1:1 (v/v) (© American Chemical Society, 1980. Reproduced with permission. Taken from reference 74).

REFERENCES Chapter III.2.

1) Schott, H., Royce, A.E., and Han, S.K., J. Colloid Interface Sci., **98**, 196 (1984).
2) Valaulikar, B.S. and Manohar, C., J. Colloid Interface Sci., **108**, 403 (1985).
3) Hans, S.K., Lee, S.M. and Schott, H., J. Colloid Interface Sci., **126**, 393 (1988).
4) Brown, W., Rymdén, R., van Stam, J., Almgren, M., and Svensk, G., J. Phys. Chem., **93**, 2512 (1989).
5) Goldfarb, J. and Sepulveda, L., J. Colloid Interface Sci., **31**, 454 (1969).
6) Mitchell, D.J., Tiddy, G.J.T., Waring, L., Bostock, T., and McDonald, M.P., J. Chem. Soc., Faraday Trans. I, **79**, 975 (1983).
7) Grieser, F., Healey, T.W., Hsu, W.P., Kratohvil, J.P., and Warr, G., Colloids Surfaces, **42**, 97 (1989).
8) Becker, P. in "Nonionic Surfactants", Schick, M.J. (Ed.), Marcel Dekker, New York (1966).
9) Crock, E.H., Fordyce, D.B., and Trebbi, G.F., J. Phys. Chem., **67**, 1987 (1963).
10) Graciaa, A., Ghoulam, M.B., Marion, G., and Lachaise, J., J. Phys. Chem., **93**, 4167 (1989).
11) Andersson B., and Olofsson, G., J. Chem. Soc., Faraday Trans. I, **84**, 4087 (1988).
12) Malliaris, A., Le Moigne, J., Sturm, J., and Zana, R., J. Phys. Chem., **89**, 2709 (1985).
13) Mandal, A.B., Nair, B.U., and Ramaswamy, D., B. Electrochem., **4**, 565 (1988).
14) Von Rybinsky, W. and Schwuger, M.J., Langmuir, **2**, 639 (1986).
15) Carrión Fité, F.J., Tenside Detergents, **22**, 225 (1985).
16) Garon, G. and Desnoyers, J.E., J. Colloid Interface Sci., **119**, 141 (1987).
17) Hsiao, L., Dunning, H.N., and Lorentz, P.B., J. Phys. Chem., **60**, 657 (1956).
18) Schick, M.J., Atlas, S.M., and Eirich, F.R., J. Phys. Chem., **66**, 1326 (1962).
19) Becker, P., J. Colloid Interface Sci., **16**, 49 (1961).
20) Rathman, J.F. and Scamehorn, J.F., Langmuir, **4**, 474 (1988).
21) Turro, N.J., Kuo, P.L., Somasundaran, P., and Wong, K., J. Phys. Chem., **90**, 288 (1986).
22) Schick, M.J. and Gilbert, A.H., J. Colloid Sci., **20**, 464 (1965).
23) Crook, E.H., Trebbi, G.F., and Fordyce, D.B., J. Phys. Chem., **68**, 3592 (1964).
24) Evans, D.F., Yamauchi, A., Roman, R., and Casassa, E.Z., J. Colloid Interface Sci., **88**, 89 (1982).
25) Ray, A., Nature, **231**, 313 (1971).
26) Paredes, S., Tribout, M., Ferreira, J., and Leanis, J., Colloid Polym. Sci., **254**, 637 (1976).
27) Klimenko, N.A., Karmazina, T.V., Yaroskenko, N.A., Bartnitskii, A.E., and Aryamova, Zh.M., Kolloidnyi Zhurnal, **46**, 1112 (1982).

28) Ray, A., and Nemethy, G., J. Phys. Chem., **75**, 809 (1971).
29) Keh, E., Partyka, S., and Zaini, S., J. Colloid Interface Sci., **129**, 363 (1989).
30) Mankowich, A.M., J. Phys. Chem., **58**, 1027 (1954).
31) Zana, R. and Weill, C., J. Phys. Lett., **46**, L-953 (1985).
32) Robson, R.J. and Dennis, E.A., J. Phys. Chem., **81**, 1075 (1977).
33) Birdi, K.S., Progr. Colloid Polym. Sci., **70**, 23 (1985).
34) Rau, H., Greiner, G., and Hämmerle, H., Ber. Bunsenges. Phys. Chem., **88**, 116 (1984).
35) Dwiggings, C.W., Bolen, R.J., and Dunning, H.N., J. Phys. Chem., **64**, 1175 (1960).
36) Kushner, L.M. and Hubbard, W.D., J. Phys. Chem., **58**, 1163 (1954).
37) Sirianni, A.F. and Coleman, R.D., Can. J. Chem., **42**, 682 (1964).
38) Bedö, Z., Berecz, E., and Lakatos, I., Colloid Polym. Sci., **265**, 715 (1987).
39) Knopf, K. and Schollmeyer, E., Tenside Detergents, **24**, 101 (1987).
40) Becker, P., J. Colloid Sci., **17**, 325 (1962).
41) Dubin, P.L., Principi, J.M., Smith, B.A., and Fallon, M.A., J. Colloid Interface Sci., **127**, 558 (1989).
42) Robson, R.J. and Dennis, E.A., Biochim. Biophys. Acta, **508**, 513 (1978).
43) Akens, R.J. and Riley, P.W., J. Colloid Interface Sci., **48**, 162 (1974).
44) Saenger, W. and Müller-Fahrnow, A., Angew. Chemie, **100**, 429 (1988).
45) Rosen, M.J., "Surfactants and Interfacial Phenomena", 2nd Ed., Wiley, New York, (1989).
46) Stubicar, N., Matejar, J., Zipper, P., and Wilfing, R. in "Surfactants in Solution", Mittal, K.L. (Ed.), Vol. 7, Plenum Press, New York (1989).
47) Sharma, B.G., Rakshit, A.K. in "Surfactants in Solution", Mittal, K.L. (Ed.), Vol. 7, Plenum Press, New York (1989).
48) Moulik, S.P., Gupta, S., and Das, A.R., Can. J. Chem., **67**, 356 (1989).
49) Coppola, L., La Mesa, C., Ranieri, G.A., and Terenzi, M., Colloid Polym. Sci., **267**, 86 (1989).
50) La Mesa, C., Sesta, B., Bonincontro, A., and Cametti, C., J. Colloid Interface Sci., **125**, 634 (1988).
51) Knopf, K. and Schollmeyer, E., Tenside Detergents, **24**, 101 (1987).
52) Kalyanasundaram, K. and Thomas, J.K. in "Micellization, Solubilization, and Microemulsions", Mittal, K.L. (Ed.), Plenum Press, New York (1977).
53) Mandal, A.B., Ray, S., Biswas, A.M., and Moulik, S.P., J. Phys. Chem., **84**, 856 (1980).
54) Grieser, F. and Drummond, C.J., J. Phys. Chem., **92**, 5580 (1988).
55) Gratzel, M., Thomas, J.K. in "Modern Fluorescence Spectroscopy", Wehry, E. (Ed.), Vol. 2, Plenum Press, New York (1976).
56) Zachariasse, K.A., van Phuc, N., and Kozankiewic, B., J. Phys. Chem., **85**, 2676 (1981).
57) Kalyanasundaran, K. and Thomas, J.K., J. Phys. Chem., **81**, 2176 (1977).
58) Law, K.Y., Photochem. Photobiol., **33**, 799 (1981).

59) Fernandez, M.S. and Fromhertz, P., J. Phys. Chem., **81**, 1755 (1977).
60) Costa, S.M.B. and Macanita, A.L. J. Phys. Chem., **84**, 2408 (1980).
61) Blatt, E., Aust. J. Chem., **40**, 851 (1987).
62) Hertz, R. and Barenholtz, Y., J. Colloid Interface Sci., **60**, 188 (1977).
63) Lianos, P. and Zana, R., Chem. Phys. Lett., **72**, 171 (1980).
64) Henderson, C.N., Selinger, B.K. and Watkins, A.R., J. Photochem., **16**, 215 (1981).
65) Zachariasse, K.A. in "Excited-State Probes in Biochemistry and Biology", Szabo, A. (Ed.), Plenum Press, New York (1986).
66) Turro, N.J. and Kuo, P.L., Langmuir, **1**, 170 (1985).
67) Crook, E.H., Fordyce, D.B., and Trebbi, G.F., J. Colloid Sci., **20**, 191 (1965).
68) Shinoda, K., Saito, H., and Arai, H., J. Colloid Interface Sci., **35**, 624 (1971).
69) Pandit, N.K. and Caronia, J., J. Colloid Interface Sci., **122**, 100 (1988).
70) Heusch, R. and Kopp, F., Progr. Colloid Polym. Sci., **77**, 77 (1988).
71) Maclay, W.N., J. Colloid Sci., **11**, 272 (1956).
72) Nakagawa, T. in "Nonionic Surfactants", Schick, M.J. (Ed.), Marcel Dekker, New York (1966).
73) Heusch, R., Tenside Detergents, **20**, 1 (1983).
74) Shinoda, K. and Arai, H., J. Phys. Chem., **68**, 3485 (1964).
75) Schick, M.J., J. Colloid Sci., **17**, 801 (1962).
76) Laughlin, R.G., J. Colloid Interface Sci., **55**, 239 (1976).
77) Nandeau, H.G. and Siggia, S. in "Nonionic Surfactants", Schick, M.J. (Ed.), Marcel Dekker, New York (1966).
78) Shinoda, K. and Arai, H., J. Colloid Sci., **20**, 93 (1965).
79) Marsden, S.S. and McBain, J.W., J. Phys. Chem., **52**, 110 (1948).
80) Almgren, M., van Stam, J., Swarup, S., and Löfroth, J.E., Langmuir, **2**, 432 (1986).
81) La Mesa, C., Ramieri, G.A., and Terenzi, M., Colloids Surfaces, **42**, 59 (1989).
82) Shinoda, K. and Kunieda, H., J. Colloid Interface Sci., **42**, 381 (1973).
83) Shinoda, K. and Arai, H., J. Colloid Interface Sci., **25**, 429 (1967).
84) Arai, H., Nippon Kagaku Zasshi, **86**, 1126 (1965).
85) Harusawa, F., Saito, T., Nakajima, H., and Fukushima, S., J. Colloid Interface Sci., **74**, 435 (1980).
86) Heusch, R., and Kopp, F., Ber. Bunsenges. Phys. Chem., **24**, 806 (1987).

PART IV: INDEXES

The **COMPOUND INDEX** is an alphabetical list of all the compounds to be found in this compilation. Each entry gives the chemical name of the compound, its general formula, and/or an alternative formula under which the compound can be found. This part of the entry is followed by the properties of the compound, arranged in alphabetical order as they appear in a particular Table and/or Figure in the compilation. The example below shows the format of a typical entry in the Compound Index.

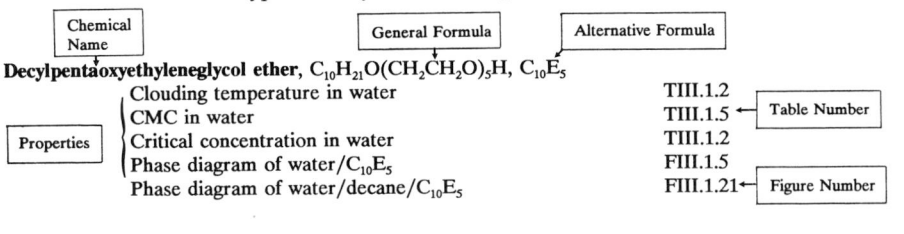

The **MOLECULAR FORMULA INDEX** gives the molecular formula of each compound in the compilation according to the Hill convention, i.e. carbon and hydrogen first, followed by the chemical name under which the compound is listed in the Compound Index.

The **GENERAL FORMULA INDEX** gives the chemical or alternative formula of the compounds in this compilation. Each formula is followed by the chemical name under which the compound can be found in the Compound Index.

The **CROSS INDEX** is an index of trivial or alternative names and abbreviations under which some compounds are known in the literature. Each name is followed by the chemical name which directs the reader to the Compound Index.

The **PROPERTY INDEX** is an alphabetical list of all the properties listed in the compilation. Most entries are arranged as follows:

Aggregation number [Property] [Description of Solution]
in aqueous solutions of CaCl$_2$

[Compounds]
- Dodecyldodecaoxyethylene glycol ether, $C_{12}H_{25}O(CH_2CH_2O)_{12}H$, $C_{12}E_{12}$ — TIII.1.10
- Dodecyloctaoxyethylene glycol ether, $C_{12}H_{25}O(CH_2CH_2O)_8H$, $C_{12}E_8$ — TIII.1.10
- Dodecyltricosaoxyethylene glycol ether, $C_{12}H_{25}O(CH_2CH_2O)_{23}H$, $C_{12}E_{23}$ — TIII.1.10

in aqueous solutions of LiCl [Description of Solution]

[Compounds]
- Nonylphenol(ethylene oxide)$_{50}$ ether, $C_9H_{19}C_6H_4O(CH_2CH_2O)_{50}H$, $C_9\phi EO_{50}$ — TIII.2.11
- Octylphenol(ethylene oxide)$_{9.5}$ ether, $C_8H_{17}C_6H_4O(CH_2CH_2O)_{9.5}H$, $C_8\phi EO_{9.5}$ — TIII.2.11

[Chemical Name] [General Formula] [Alternative Formula]

Such an entry gives an alphabetical list of (aqueous solutions, each of which is followed by the compounds, listed in alphabetical order, for which the property in question was determined. Each compound is indicated by its chemical name, its general formula, and/or an alternative formula under which it appears in a particular Table and/or Figure in the compilation.

COMPOUND INDEX

Cyclohexane (continued)
    Phase inversion temperature
        in dodecylphenol(ethylene oxide)$_{9.4}$ ether ($C_{12}\phi EO_{9.4}$)
          - water - cyclohexane   TIII.2.25
        in dodecylphenol(ethylene oxide)$_{9.7}$ ether ($C_{12}\phi EO_{9.7}$)
          - water - cyclohexane   TIII.2.25
        in hexadecylphenol(ethylene oxide)$_{12.4}$ ether ($C_{16}\phi EO_{12.4}$)
          - water - cyclohexane   TIII.2.25
        in hexylphenol(ethylene oxide)$_{7.5}$ ether ($C_6\phi EO_{7.5}$)
          - water - cyclohexane   TIII.2.25
        in nonylphenol(ethylene oxide)$_6$ ether ($C_9\phi EO_6$) mixture
          with nonylphenol(ethylene oxide)$_8$ ether ($C_9\phi EO_8$)
          - water - cyclohexane   FIII.2.31
        in nonylphenol(ethylene oxide)$_{6.2}$ ether ($C_9\phi EO_{6.2}$)
          - water - cyclohexane   FIII.2.27
                                          & FIII.2.28
        in nonylphenol(ethylene oxide)$_{6.2}$ ether ($C_9\phi EO_{6.2}$
          - water - heptane - cyclohexane   FIII.2.29
        in nonylphenol(ethylene oxide)$_{6.2}$ ether ($C_9\phi EO_{6.2}$) mixture
          with nonylphenol(ethylene oxide)$_{9.6}$ ether ($C_9\phi EO_{9.6}$)
          - water - cylohexane   FIII.2.28
        in nonylphenol(ethylene oxide)$_{7.4}$ ether ($C_9\phi EO_{7.4}$
          - water - heptane - cyclohexane   FIII.2.29
        in nonylphenol(ethylene oxide)$_8$ ether ($C_9\phi EO_8$)
          - water - cyclohexane   FIII.2.31
        in nonylphenol(ethylene oxide)$_{8.1}$ ether ($C_9\phi EO_{8.1}$)
          - water - cyclohexane   TIII.2.25
        in nonylphenol(ethylene oxide)$_{8.1}$ ether ($C_9\phi EO_{8.1}$)
          - water - heptane - cyclohexane   FIII.2.29
        in nonylphenol(ethylene oxide)$_{8.6}$ ether ($C_9\phi EO_{8.6}$)
          - water - cyclohexane   TIII.2.25
        in nonylphenol(ethylene oxide)$_{8.6}$ ether ($C_9\phi EO_{8.6}$) mixture
          with calcium bis(dodecylbenzenesulphonate)
          (($C_{12}H_{25}C_6H_5SO_3)_2Ca$) - water - cyclohexane   TIII.2.25
        in nonylphenol(ethylene oxide)$_{9.6}$ ether ($C_9\phi EO_{9.6}$)
          - water - cyclohexane   FIII.2.22
                                          & FIII.2.24
        in nonylphenol(ethylene oxide)$_{9.6}$ ether ($C_9\phi EO_{9.6}$)
          - water - heptane - cylohexane   FIII.2.29
        in nonylphenol(ethylene oxide)$_{14}$ ether ($C_9\phi EO_{14}$)
          - water - cyclohexane   FIII.2.27
        in nonylphenol(ethylene oxide)$_{15.8}$ ether ($C_9\phi EO_{15.8}$)
          - water - heptane - cyclohexane   FIII.2.29
        in nonylphenol(ethylene oxide)$_{17.7}$ ether ($C_9\phi EO_{17.7}$)
          - water - cyclohexane   FIII.2.27
        in octylphenol(ethylene oxide)$_{8.4}$ ether ($C_8\phi EO_{8.4}$)
          - water - cyclohexane   TIII.2.25
        in octylphenol(ethylene oxide)$_{8.5}$ ether ($C_8\phi EO_{8.5}$)
          - water - cyclohexane   TIII.2.25
    Slow relaxation rate of N-hexadecylpyridinium chloride
        ($C_{16}NC_5H_5Cl$) in water - pentanol; effect of =   FII.2.6
Cyclohexane (to be continued)

410

434

Octylphenol(ethylene oxide)$_{9.5}$ ether (continued)
    Clouding temperature
        in aqueous solutions of Na$_2$SO$_4$ FIII.2.4
        in aqueous solutions of NH$_4^+$ TIII.2.1
        in aqueous solutions of NH$_4$Br   FIII.2.3
        in aqueous solutions of NH$_4$Cl   FIII.2.3
        in aqueous solutions of (NH$_4$)$_2$SO$_4$ FIII.2.3
        in aqueous solutions of Ni^{2+} TIII.2.1
        in aqueous solutions of Pb^{2+} TIII.2.1
        in aqueous solutions of PO$_4^{3-}$ TIII.2.1
        in aqueous solutions of SCN$^-$ TIII.2.1
        in aqueous solutions of SO$_4^{2-}$ TIII.2.1
        in aqueous solutions of Zn^{2+} TIII.2.1
        in aqueous solutions of Zn(NO$_3$)$_2$ FIII.2.3
        in water   TIII.2.1
        in water/acetamide   TIII.2.1
        in water/acetone   FIII.2.5
        in water/aniline   FIII.2.5
        in water/benzene   FIII.2.5
        in water/butyl acetate   FIII.2.5
        in water/cyclohexane   FIII.2.5
        in water/decane   FIII.2.5
        in water/1,1-diethylurea   TIII.2.1
        in water/1,3-dimethylthiourea   TIII.2.1
        in water/1,1-dimethylurea   TIII.2.1
        in water/1,3-dimethylurea   TIII.2.1
        in water/dodecane   FIII.2.5
        in water/dodecanol/nitrobenzene   FIII.2.5
        in water/ethylene dichloride   FIII.2.5
        in water/2-ethylhexene   FIII.2.5
        in water/ethylurea   TIII.2.1
        in water/guanidinium chloride   TIII.2.1
        in water/hexadecane   FIII.2.5
        in water/hexane   FIII.2.5
        in water/methylthiourea   TIII.2.1
        in water/methylurea   TIII.2.1
        in water/nitrobenzene/dodecanol   FIII.2.5
        in water/octene   FIII.2.5
        in water/oleic acid/phenol   FIII.2.5
        in water/phenol/oleic acid   FIII.2.5
        in water/phenylurea   TIII.2.1
        in water/sodium decylsulphate (C$_{10}$H$_{21}$OSO$_3$Na)   TIII.2.1
        in water/sodium dodecylbenzenesulphonate (C$_{12}$H$_{25}$C$_6$H$_4$SO$_3$Na)   TIII.2.1
        in water/sodium dodecylsulphate (C$_{12}$H$_{25}$OSO$_3$Na)   TIII.2.1
        in water/tetradecene   FIII.2.5
        in water/tetradecylmercaptan   FIII.2.5
        in water/tetramethylurea   TIII.2.1
        in water/thiourea   TIII.2.1
        in water/urea   TIII.2.1
    CMC
        in aqueous solutions of KBr   TIII.2.2
        in aqueous solutions of KCl   TIII.2.2
        in aqueous solutions of KI   TIII.2.2
Octylphenol(ethylene oxide)$_{9.5}$ ether (to be continued)

450

MOLECULAR FORMULA INDEX

H_4BrN
  Ammonium bromide
H_4ClN
  Ammonium chloride
H_4N_2
  Hydrazine
$H_8N_2O_4S$
  Ammonium sulphate
$CDCl_3$
  Deuterochloroform
CD_4O
  Deuteromethanol
CKNS
  Potassium thiocyanate
CNNaS
  Sodium thiocyanate
CNa_2O_3
  Sodium carbonate
CH_2O_2
  Formic acid
CH_3NO
  Formamide
CH_4N_2O
  Urea
CH_4N_2S
  Thiourea
CH_6ClN_3
  Guanidinium chloride
C_2Cl_4
  Tetrachloroethene
$C_2Cl_4F_2$
  Difluorotetrachloroethane
$C_2H_3NaO_2$
  Sodium acetate
$C_2H_4Cl_2$
  Ethylene dichloride
$(C_2H_4)_n$
  Poly(ethylene glycol)
C_2H_5NO
  Acetamide
$C_2H_6N_2O$
  Methylurea
$C_2H_6N_2S$
  Methylthiourea
C_2H_6O
  Ethanol
$C_2H_6O_2$
  Ethylene glycol
C_2H_6OS
  2-Mercaptoethanol
C_2H_7NO
  2-Aminoethanol
$C_2H_8N_2$
  Ethylenediamine

$C_2H_8O_2O_3$
  Ethylammonium nitrate

C_3H_6O
  Acetone

$C_3H_8N_2O$
  1,1-Dimethylurea
  1,3-Dimethylurea
  Ethylurea

$C_3H_8N_2S$
  1,3-Dimethylthiourea

C_3H_8O
  Propanol

$C_3H_8O_2$
  Methylmonooxyethylene glycol ether
  1,2-Propanediol
  1,3-Propanediol

$C_3H_8O_3$
  Glycerol

C_3H_9NO
  1-Amino-2-propanol

$C_4H_8O_2$
  Dioxane

$C_4H_{10}O$
  Butanol
  2-Methyl-2-propanol

$C_4H_{10}O_2$
  1,3-Butanediol
  1,4-Butanediol
  Dimethylmonooxyethylene glycol diether
  Ethylmonooxyethylene glycol ether

$C_4H_{12}ClN$
  Tetramethylammonium chloride

C_5H_{12}
  Pentane

$C_5H_{12}N_2O$
  1,1-Diethylurea
  Tetramethylurea

$C_5H_{12}O$
  Pentanol

$C_5H_{12}O_2$
  Propylmonooxyethylene glycol ether

C_6H_5Br
  Bromobenzene

$C_6H_5NO_2$
  Nitrobenzene

$C_6H_5Na_3O_7$
  Sodium citrate

C_6H_6
  Benzene

C_6H_6O
  Phenol

C_6H_7N
  Aniline

$C_6H_8O_7$
  Citric acid

C_6H_{12}
  Cyclohexane
  Hexene
$C_6H_{12}O_2$
  Butyl acetate
$C_6H_{13}NaO_3S$
  Sodium hexanesulphonate
$C_6H_{13}NaO_4S$
  Sodium hexylsulphate
C_6H_{14}
  Hexane
$C_6H_{14}O$
  Hexanol
$C_6H_{14}O_2$
  Butylmonooxyethylene glycol ether
$C_6H_{14}O_3$
  Ethyldioxyethylene glycol ether
$C_7H_7NaO_3S$
  Sodium 4-methylbenzenesulphonate
C_7H_8
  Toluene
$C_7H_8N_2O$
  Phenylurea
C_7H_8O
  Benzyl alcohol
  p-Cresol
$C_7H_{15}NaO_4S$
  Sodium heptylsulphate
C_7H_{16}
  Heptane
$C_7H_{16}O$
  Heptanol
C_8H_7N
  Indole
C_8H_9ClO
  4-Chloro-3,5-xylenol
C_8H_{10}
  Xylene
  m-Xylene
$C_8H_{11}N$
  N,N-Dimethylaniline
$C_8F_{15}NaO_2$
  Sodium pentadecafluorooctanoate
C_8H_{16}
  2-Ethylhexene
  Octene
$C_8H_{17}NaO_3S$
  Sodium octanesulphonate
$C_8H_{17}NaO_4S$
  Sodium octylsulphate
C_8H_{18}
  Isooctane
C_8H_{18}
  Octane

$C_8H_{18}O$
Octanol

$C_8H_{18}O_2$
Hexylmonooxyethylene glycol ether

$C_8H_{18}O_4$
Ethyltrioxyethylene glycol ether

$C_8H_{20}BrN$
Tetraethylammonium bromide

$C_9H_{19}NaO_4S$
Sodium nonylsulphate

C_9H_{20}
Nonane

$C_9H_{20}O_3$
Pentyldioxyethylene glycol ether

$C_{10}-C_{14}LAS$
Commercial LAS

$C_{10}H_{14}$
Butylbenzene

$C_{10}H_{21}NaO_3S$
Sodium decanesulphonate

$C_{10}H_{21}NaO_4S$
Sodium decylsulphate
Sodium 2-decylsulphate
Sodium 2-hydroxydecanesulphonate

$C_{10}H_{22}$
Decane

$C_{10}H_{22}O$
Decanol

$C_{10}H_{22}O_2$
Octylmonooxyethylene glycol ether

$C_{10}H_{22}O_3$
Hexyldioxyethylene glycol ether

$C_{10}H_{22}O_5$
Ethyltetraoxyethylene glycol ether

$C_{11}H_{18}BrN$
N-Hexylpyridinium bromide

$C_{11}H_{23}NaO_3S$
Secondary sodium undecanesulphonate (isomeric mixture)
Sodium undecanesulphonate

$C_{11}H_{23}NaO_4S$
Sodium 2-hydroxyundecanesulphonate
Sodium undecylsulphate

$C_{11}H_{26}BrN$
Octyltrimethylammonium bromide

$C_{12}H_{17}NaO_3S$
Sodium 4-(hexyl)benzenesulphonate

$C_{12}H_{23}NaO_3S$
Sodium 2-dodecenesulphonate

$C_{12}H_{23}NaO_4S$
Sodium 2-oxododecanesulphonate
Sodium 3-oxododecanesulphonate

$C_{12}H_{24}$
Dodecene

$C_{12}H_{25}KO_4S$
Potassium dodecylsulphate

$C_{12}H_{25}NaO_3S$
  Secondary sodium dodecanesulphonate (isomeric mixture)
  Sodium dodecanesulphonate
  Sodium 2-dodecanesulphonate
  Sodium 3-dodecanesulphonate
  Sodium 4-dodecanesulphonate
  Sodium 5-dodecanesulphonate
  Sodium 6-dodecanesulphonate
$C_{12}H_{25}NaO_4S$
  Sodium dodecylsulphate
  Sodium 1-hydroxydodecanesulphonate
  Sodium 1-hydroxy-2-dodecanesulphonate
  Sodium 2-hydroxydodecanesulphonate
  Sodium 3-hydroxydodecanesulphonate
  Sodium 2-methylundecylsulphate
$C_{12}H_{26}$
  Dodecane
$C_{12}H_{26}O$
  Dodecanol
$C_{12}H_{26}O_2$
  Decylmonooxyethylene glycol ether
$C_{12}H_{26}O_3$
  Octyldioxyethylene glycol ether
$C_{12}H_{26}O_4$
  Hexyltrioxyethylene glycol ether
$C_{12}H_{26}O_6$
  Ethylpentaoxyethylene glycol ether
$C_{12}H_{26}O_7$
  Hexaoxyethylene glycol
$C_{13}H_{19}NaO_3S$
  Sodium 4-(heptyl)benzenesulphonate
$C_{13}H_{22}BrN$
  N-Octylpyridinium bromide
$C_{13}H_{25}NaO_4S$
  Sodium 2-oxotridecanesulphonate
$C_{13}H_{25}NaO_5S$
  Sodium 1-carbomethoxyundecanesulphonate
$C_{13}H_{27}NaO_3S$
  Secondary sodium tridecanesulphonate (isomeric mixture)
  Sodium tridecanesulphonate
$C_{13}H_{27}NaO_4S$
  Sodium 2-hydroxytridecanesulphonate
  Sodium 3-methoxydodecanesulphonate
  Sodium tridecylsulphate
$C_{13}H_{30}BrN$
  Decyltrimethylammonium bromide
$C_{14}H_{10}$
  Anthracene
$C_{14}H_{21}NaO_3S$
  Sodium 4-(2-ethyl-hexyl)benzenesulphonate
  Sodium 4-(octyl)benzenesulphonate
  Sodium 4-(2-octyl)benzenesulphonate
  Sodium 2-tetradecenesulphonate
$C_{14}H_{21}NaO_4S$
  Sodium 3-oxotetradecanesulphonate

$C_{14}H_{28}$
  Tetradecene
$C_{14}H_{29}KO_4S$
  Potassium tetradecylsulphate
$C_{14}H_{29}NaO_3S$
  Secondary sodium tetradecanesulphonate (isomeric mixture)
  Sodium tetradecanesulphonate
  Sodium 5-tetradecanesulphonate
$C_{14}H_{29}NaO_4S$
  Sodium 3-ethoxydodecanesulphonate
  Sodium 1-hydroxytetradecanesulphonate
  Sodium 1-hydroxy-2-tetradecanesulphonate
  Sodium 3-hydroxytetradecanesulphonate
  Sodium 2-methyltridecylsulphate
  Sodium tetradecylsulphate
  Sodium 2-tetradecylsulphate
  Sodium 3-tetradecylsulphate
  Sodium 4-tetradecylsulphate
  Sodium 5-tetradecylsulphate
  Sodium 7-tetradecylsulphate
$C_{14}H_{30}$
  Tetradecane
$C_{14}H_{30}O$
  Tetradecanol
$C_{14}H_{30}O_2$
  Dodecylmonooxyethylene glycol ether
$C_{14}H_{30}O_3$
  Decyldioxyethylene glycol ether
$C_{14}H_{30}O_4$
  Octyltrioxyethylene glycol ether
$C_{14}H_{30}O_5$
  Hexyltetraoxyethylene glycol ether
$C_{14}H_{30}O_7$
  Ethylhexaoxyethylene glycol ether
$C_{14}H_{30}S$
  Tetradecylmercaptan
$C_{14}H_{31}OP$
  Dodecyldimethyl phosphoxide
$C_{15}H_{12}$
  2-Methylanthracene
  9-Methylanthracene
$C_{15}H_{23}NaO_3S$
  Sodium 4-(nonyl)benzenesulphonate
  Sodium 4-(3-nonyl)benzenesulphonate
$C_{15}H_{26}ClN$
  N-Decylpyridinium chloride
$C_{15}H_{29}NaO_3S$
  Sodium 2-pentadecenesulphonate
$C_{15}H_{29}NaO_4S$
  Sodium 2-oxopentadecanesulphonate
$C_{15}H_{29}NaO_5S$
  Sodium 1-carbomethoxytridecanesulphonate

$C_{15}H_{31}NaO_3S$
 Secondary sodium pentadecanesulphonate (isomeric mixture)
 Sodium pentadecanesulphonate
 Sodium 8-pentadecanesulphonate
$C_{15}H_{31}NaO_4S$
 Sodium 2-hydroxypentadecanesulphonate
 Sodium 3-hydroxypentadecanesulphonate
 Sodium pentadecylsulphate
 Sodium 2-pentadecylsulphate
 Sodium 3-pentadecylsulphate
 Sodium 5-pentadecylsulphate
 Sodium 8-pentadecylsulphate
 Sodium 3-(propoxy)dodecanesulphonate
 Sodium 3-(2-propoxy)dodecanesulphonate
$C_{15}H_{32}$
 Pentadecane
$C_{15}H_{34}BrN$
 Dodecyltrimethylammonium bromide
$C_{15}H_{34}ClN$
 Dodecyltrimethylammonium chloride
$C_{15}H_{35}NO$
 Dodecyltrimethylammonium hydroxide
$C_{16}H_{10}$
 Pyrene
$C_{16}H_{12}NNaO_3S$
 Sodium 1-anilino-8-naphthalenesulphonate
$C_{16}H_{23}NaO_6S$
 Sodium phenoxysulphonate ethanoic acid octyl ester
$C_{16}H_{25}NaO_3S$
 Sodium decylbenzenesulphonate (technical grade, purified)
 Sodium decylbenzenesulphonate (isomeric mixture)
 Sodium decylbenzenesulphonate (inner isomeric mixture)
 Sodium 2-(decyl)benzenesulphonate
 Sodium 4-(decyl)benzenesulphonate
 Sodium 4-(2-decyl)benzenesulphonate
 Sodium 4-(3-decyl)benzenesulphonate
 Sodium 4-(5-decyl)benzenesulphonate
 Sodium 4-(2-propyl-heptyl)benzenesulphonate
$C_{16}H_{26}O_2$
 Octylphenol(ethylene oxide)$_1$ ether
$C_{16}H_{28}BrN$
 N-Undecylpyridinium bromide
$C_{16}H_{30}Na_2O_5S$
 Disodium pentadecanesulphonate-1-carboxylate
$C_{16}H_{31}NaO_3S$
 Sodium 2-hexadecenesulphonate
$C_{16}H_{31}NaO_4S$
 Sodium 3-oxohexadecanesulphonate
$C_{16}H_{32}O_2$
 Palmitic acid
$C_{16}H_{33}NaO_3S$
 Secondary sodium hexadecanesulphonate (isomeric mixture)
 Sodium hexadecanesulphonate
 Sodium 7-hexadecanesulphonate

$C_{16}H_{33}NaO_4S$
  Sodium 3(butoxy)dodecanesulphonate
  Sodium hexadecylsulphate
  Sodium 4-hexadecylsulphate
  Sodium 6-hexadecylsulphate
  Sodium 8-hexadecylsulphate
  Sodium 1-hydroxyhexadecanesulphonate
  Sodium 1-hydroxy-2-hexadecanesulphonate
  Sodium 3-hydroxyhexadecanesulphonate
  Sodium 2-methylpentadecylsulphate
$C_{16}H_{33}NaO_5S$
  Sodium 3-hydroxyethoxytetradecanesulphonate
$C_{16}H_{34}$
  Hexadecane
$C_{16}H_{34}MgO_6S_2$
  Magnesium bis(octanesulphonate)
$C_{16}H_{34}NNaO_3S$
  Sodium 3-(dimethylamino)tetradecanesulphonate
$C_{16}H_{34}O$
  Hexadecanol
$C_{16}H_{34}O_2$
  Tetradecylmonooxyethylene glycol ether
$C_{16}H_{34}O_3$
  Dodecyldioxyethylene glycol ether
$C_{16}H_{34}O_4$
  Decyltrioxyethylene glycol ether
$C_{16}H_{34}O_5$
  Octyltetraoxyethylene glycol ether
$C_{16}H_{34}O_6$
  Hexylpentaoxyethylene glycol ether
$C_{16}H_{34}O_7$
  Butylhexaoxyethylene glycol ether
$C_{16}H_{34}O_8$
  Ethylheptaoxyethylene glycol ether
$C_{16}H_{36}BrN$
  Dodecylethyldimethylammonium bromide
$C_{16}H_{36}NBr$
  Tetrabutylammonium bromide
$C_{17}H_{11}O$
  1-Pyrene-3-carboxaldehyde
$C_{17}H_{25}NaO_5S$
  Sodium decyl-4-sulphobenzoate
  Sodium phenylsulphonate ethanoic acid nonyl ester
  Sodium 4-(undecanoic acid)benzenesulphonate ester
$C_{17}H_{26}NNaO_4S$
  Sodium undecanoyl-4-aminobenzenesulphonate
$C_{17}H_{27}NaO_3S$
  Sodium undecylbenzenesulphonate (isomeric mixture)
  Sodium 4-(undecyl)benzenesulphonate
  Sodium 4-(2-undecyl)benzenesulphonate
$C_{17}H_{27}NaO_3S_2$
  Sodium 4-(thioundecyl)benzenesulphonate
$C_{17}H_{27}NaO_4S$
  Sodium 4-(undecoxy)benzenesulphonate

$C_{17}H_{27}NaO_4S_2$
Sodium 4-(undecylsulphoxy)benzenesulphonate
$C_{17}H_{27}NaO_5S_2$
Sodium 4-(undecylsulphonyl)benzenesulphonate

$C_{17}H_{28}NNaO_3S$
Sodium 4-(undecylamino)benzenesulphonate
$C_{17}H_{30}BrN$
N-Dodecylpyridinium bromide
$C_{17}H_{30}ClN$
N-Dodecylpyridinium chloride
$C_{17}H_{30}IN$
N-Dodecylpyridinium iodide
$C_{17}H_{33}NaO_4S$
Sodium 2-oxoheptadecanesulphonate
$C_{17}H_{33}NaO_5S$
Sodium 1-carbomethoxypentadecanesulphonate
$C_{17}H_{35}NaO_3S$
Secondary sodium heptadecanesulphonate (isomeric mixture)
Sodium heptadecanesulphonate
$C_{17}H_{35}NaO_4S$
Sodium heptadecylsulphate
Sodium 2-heptadecylsulphate
Sodium 9-heptadecylsulphate
Sodium 2-hydroxyheptadecanesulphonate
Sodium 3-(methoxy)hexadecanesulphonate
$C_{17}H_{36}NNaO_3S$
Sodium 3-(propylamino)tetradecanesulphonate
$C_{17}H_{36}O_5$
Nonyltetraoxyethylene glycol ether
$C_{17}H_{38}BrN$
Tetradecyltrimethylammonium bromide
$C_{17}H_{38}ClN$
Tetradecyltrimethylammonium chloride
$C_{17}H_{38}N_2O_3$
Tetradecyltrimethylammonium nitrate
$C_{17}H_{39}NO$
Tetradecyltrimethylammonium hydroxide
$C_{18}H_{14}O_2$
Benzyl-1-naphthoate
$C_{18}H_{23}NaO_3S$
Sodium octylnaphthalenesulphonate
$C_{18}H_{28}IN$
N-Methyl-4-(dodec-5-ynyl)pyridinium iodide
$C_{18}H_{29}CsO_3S$
Cesium 4-(dodecyl)benzenesulphonate
$C_{18}H_{29}KO_3S$
Potassium 4-(dodecyl)benzenesulphonate
$C_{18}H_{29}LiO_3S$
Lithium 4-(dodecyl)benzenesulphonate
$C_{18}H_{29}NaO_3S$
Commercial C_{12}LAS based on propylene tetramer
Sodium 4-(2-(butyl)octyl)benzenesulphonate
Sodium dodecylbenzenesulphonate (isomeric mixture)
$C_{18}H_{29}NaO_3S$ (to be continued)

$C_{18}H_{29}NaO_3S$ (continued)
  Sodium dodecylbenzenesulphonate (inner isomeric mixture)
  Sodium dodecylbenzenesulphonate (80% para, 20% ortho isomers)
  Sodium dodecylbenzenesulphonate (technical grade, purified)
  Sodium 4-(dodecyl)benzenesulphonate
  Sodium 4-(2-dodecyl)benzenesulphonate
  Sodium 4-(3-dodecyl)benzenesulphonate
  Sodium 4-(4-dodecyl)benzenesulphonate
  Sodium 4-(6-dodecyl)benzenesulphonate
$C_{18}H_{29}O_3RbS$
  Rubidium 4-(dodecyl)benzenesulphonate
$C_{18}H_{29}NaO_4S$
  Sodium 4-(phenoxy)dodecanesulphonate
$C_{18}H_{30}O_3$
  Octylphenol(ethylene oxide)$_2$ ether
$C_{18}H_{32}BrN$
  N-Dodecyl-4-methoxypyridinium bromide
  N-Tridecylpyridinium bromide
$C_{18}H_{32}BrNO$
  N-Methyl-4-dodecyloxypyridinium bromide
$C_{18}H_{32}ClNO$
  N-Dodecyl-4-methoxypyridinium chloride
$C_{18}H_{32}IN$
  N-Methyl-4-(9,9-dimethyldecyl)pyridinium iodide
  N-Methyl-4-dodecylpyridinium iodide
  N-Methyl-4-(2-dodecyl)pyridinium iodide
  N-Methyl-4-(8-ethyldecyl)pyridinium iodide
$C_{18}H_{34}Na_2O_5S$
  Disodium heptadecanesulphonate-1-carboxylate
$C_{18}H_{34}Na_2O_8S_2$
  Disodium pentadecanesulphonate-1-carboxyethanesulphonate
$C_{18}H_{34}O_2$
  Oleic acid
$C_{18}H_{35}NaO_3S$
  Sodium 2-octadecenesulphonate
$C_{18}H_{35}NaO_4S$
  Sodium 1-carbethoxypentadecanesulphonate
  Sodium 3-oxooctadecanesulphonate
$C_{18}H_{36}NaO_4S$
  Sodium 3-(morpholino)tetradecanesulphonate
$C_{18}H_{37}NaO_3S$
  Sodium octadecanesulphonate
  Sodium 9-octadecanesulphonate
$C_{18}H_{37}NaO_4S$
  Sodium 3-(hexoxy)dodecanesulphonate
  Sodium 1-hydroxyoctadecanesulphonate
  Sodium 1-hydroxy-2-octadecanesulphonate
  Sodium 3-hydroxyoctadecanesulphonate
  Sodium 2-methylheptadecylsulphate
  Sodium octadecylsulphate
  Sodium 2-octadecylsulphate
  Sodium 4-octadecylsulphate
  Sodium 6-octadecylsulphate
$C_{18}H_{37}NaO_6S$
  Sodium 3-(hydroxyethoxyethoxy)tetradecanesulphonate

$C_{18}H_{38}NNaO_3S$
    Sodium 3-(butylamino)tetradecanesulphonate
$C_{18}H_{38}O$
    Octadecanol
$C_{18}H_{38}O_2$
    Hexadecylmonooxyethylene glycol ether

$C_{18}H_{38}O_3$
    Tetradecyldioxyethylene glycol ether
$C_{18}H_{38}O_4$
    Dodecyltrioxyethylene glycol ether
$C_{18}H_{38}O_5$
    Decyltetraoxyethylene glycol ether
$C_{18}H_{38}O_6$
    Octylpentaoxyethylene glycol ether
$C_{18}H_{38}O_7$
    Hexylhexaoxyethylene glycol ether
$C_{18}H_{38}O_9$
    Ethyloctaoxyethylene glycol ether
$C_{18}H_{40}BrN$
    Dodecylbutyldimethylammonium bromide
$C_{19}H_{31}NaO_3S$
    Sodium 3-methyl-4-(3-dodecyl)benzenesulphonate
    Sodium tridecylbenzenesulphonate (isomeric mixture)
    Sodium 4-(tridecyl)benzenesulphonate
    Sodium 4-(2-tridecyl)benzenesulphonate
$C_{19}H_{33}BrD_9N$
    Hexadecyl-nona-deuteriotrimethylammonium bromide
$C_{19}H_{34}BrN$
    N-Tetradecylpyridinium bromide
$C_{19}H_{34}ClN$
    N-Tetradecylpyridinium chloride
$C_{19}H_{34}IN$
    N-Ethyl-4-dodecylpyridinium iodide
    N-Tetradecylpyridinium iodide
$C_{19}H_{37}NaO_5S$
    Sodium 1-carbomethoxyheptadecanesulphonate
    Sodium 1-carbopropoxypentadecanesulphonate
$C_{19}H_{38}NaO_5S$
    Sodium 3-(piperidino)tetradecanesulphonate
$C_{19}H_{39}NaO_3S$
    Sodium 10-nonadecanesulphonate
$C_{19}H_{39}NaO_4S$
    Sodium 5-nonadecylsulphate
    Sodium 10-nonadecylsulphate
    Sodium 3-propoxyhexadecanesulphonate
$C_{19}H_{40}O_5$
    Undecyltetraoxyethylene glycol ether
$C_{19}H_{40}O_6$
    Nonylpentaoxyethylene glycol ether
$C_{19}H_{42}BrN$
    Hexadecyltrimethylammonium bromide
$C_{19}H_{42}ClN$
    Hexadecyltrimethylammonium chloride

$C_{19}H_{42}N_2O_3$
  Hexadeyltrimethylammonium nitrate
$C_{19}H_{43}NO$
  Hexadeyltrimethylammonium hydroxide
$C_{19.96}H_{33.92}O_{3.98}$
  Octylphenol(ethylene oxide)$_{2.98}$ ether
$C_{20}H_{12}$
  Perylene
$C_{20}H_{27}NaO_3S$
  Sodium decylnaphthalenesulphonate
$C_{20}H_{30}Cl_3NaO_4S$
  Sodium 3-(trichlorophenoxy)tetradecanesulphonate
$C_{20}H_{33}NaO_3S$
  Sodium 2,5-diheptylbenzenesulphonate
  Sodium 2,3-dimethyl-4-(dodecyl)benzenesulphonate
  Sodium 2,3-dimethyl-4-(3-dodecyl)benzenesulphonate
  Sodium 2-(dodecyl)-5-ethylbenzenesulphonate
  Sodium 2-ethyl-5-(dodecyl)benzenesulphonate
  Sodium 2-methyl-5-(tridecyl)benzenesulphonate
  Sodium 4-(2-pentyl-nonyl)benzenesulphonate
  Sodium 2-propyl-5-(undecyl)benzenesulphonate
  Sodium tetradecylbenzenesulphonate (technical grade, purified)
  Sodium tetradecylbenzenesulphonate (isomeric mixture)
  Sodium 4-(tetradecyl)benzenesulphonate
  Sodium 2-(undecyl)-5-propylbenzenesulphonate
$C_{20}H_{33}NaO_4S$
  Sodium 3-(phenoxy)tetradecanesulphonate
$C_{20}H_{34}O_4$
  Octylphenol(ethylene oxide)$_3$ ether
$C_{20}H_{36}BrNO$
  N-Tetradecyl-4-methoxypyridinium bromide
$C_{20}H_{36}IN$
  N-Propyl-4-(dodecyl)pyridinium iodide
  N-(2-Propyl)-4-(dodecyl)pyridinium iodide
$C_{20}H_{37}NO_3S$
  Dimethylammonium 4-(6-dodecyl)benzenesulphonate
$C_{20}H_{38}O_2$
  Ethyl oleate
$C_{20}H_{38}Na_2O_8S_2$
  Disodium heptadecanesulphonate-1-carboxyethanesulphonate
$C_{20}H_{39}NaO_5S$
  Sodium 1-carbethoxyheptadecanesulphonate
$C_{20}H_{41}NaO_3S$
  Sodium eicosanesulphonate
$C_{20}H_{41}NaO_4S$
  Sodium 3-(butoxy)hexadecanesulphonate
  Sodium eicosylsulphate
  Sodium 3-(2-(ethyl)hexoxy)dodecanesulphonate
  Sodium 3-(octoxy)dodecanesulphonate
$C_{20}H_{42}CuO_6S_2$
  Copper bis(decanesulphonate)
$C_{20}H_{42}MgO_6S_2$
  Magnesium bis(decanesulphonate)
$C_{20}H_{42}O_2$
  Octadecylmonooxyethylene glycol ether

$C_{20}H_{42}O_3$
  Hexadecyldioxyethylene glycol ether
$C_{20}H_{42}O_4$
  Tetradecyltrioxyethylene glycol ether
$C_{20}H_{42}O_5$
  Dodecyltetraoxyethylene glycol ether
$C_{20}H_{42}O_6$
  Decylpentaoxyethylene glycol ether
$C_{20}H_{42}O_7$
  Octylhexaoxyethylene glycol ether
$C_{20}H_{42}O_{10}$
  Ethylnonaoxyethylene glycol ether
$C_{20}H_{44}BrN$
  Tetradecyltriethylammonium bromide
$C_{21}H_{35}NaO_3S$
  Sodium pentadecylbenzenesulphonate (isomeric mixture)
  Sodium 4-(pentadecyl)benzenesulphonate
  Sodium 4-(2-pentadecyl)benzenesulphonate
$C_{21}H_{38}BF_4N$
  N-Hexadecylpyridinium tetrafluoroborate
$C_{21}H_{38}BrN$
  N-Hexadecylpyridinium bromide
$C_{21}H_{38}ClN$
  N-Hexadecylpyridinium chloride
$C_{21}H_{38}IN$
  N-Butyl-4-(dodecyl)pyridinium iodide
  N-Hexadecylpyridinium iodide
$C_{21}H_{38}N_2O_3$
  N-Hexadecylpyridinium nitrate
$C_{21}H_{39}NO_4S$
  N-Hexadecylpyridinium bisulphate
$C_{21}H_{40}NO_4P$
  N-Hexadecylpyridinium phosphate
$C_{21}H_{41}NaO_5S$
  Sodium 1-carbopropoxyheptadecanesulphonate
  Sodium 1-carbo(2-propoxy)heptadecanesulphonate
$C_{21}H_{44}O_6$
  Undecylpentaoxyethylene glycol ether
$C_{21}H_{44}O_7$
  Nonylhexaoxyethylene glycol ether
$C_{21}H_{46}BrN$
  Octadecyltrimethylammonium bromide
$C_{21}H_{46}ClN$
  Octadecyltrimethylammonium chloride
$C_{22}H_{16}O_2$
  Benzyl-9-anthroate
$C_{22}H_{31}NaO_3S$
  Sodium dodecylnaphthalenesulphonate
$C_{22}H_{37}NaO_3S$
  Sodium hexadecylbenzenesulphonate (technical grade, purified)
  Sodium 4-(hexadecyl)benzenesulphonate
  Sodium 4-(8-hexadecyl)benzenesulphonate
$C_{22}H_{38}O_5$
  Octylphenol(ethylene oxide)$_4$ ether

$C_{22}H_{41}NNaO_3S$
Diethylammonium 4-(6-dodecyl)benzenesulphonate
$C_{22}H_{45}NaO_3S$
Sodium docosanesulphonate
$C_{22}H_{45}NaO_4S$
Sodium docosylsulphate
$C_{22}H_{46}O_3$
Octadecyldioxyethylene glycol ether
$C_{22}H_{46}O_4$
Hexadecyltrioxyethylene glycol ether
$C_{22}H_{46}O_5$
Tetradecyltetraoxyethylene glycol ether
$C_{22}H_{46}O_6$
Dodecylpentaoxyethylene glycol ether
$C_{22}H_{46}O_7$
Decylhexaoxyethylene glycol ether
$C_{22}H_{46}O_{11}$
Ethyldecaoxyethylene glycol ether
$C_{22}H_{48}BrN$
Didecyldimethylammonium bromide
Dodecyloctyldimethylammonium bromide
$C_{22.14}H_{38.28}O_{5.07}$
Octylphenol(ethylene oxide)$_{4.07}$ ether
$C_{23}H_{39}NaO_3S$
Sodium 4-(2-heptadecyl)benzenesulphonate
$C_{23}H_{42}BrN$
N-Octadecylpyridinium bromide
$C_{23}H_{42}IN$
N-Octadecylpyridinium iodide
$C_{23}H_{48}O_6$
Tridecylpentaoxyethylene glycol ether
$C_{23}H_{48}O_7$
Undecylhexaoxyethylene glycol ether
$C_{23}H_{50}BrN$
Tetradecyltripropylammonium bromide
$C_{23}H_{50}ClN$
Eicosyltrimethylammonium chloride
$C_{24}H_{16}O_2$
Benzyl-1-pyrenoate
$C_{24}H_{35}NaO_3S$
Sodium tetradecylnaphthalenesulphonate
$C_{24}H_{39}O_5Na$
Sodium cholate
$C_{24}H_{41}NaO_3S$
Sodium 4-(octadecyl)benzenesulphonate
$C_{24}H_{42}O_6$
Octylphenol(ethylene oxide)$_5$ ether
$C_{24}H_{45}NaO_3S$
Dipropylammonium 4-(6-dodecyl)benzenesulphonate
$C_{24}H_{50}BaO_8S_2$
Barium bis(dodecylsulphate)
$C_{24}H_{50}CaO_8S_2$
Calcium bis(dodecylsulphate)
$C_{24}H_{50}CuO_6S_2$
Copper bis(dodecanesulphonate)

$C_{24}H_{50}MgO_6S_2$
  Magnesium bis(dodecanesulphonate)
$C_{24}H_{50}MnO_8S_2$
  Manganese bis(dodecylsulphate)
$C_{24}H_{50}O_4$
  Octadecyltrioxyethylene glycol ether
$C_{24}H_{50}O_5$
  Hexadecyltetraoxyethylene glycol ether
$C_{24}H_{50}O_6$
  Tetradecylpentaoxyethylene glycol ether
$C_{24}H_{50}O_7$
  Dodecylhexaoxyethylene glycol ether
$C_{24}H_{50}O_8PbS_2$
  Lead bis(dodecylsulphate)
$C_{24}H_{50}O_8S_2Sr$
  Strontium bis(dodecylsulphate)
$C_{24}H_{50}O_8S_2Zn$
  Zinc bis(dodecylsulphate)
$C_{24}H_{50}O_9$
  Octyloctaoxyethylene glycol ether
$C_{24}H_{52}BrN$
  Dodecyldecyldimethylammonium bromide
$C_{24}H_{58}CuO_{12}S_2$
  Copper bis(dodecylsulphate) tetrahydrate
$C_{24}H_{62}CoO_{14}S_2$
  Cobalt bis(dodecylsulphate) hexahydrate
$C_{24}H_{62}MgO_{14}S_2$
  Magnesium bis(dodecylsulphate) hexahydrate
$C_{24.02}H_{42.04}O_{6.01}$
  Octylphenol(ethylene oxide)$_{5.01}$ ether
$C_{25}H_{44}O_6$
  Nonylphenol(ethylene oxide)$_5$ ether
$C_{25}H_{52}O_7$
  Tridecylhexaoxyethylene glycol ether
$C_{25}H_{52}O_9$
  Nonyloctaoxyethylene glycol ether
$C_{26}H_{39}NaO_3S$
  Sodium hexadecylnaphthalenesulphonate
$C_{26}H_{40}O_3$
  4-Heptadecyl-7-hydroxycoumarin
$C_{26}H_{45}NaO_3S$
  Sodium 4-(eicosyl)benzenesulphonate
$C_{26}H_{46}O_7$
  Octylphenol(ethylene oxide)$_6$ ether
$C_{26}H_{54}O_5$
  Octadecyltetraoxyethylene glycol ether
$C_{26}H_{54}O_6$
  Hexadecylpentaoxyethylene glycol ether
$C_{26}H_{54}O_7$
  Tetradecylhexaoxyethylene glycol ether
$C_{26}H_{54}O_8$
  Dodecylheptaoxyethylene glycol ether
$C_{26}H_{54}O_9$
  Decyloctaoxyethylene glycol ether

$C_{26}H_{54}O_{10}$
Octylnonaoxyethylene glycol ether

$C_{26}H_{56}BrN$
Didodecyldimethylammonium bromide
Hexadecyloctyldimethylammonium bromide
Tetradecyltributylammonium bromide

$C_{26.06}H_{46.12}O_{7.03}$
Octylphenol(ethylene oxide)$_{6.03}$ ether

$C_{27}H_{48}O_7$
Nonylphenol(ethylene oxide)$_6$ ether

$C_{27}H_{48}O_{8.5}$
Hexylphenol(ethylene oxide)$_{7.5}$ ether

$C_{27}H_{56}O_7$
Pentadecylhexaoxyethylene glycol ether

$C_{27}H_{56}O_9$
Undecyloctaoxyethylene glycol ether

$C_{27.4}H_{48.8}O_{7.2}$
Nonylphenol(ethylene oxide)$_{6.2}$ ether

$C_{28}H_{45}NO_3S$
N-Hexadecylpyridinium tosylate

$C_{28}H_{50}O_8$
Octylphenol(ethylene oxide)$_7$ ether

$C_{28}H_{58}BaO_8S_2$
Barium bis(tetradecylsulphate)

$C_{28}H_{58}CaO_8S_2$
Calcium bis(tetradecylsulphate)

$C_{28}H_{58}CoO_8S_2$
Cobalt bis(tetradecylsulphate)

$C_{28}H_{58}CuO_6S_2$
Copper bis(tetradecanesulphonate)

$C_{28}H_{58}CuO_8S_2$
Copper bis(tetradecylsulphate)

$C_{28}H_{58}MgO_8S_2$
Magnesium bis(tetradecylsulphate)

$C_{28}H_{58}MnO_8S_2$
Manganese bis(tetradecylsulphate)

$C_{28}H_{58}O_8S_2Zn$
Zinc bis(tetradecylsulphate)

$C_{28}H_{58}O_7$
Hexadecylhexaoxyethylene glycol ether

$C_{28}H_{58}O_8$
Tetradecylheptaoxyethylene glycol ether

$C_{28}H_{58}O_9$
Dodecyloctaoxyethylene glycol ether

$C_{28}H_{58}O_{10}$
Decylnonaoxyethylene glycol ether

$C_{28}H_{62}CaO_6S_2$
Calcium bis(octylbenzenesulphonate)

$C_{28.1}H_{50.2}O_{8.05}$
Octylphenol(ethylene oxide)$_{7.05}$ ether

$C_{28.6}H_{51.2}O_{6.3}$
Dodecylphenol(ethylene oxide)$_{5.3}$ ether

$C_{28.6}H_{51.2}O_{8.3}$
Octylphenol(ethylene oxide)$_{7.3}$ ether

$C_{29}H_{52}O_8$
Nonylphenol(ethylene oxide)$_7$ ether
$C_{29}H_{52}O_{8.5}$
Octylphenol(ethylene oxide)$_{7.5}$ ether
$C_{29}H_{59}NaO_4S$
Sodium 15-nonacosylsulphate
$C_{29}H_{60}O_9$
Tridecyloctaoxyethylene glycol ether
$C_{29.4}H_{51.8}O_{8.7}$
Octylphenol(ethylene oxide)$_{7.7}$ ether
$C_{29.8}H_{53.6}O_{8.4}$
Nonylphenol(ethylene oxide)$_{7.4}$ ether
$C_{30}H_{54}O_{8.5}$
Nonylphenol(ethylene oxide)$_{7.5}$ ether
$C_{30}H_{54}O_9$
Octylphenol(ethylene oxide)$_8$ ether
$C_{30}H_{58}CaO_6S_2$
Calcium bis(2-pentadecenesulphonate)
$C_{30}H_{62}$
2,6,10,15,19,23-Hexamethyltetracosane
$C_{30}H_{62}CaO_8S_2$
Calcium bis(3-hydroxypentadecanesulphonate)
$C_{30}H_{62}O_8$
Hexadecylheptaoxyethylene glycol ether
$C_{30}H_{62}O_9$
Tetradecyloctaoxyethylene glycol ether
$C_{30}H_{62}O_{10}$
Dodecylnonaoxyethylene glycol ether
$C_{30}H_{62}O_{11}$
Decyldecaoxyethylene glycol ether
$C_{30}H_{64}BrN$
Ditetradecyldimethylammonium bromide
$C_{30.06}H_{54.12}O_{9.03}$
Octylphenol(ethylene oxide)$_{8.03}$ ether
$C_{30.8}H_{55.6}O_{9.4}$
Octylphenol(ethylene oxide)$_{8.4}$ ether
$C_{31}H_{40}O_4$
2-(9-Anthroyloxy)palmitic acid
$C_{31}H_{56}O_9$
Nonylphenol(ethylene oxide)$_8$ ether
$C_{31}H_{56}O_{9.5}$
Octylphenol(ethylene oxide)$_{8.5}$ ether
$C_{31}H_{64}O_9$
Pentadecyloctaoxyethylene glycol ether
$C_{31.2}H_{56.4}O_{9.1}$
Nonylphenol(ethylene oxide)$_{8.1}$ ether
$C_{31.6}H_{57.2}O_{9.8}$
Octylphenol(ethylene oxide)$_{8.8}$ ether
$C_{32}H_{56}N_2O_6S_2$
Methylviologen bis(decanesulphonate)
$C_{32}H_{58}O_{9.5}$
Nonylphenol(ethylene oxide)$_{8.5}$ ether
$C_{32}H_{58}O_{10}$
Octylphenol(ethylene oxide)$_9$ ether

$C_{32}H_{66}O_9$
   Hexadecyloctaoxyethylene glycol ether
$C_{32}H_{66}O_{11}$
   Dodecyldecaoxyethylene glycol ether
$C_{32}H_{66}O_{12}$
   Decylundecaoxyethylene glycol ether
$C_{32}H_{66}O_{13}$
   Octyldodecaoxyethylene glycol ether
$C_{32.12}H_{58.24}O_{10.06}$
   Octylphenol(ethylene oxide)$_{9.06}$ ether
$C_{32.2}H_{58.4}O_{9.6}$
   Nonylphenol(ethylene oxide)$_{8.6}$ ether
$C_{33}H_{44}O_4$
   6-(9-Anthroyloxy)stearic acid
   6-(9-Anthroyloxy)stearic acid
   12-(9-Anthroyloxy)stearic acid
$C_{33}H_{60}O_{10}$
   Nonylphenol(ethylene oxide)$_9$ ether
$C_{33}H_{60}O_{10.5}$
   Octylphenol(ethylene oxide)$_{9.5}$ ether
$C_{33}H_{68}O_{11}$
   Tridecyldecaoxyethylene glycol ether
$C_{33.2}H_{60.4}O_{10.6}$
   Octylphenol(ethylene oxide)$_{9.6}$ ether
$C_{33.4}H_{60.8}O_{10.7}$
   Octylphenol(ethylene oxide)$_{9.7}$ ether
$C_{33.86}H_{61.72}O_{10.93}$
   Octylphenol(ethylene oxide)$_{9.93}$ ether
$C_{34}H_{56}CaN_2O_6S$
   Calcium bis(N-undecyl-4-aminobenzenesulphonate)
$C_{34}H_{62}O_9$
   Dodecylphenol(ethylene oxide)$_8$ ether
$C_{34}H_{62}O_{10}$
   Decylphenol(ethylene oxide)$_9$ ether
$C_{34}H_{62}O_{10.5}$
   Nonylphenol(ethylene oxide)$_{9.5}$ ether
$C_{34}H_{62}O_{11}$
   Octylphenol(ethylene oxide)$_{10}$ ether
$C_{34}H_{70}O_9$
   Octadecyloctaoxyethylene glycol ether
$C_{34}H_{70}O_{10}$
   Hexadecylnonaoxyethylene glycol ether
$C_{34}H_{70}O_{12}$
   Dodecylundecaoxyethylene glycol ether
$C_{34}H_{72}BrN$
   Dihexadecyldimethylammonium bromide
$C_{34.2}H_{62.4}O_{10.6}$
   Nonylphenol(ethylene oxide)$_{9.6}$ ether
$C_{34.4}H_{62.8}O_{9.2}$
   Dodecylphenol(ethylene oxide)$_{8.2}$ ether
$C_{35}H_{24}$
   1,3-Di(1-pyrenyl)propane
$C_{35}H_{64}O_{11}$
   Nonylphenol(ethylene oxide)$_{10}$ ether

$C_{35.6}H_{65.2}O_{11.3}$
  Nonylphenol(ethylene oxide)$_{10.3}$ ether
$C_{36}H_{58}CaO_6S_2$
  Calcium bis(dodecylbenzenesulphonate)
$C_{36}H_{60}O_{30}$
  α-Cyclodextrin
$C_{36}H_{64}N_2O_6S_2$
  Methylviologen bis(dodecanesulphonate)
$C_{36}H_{64}N_2O_8S_2$
  Methylviologen bis(dodecylsulphate)
$C_{36}H_{66}O_{10}$
  Dodecylphenol(ethylene oxide)$_9$ ether
$C_{36}H_{66}O_{11.5}$
  Nonylphenol(ethylene oxide)$_{10.5}$ ether
$C_{36}H_{72}NO_8P$
  Dimyristoylphosphatidyl chloline
$C_{36}H_{74}O_{13}$
  Dodecyldodecaoxyethylene glycol ether
$C_{36.8}H_{57.6}O_{10.4}$
  Dodecylphenol(ethylene oxide)$_{9.4}$ ether
$C_{37}H_{68}O_{12}$
  Nonylphenol(ethylene oxide)$_{11}$ ether
$C_{37}H_{76}O_{13}$
  Tridecyldodecaoxyethylene glycol ether
$C_{37.4}H_{68.8}O_{107}$
  Dodecylphenol(ethylene oxide)$_{9.7}$ ether
$C_{38}H_{84}N_2O_3$
  Bis(hexadecyltrimethylammonium)sulphate
$C_{39}H_{84}NO_3$
  Bis(hexadecyltrimethylammonium)carbonate
$C_{39}H_{72}O_{13.5}$
  Octylphenol(ethylene oxide)$_{12.5}$ ether
$C_{40}H_{72}N_2O_6S_2$
  1,1'-(1,ω-Ethanediyl)bispyridinium bis(tetradecanesulphonate)
  Methylviologen bis(tetradecanesulphonate)
$C_{40}H_{74}O_{13.5}$
  Nonylphenol(ethylene oxide)$_{12.5}$ ether
$C_{40}H_{74}O_{14}$
  Octylphenol(ethylene oxide)$_{13}$ ether
$C_{40}H_{80}NO_8P$
  Dipalmitoylphosphatidyl choline
$C_{40}H_{82}O_{13}$
  Hexadecyldodecaoxyethylene glycol ether
$C_{40}H_{82}O_{15}$
  Dodecyltetradecaoxyethylene glycol ether
$C_{41}H_{29}NO$
  2,6-Diphenyl-4-(2,4,6-triphenyl-1-pyridino)phenoxide
$C_{41}H_{76}O_{14}$
  Nonylphenol(ethylene oxide)$_{13}$ ether
$C_{41.6}H_{85.2}O_{15.8}$
  Dodecylpoly(oxyethylene glycol)$_{14.8}$ ether
$C_{42}H_{38}$
  1,10-Di(1-pyrenyl)decane
$C_{42}H_{76}N_2O_6S_2$
  1,1'-(1,ω-Butanediyl)bispyridinium bis(tetradecanesulphonate)

$C_{42}H_{86}O_{16}$
  Dodecylpentadecaoxyethylene glycol ether
$C_{42}H_{86}O_{17}$
  Decylhexadecaoxyethylene glycol ether
$C_{42}H_{87}AlO_{12}S_3$
  Aluminium tris(tetradecylsulphate)
$C_{43}H_{80}O_{15}$
  Nonylphenol(ethylene oxide)$_{14}$ ether
$C_{43}H_{88}O_{16}$
  Tridecylpentadecaoxyethylene glycol ether
$C_{44}H_{80}N_2O_6S_2$
  1,1'-(1,ω-Hexanediyl)bispyridinium bis(tetradecanesulphonate)
  Methylviologen bis(hexadecanesulphonate)
$C_{45}H_{84}O_{15}$
  Nonylphenol(ethylene oxide)$_{15}$ ether
$C_{46}H_{84}N_2O_6S_2$
  1,1'-(1,ω-Octanediyl)bispyridinium bis(tetradecanesulphonate)
$C_{46}H_{86}O_{17}$
  Octylphenol(ethylene oxide)$_{16}$ ether
$C_{46}H_{94}O_{16}$
  Hexadecylpentadecaoxyethylene glycol ether
$C_{46.6}H_{87.2}O_{16.8}$
  Nonylphenol(ethylene oxide)$_{15.8}$ ether
$C_{46.8}H_{87.6}O_{13.4}$
  Hexadecylphenol(ethylene oxide)$_{12.4}$ ether
$C_{48}H_{88}N_2O_6S_2$
  1,1'-(1,ω-Decanediyl)bispyridinium bis(tetradecanesulphonate)
  Methylviologen bis(octadecanesulphonate)
$C_{48}H_{90}O_{18}$
  Octylphenol(ethylene oxide)$_{17}$ ether
$C_{48}H_{98}O_{19}$
  Dodecyloctadecaoxyethylene glycol ether
$C_{50}H_{92}N_2O_6S_2$
  1,1'-(1,ω-Dodecanediyl)bispyridinium bis(tetradecanesulphonate)
$C_{50.4}H_{94.8}O_{18.7}$
  Nonylphenol(ethylene oxide)$_{17.7}$ ether
$C_{50.6}H_{102.6}O_{20.3}$
  Dodecylpoly(oxyethylene glycol)$_{19.3}$ ether
$C_{51}H_{96}O_{19}$
  Nonylphenol(ethylene oxide)$_{18}$ ether
$C_{52}H_{96}N_2O_6S_2$
  1,1'-(1,ω-Tetradecanediyl)bispyridinium bis(tetradecanesulphonate)
$C_{54}H_{102}O_{21}$
  Octylphenol(ethylene oxide)$_{20}$ ether
$C_{55}H_{104}O_{21}$
  Nonylphenol(ethylene oxide)$_{20}$ ether
$C_{57}H_{116}O_{23}$
  Tridecyldocosaoxyethylene glycol ether
$C_{58}H_{118}O_{22}$
  Hexadecylheneicosaoxyethylene glycol ether
$C_{58}H_{118}O_{24}$
  Dodecyltricosaoxyethylene glycol ether
$C_{59.8}H_{121.6}O_{24.9}$
  Dodecylpoly(oxyethylene glycol)$_{23.9}$ ether

$C_{72}H_{146}O_{31}$
    Dodecyltriacontaoxyethylene glycol ether
$C_{74}H_{142}O_{31}$
    Octylphenol(ethylene oxide)$_{30}$ ether
$C_{74}H_{150}O_{32}$
    Dodecylhentriacontaoxyethylene glycol ether
$C_{75}H_{144}O_{31}$
    Nonylphenol(ethylene oxide)$_{30}$ ether
$C_{76}H_{154}O_{31}$
    Hexadecyltriacontaoxyethylene glycol ether
$C_{77}H_{148}O_{32}$
    Nonylphenol(ethylene oxide)$_{31}$ ether
$C_{94}H_{184}O_{41}$
    Octylphenol(ethylene oxide)$_{40}$ ether
$C_{95}H_{184}O_{41}$
    Nonylphenol(ethylene oxide)$_{40}$ ether
$C_{115}H_{224}O_{51}$
    Nonylphenol(ethylene oxide)$_{50}$ ether
$C_{215}H_{424}O_{101}$
    Nonylphenol(ethylene oxide)$_{100}$ ether

GENERAL FORMULA INDEX

Anionic Surfactants
   Sodium alkylsulphates
   Alkylsulphates, other counterions
   Sodium alkanesulphonates
   Alkanesulponates, other counterions
   Sodium alkylarenesulphonates
   Alkylarenesulphonates, other counterions
   Sodium alkylnaphtalenesulphonates

Cationic Surfactants
   Alkyltrimethylammonium salts
   Alkylpyridinium salts

Nonionic Surfactants
   Alkylpolyoxyethylene compounds
   Alkylphenolethoxylate compounds
   Phosphoxides

Alcohols
Alkanes - Cycloalkanes - Alkenes
Deuterated Compounds
Fluorinated Compounds
Miscellaneous Inorganic Compounds
Miscellaneous Organic Compounds

Anionic Surfactants

Sodium alkylsulphates

$C_6H_{13}OSO_3Na$
  Sodium hexylsulphate
$C_7H_{15}OSO_3Na$
  Sodium heptylsulphate
$C_8H_{17}OSO_3Na$
  Sodium octylsulphate
$C_9H_{19}OSO_3Na$
  Sodium nonylsulphate
$C_{10}H_{21}OSO_3Na$
  Sodium decylsulphate
$C_8H_{17}CH(CH_3)OSO_3Na$
  Sodium 2-decylsulphate
$C_{11}H_{23}OSO_3Na$
  Sodium undecylsulphate
$C_{12}H_{25}OSO_3Na$
  Sodium dodecylsulphate
$C_9H_{19}CH(CH_3)CH_2OSO_3Na$
  Sodium 2-methylundecylsulphate
$C_{13}H_{27}OSO_3Na$
  Sodium tridecylsulphate
$C_{14}H_{29}OSO_3Na$
  Sodium tetradecylsulphate
$C_{12}H_{25}CH(CH_3)OSO_3Na$
  Sodium 2-tetradecylsulphate
$C_{11}H_{23}CH(C_2H_5)OSO_3Na$
  Sodium 3-tetradecylsulphate
$C_{10}H_{21}CH(C_3H_7)OSO_3Na$
  Sodium 4-tetradecylsulphate
$C_9H_{19}CH(C_4H_9)OSO_3Na$
  Sodium 5-tetradecylsulphate
$C_7H_{15}CH(C_6H_{13})OSO_3Na$
  Sodium 7-tetradecylsulphate
$C_{11}H_{23}CH(CH_3)CH_2OSO_3Na$
  Sodium 2-methyltridecylsulphate
$C_{15}H_{31}OSO_3Na$
  Sodium pentadecylsulphate
$C_{13}H_{27}CH(CH_3)OSO_3Na$
  Sodium 2-pentadecylsulphate
$C_{12}H_{25}CH(C_2H_5)OSO_3Na$
  Sodium 3-pentadecylsulphate
$C_{10}H_{21}CH(C_4H_9)OSO_3Na$
  Sodium 5-pentadecylsulphate
$C_7H_{15}CH(C_7H_{15})OSO_3Na$
  Sodium 8-pentadecylsulphate
$C_{16}H_{33}OSO_3Na$
  Sodium hexadecylsulphate
$C_{12}H_{25}CH(C_3H_7)OSO_3Na$
  Sodium 4-hexadecylsulphate
$C_{10}H_{21}CH(C_5H_{11})OSO_3Na$
  Sodium 6-hexadecylsulphate

$C_8H_{17}CH(C_7H_{15})OSO_3Na$
Sodium 8-hexadecylsulphate
$C_{13}H_{27}CH(CH_3)CH_2OSO_3Na$
Sodium 2-methylpentadecylsulphate
$C_{17}H_{35}OSO_3Na$
Sodium heptadecylsulphate
$C_{15}H_{31}CH(CH_3)OSO_3Na$
Sodium 2-heptadecylsulphate
$C_8H_{17}CH(C_8H_{17})OSO_3Na$
Sodium 9-heptadecylsulphate
$C_{18}H_{37}OSO_3Na$
Sodium octadecylsulphate
$C_{16}H_{33}CH(CH_3)OSO_3Na$
Sodium 2-octadecylsulphate
$C_{14}H_{29}CH(C_3H_7)OSO_3Na$
Sodium 4-octadecylsulphate
$C_{12}H_{25}CH(C_5H_{11})OSO_3Na$
Sodium 6-octadecylsulphate
$C_{15}H_{31}CH(CH_3)CH_2OSO_3Na$
Sodium 2-methylheptadecylsulphate
$C_{14}H_{29}CH(C_4H_9)OSO_3Na$
Sodium 5-nonadecylsulphate
$C_9H_{19}CH(C_9H_{19})OSO_3Na$
Sodium 10-nonadecylsulphate
$C_{20}H_{41}OSO_3Na$
Sodium eicosylsulphate
$C_{22}H_{45}OSO_3Na$
Sodium docosylsulphate
$C_{14}H_{29}CH(C_{14}H_{29})OSO_3Na$
Sodium 15-nonacosylsulphate

Alkylsulphates, other counterions

$C_{12}H_{25}OSO_3K$
Potassium dodecylsulphate
$C_{14}H_{29}OSO_3K$
Potassium tetradecylsulphate
$(C_{12}H_{25}OSO_3)_2Ba$
Barium bis(dodecylsulphate)
$(C_{12}H_{25}OSO_3)_2Ca$
Calcium bis(dodecylsulphate)
$(C_{12}H_{25}OSO_3)_2Co.6H_2O$
Cobalt bis(dodecylsulphate) hexahydrate
$(C_{12}H_{25}OSO_3)_2Cu.4H_2O$
Copper bis(dodecylsulphate) tetrahydrate
$(C_{12}H_{25}OSO_3)_2Mg.6H_2O$
Magnesium bis(dodecylsulphate) hexahydrate
$(C_{12}H_{25}OSO_3)_2Mn$
Manganese bis(dodecylsulphate)
$(C_{12}H_{25}OSO_3)_2MV$
Methylviologen bis(dodecylsulphate)
$(C_{12}H_{25}OSO_3)_2Pb$
Lead bis(dodecylsulphate)

(C₁₂H₂₅OSO₃)₂Sr
  Strontium bis(dodecylsulphate)
(C₁₂H₂₅OSO₃)₂Zn
  Zinc bis(dodecylsulphate)
(C₁₄H₂₉OSO₃)₃Al
  Aluminium tris(tetradecylsulphate)
(C₁₄H₂₉OSO₃)₂Ba
  Barium bis(tetradecylsulphate)
(C₁₄H₂₉OSO₃)₂Ca
  Calcium bis(tetradecylsulphate)
(C₁₄H₂₉OSO₃)₂Co
  Cobalt bis(tetradecylsulphate)
(C₁₄H₂₉OSO₃)₂Cu
  Copper bis(tetradecylsulphate)
(C₁₄H₂₉OSO₃)₂Mg
  Magnesium bis(tetradecylsulphate)
(C₁₄H₂₉OSO₃)₂Mn
  Manganese bis(tetradecylsulphate)
(C₁₄H₂₉OSO₃)₂Zn
  Zinc bis(tetradecylsulphate)

Sodium alkanesulphonates

C₆H₁₃SO₃Na
  Sodium hexanesulphonate
C₈H₁₇SO₃Na
  Sodium octanesulphonate
C₁₀H₂₁SO₃Na
  Sodium decanesulphonate
C₈H₁₇CHOHCH₂SO₃Na
  Sodium 2-hydroxydecanesulphonate
C₁₁H₂₃SO₃Na
  Sodium undecanesulphonate
C₁₀H₂₂CH(COOCH₃SO₃Na)
  Sodium 1-carbomethoxyundecanesulphonate
C₉H₁₉CHOHCH₂SO₃Na
  Sodium 2-hydroxyundecanesulphonate
C₁₂H₂₅SO₃Na
  Sodium dodecanesulphonate
C₁₀H₂₁CH(CH₃)SO₃Na
  Sodium 2-dodecanesulphonate
C₉H₁₉CH(C₂H₅)SO₃Na
  Sodium 3-dodecanesulphonate
C₈H₁₇CH(C₃H₇)SO₃Na
  Sodium 4-dodecanesulphonate
C₇H₁₅CH(C₄H₉)SO₃Na
  Sodium 5-dodecanesulphonate
C₆H₁₃CH(C₅H₁₁)SO₃Na
  Sodium 6-dodecanesulphonate
C₉H₁₉CH=CHCH₂SO₃Na
  Sodium 2-dodecenesulphonate
C₉H₁₉CH(OC₄H₉)C₂H₄SO₃Na
  Sodium 3-(butoxy)dodecane-1-sulphonate

C$_9$H$_{19}$CH(OC$_2$H$_5$)C$_2$H$_4$SO$_3$Na
Sodium 3-ethoxydodecanesulphonate

C$_9$H$_{19}$CH(OCH$_2$CH(C$_2$H$_5$)C$_4$H$_9$)C$_2$H$_4$SO$_3$Na
Sodium 3-((2-ethyl)hexoxy)dodecanesulphonate

C$_9$H$_{19}$CH(OC$_6$H$_{13}$)C$_2$H$_4$SO$_3$Na
Sodium 3-(hexoxy)dodecanesulphonate

C$_{11}$H$_{23}$CHOHSO$_3$Na
Sodium 1-hydroxydodecanesulphonate

C$_{10}$H$_{21}$CHOHCH$_2$SO$_3$Na
Sodium 2-hydroxydodecanesulphonate

C$_9$H$_{19}$CHOHCH$_2$CH$_2$SO$_3$Na
Sodium 3-hydroxydodecanesulphonate

C$_{10}$H$_{21}$CH(CH$_2$OH)SO$_3$Na
Sodium 1-hydroxy-2-dodecanesulphonate

C$_9$H$_{19}$CH(OH)C$_2$H$_4$SO$_3$Na
Sodium 3-hydroxydodecanesulphonate

C$_9$H$_{19}$CH(OCH$_3$)C$_2$H$_4$SO$_3$Na
Sodium 3-methoxydodecanesulphonate

C$_9$H$_{19}$CH(OC$_8$H$_{17}$)C$_2$H$_4$SO$_3$Na
Sodium 3-(octoxy)dodecanesulphonate

C$_{10}$H$_{21}$COCH$_2$SO$_3$Na
Sodium 2-oxododecanesulphonate

C$_9$H$_{19}$COCH$_2$CH$_2$SO$_3$Na
Sodium 3-oxododecanesulphonate

C$_9$H$_{19}$CH(OC$_6$H$_5$)C$_2$H$_4$SO$_3$Na
Sodium 3-(phenoxy)dodecanesulphonate

C$_9$H$_{19}$CH(OCH(CH$_3$)$_2$)C$_2$H$_4$SO$_3$Na
Sodium 3-(i-propoxy)dodecanesulphonate

C$_9$H$_{19}$CH(OC$_3$H$_7$)C$_2$H$_4$SO$_3$Na
Sodium 3-(n-propoxy)dodecanesulphonate

C$_{13}$H$_{27}$SO$_3$Na
Sodium tridecanesulphonate

C$_{12}$H$_{25}$CH(COOCH$_3$SO$_3$Na)
Sodium 1-carbomethoxytridecanesulphonate

C$_{11}$H$_{23}$CHOHCH$_2$SO$_3$Na
Sodium 2-hydroxytridecanesulphonate

C$_{11}$H$_{23}$COCH$_2$SO$_3$Na
Sodium 2-oxotridecanesulphonate

C$_{14}$H$_{29}$SO$_3$Na
Sodium tetradecanesulphonate

C$_{11}$H$_{23}$CH=CHCH$_2$SO$_3$Na
Sodium 2-tetradecenesulphonate

C$_9$H$_{19}$CH(C$_4$H$_9$)SO$_3$Na
Sodium 5-tetradecanesulphonate

C$_{11}$H$_{23}$CH(NHC$_4$H$_9$)C$_2$H$_4$SO$_3$Na
Sodium 3-(butylamino)tetradecane-1-sulphonate

C$_{11}$H$_{23}$CH(N(CH$_3$)$_2$)C$_2$H$_4$SO$_3$Na
Sodium 3-(dimethylamino)tetradecanesulphonate

C$_{11}$H$_{23}$CH(OC$_2$H$_4$OC$_2$H$_4$OH)C$_2$H$_4$SO$_3$Na
Sodium 3-(hydroxyethoxyethoxy)tetradecanesulphonate

C$_{11}$H$_{23}$CH(OC$_2$H$_4$OH)C$_2$H$_4$SO$_3$Na
Sodium 3-(hydroxyethoxy)tetradecanesulphonate

C$_{13}$H$_{27}$CHOHSO$_3$Na
Sodium 1-hydroxytetradecanesulphonate

$C_{12}H_{25}CH(CH_2OH)SO_3Na$
   Sodium 1-hydroxy-2-tetradecanesulphonate
$C_{11}H_{23}CHOHCH_2CH_2SO_3Na$
   Sodium 3-hydroxytetradecanesulphonate
$C_{11}H_{23}CH(NC_4H_8O)C_2H_4SO_3Na$
   Sodium 3-(morpholino)tetradecanesulphonate
$C_{11}H_{23}COCH_2CH_2SO_3Na$
   Sodium 3-oxotetradecanesulphonate
$C_{11}H_{23}CH(OC_6H_5)C_2H_4SO_3Na$
   Sodium 3-(phenoxy)tetradecanesulphonate
$C_{11}H_{23}CH(NC_5H_{10})C_2H_4SO_3Na$
   Sodium 3-(piperidino)tetradecanesulphonate
$C_{11}H_{23}CH(NHC_3H_7)C_2H_4SO_3Na$
   Sodium 3-(propylamino)tetradecanesulphonate
$C_{11}H_{23}CH(OC_6H_2Cl_3)C_2H_4SO_3Na$
   Sodium 3-(trichlorophenoxy)tetradecanesulphonate
$C_{15}H_{31}SO_3Na$
   Sodium pentadecanesulphonate
$C_7H_{15}CH(C_7H_{15})SO_3Na$
   Sodium 8-pentadecanesulphonate
$C_{12}H_{25}CH=CHCH_2SO_3Na$
   Sodium 2-pentadecenesulphonate
$C_{14}H_{29}CH(SO_3Na)CO(OC_2H_5)$
   Sodium 1-carbethoxypentadecane-1-sulphonate
$C_{14}H_{29}CH(COOCH_3SO_3Na)$
   Sodium 1-carbomethoxypentadecanesulphonate
$C_{13}H_{27}CHOHCH_2SO_3Na$
   Sodium 2-hydroxypentadecanesulphonate
$C_{12}H_{25}CH(OH)C_2H_4SO_3Na$
   Sodium 3-hydroxypentadecanesulphonate
$C_{14}H_{29}CH(COOC_3H_7SO_3Na)$
   Sodium 1-carbo-n-propoxypentadecanesulphonate
$C_{13}H_{27}COCH_2SO_3Na$
   Sodium 2-oxopentadecanesulphonate
$C_{16}H_{33}SO_3Na$
   Sodium hexadecanesulphonate
$C_{13}H_{27}CH=CHCH_2SO_3Na$
   Sodium 2-hexadecenesulphonate
$C_9H_{19}CH(C_6H_{13})SO_3Na$
   Sodium 7-hexadecanesulphonate
$C_{13}H_{27}CH(OC_4H_9)C_2H_4SO_3Na$
   Sodium 3-(butoxy)hexadecane-1-sulphonate
$C_{15}H_{31}CHOHSO_3Na$
   Sodium 1-hydroxyhexadecanesulphonate
$C_{13}H_{27}CHOHCH_2CH_2SO_3Na$
   Sodium 3-hydroxyhexadecanesulphonate
$C_{14}H_{29}CH(CH_2OH)SO_3Na$
   Sodium 1-hydroxy-2-hexadecanesulphonate
$C_{13}H_{27}CH(OCH_3)C_2H_4SO_3Na$
   Sodium 3-methoxyhexadecanesulphonate
$C_{13}H_{27}COCH_2CH_2SO_3Na$
   Sodium 3-oxohexadecanesulphonate
$C_{13}H_{27}CH(OC_3H_7)C_2H_4SO_3Na$
   Sodium 3-(n-propoxy)hexadecanesulphonate

$C_{17}H_{35}SO_3Na$
  Sodium heptadecanesulphonate
$C_{16}H_{33}CH(COOC_2H_5SO_3Na)$
  Sodium 1-carbethoxyheptadecanesulphonate
$C_{16}H_{33}CH(COOCH_3SO_3Na)$
  Sodium 1-carbomethoxyheptadecanesulphonate
$C_{16}H_{33}CH(COOC_3H_7SO_3Na)$
  Sodium 1-carbo-i-propoxyheptadecanesulphonate
$C_{16}H_{33}CH(COOC_3H_7SO_3Na)$
  Sodium 1-carbo-n-propoxyheptadecanesulphonate
$C_{15}H_{31}CHOHCH_2SO_3Na$
  Sodium 2-hydroxyheptadecanesulphonate
$C_{15}H_{31}COCH_2SO_3Na$
  Sodium 2-oxoheptadecanesulphonate
$C_{18}H_{37}SO_3Na$
  Sodium octadecanesulphonate
$C_9H_{19}CH(C_8H_{17})SO_3Na$
  Sodium 9-octadecanesulphonate
$C_{15}H_{31}CH=CHCH_2SO_3Na$
  Sodium 2-octadecenesulphonate
$C_{17}H_{35}CHOHSO_3Na$
  Sodium 1-hydroxyoctadecanesulphonate
$C_{15}H_{31}CHOHCH_2CH_2SO_3Na$
  Sodium 3-hydroxyoctadecanesulphonate
$C_{16}H_{33}CH(CH_2OH)SO_3Na$
  Sodium 1-hydroxy-2-octadecanesulphonate
$C_{15}H_{31}COCH_2CH_2SO_3Na$
  Sodium 3-oxooctadecanesulphonate
$C_9H_{19}CH(C_9H_{19})SO_3Na$
  Sodium 10-nonadecanesulphonate
$C_{20}H_{41}SO_3Na$
  Sodium eicosanesulphonate
$C_{22}H_{45}SO_3Na$
  Sodium docosanesulphonate

                    Alkane sulphonates, other counterions

$(C_8H_{17}SO_3)_2Mg$
  Magnesium bisoctanesulphonate
$(C_{10}H_{21}SO_3)_2Cu$
  Copper bisdecanesulphonate
$(C_{10}H_{21}SO_3)_2Mg$
  Magnesium bisdecanesulphonate
$(C_{10}H_{21}SO_3)_2MV$
  Methylviologen bisdecanesulphonate
$(C_{12}H_{25}SO_3)_2Cu$
  Copper bisdodecanesulphonate
$(C_{12}H_{25}SO_3)_2Mg$
  Magnesium bisdodecanesulphonate
$(C_{12}H_{25}SO_3)_2MV$
  Methylviologen bisdodecanesulphonate
$C_4BP(C_{14})_2$
  1,1'-(1-ω-butanediyl)bispyridinium bistetradecanesulphonate

$(C_{14}H_{29}SO_3)_2Cu$
Copper bistetradecanesulphonate

$C_{10}BP(C_{14})_2$
1,1'-(1-ω-decanediyl)bispyridinium bistetradecanesulphonate

$C_{12}BP(C_{14})_2$
1,1'-(1-ω-dodecanediyl)bispyridinium bistetradecanesulphonate

$C_2BP(C_{14})_2$
1,1'-(1-ω-ethanediyl)bispyridinium bistetradecanesulphonate

$C_6BP(C_{14})_2$
1,1'-(1-ω-hexanediyl)bispyridinium bistetradecanesulphonate

$(C_{14}H_{29}SO_3)_2MV$
Methylviologen bistetradecanesulphonate

$C_8BP(C_{14})_2$
1,1'-(1-ω-octanediyl)bispyridinium bistetradecanesulphonate

$C_{14}BP(C_{14})_2$
1,1'-(1-ω-tetradecanediyl)bispyridinium bistetradecanesulphonate

$(C_{12}H_{25}CH(OH)C_2H_4SO_3)_2Ca$
Calcium bis(3-hydroxypentadecanesulphonate)

$(C_{12}H_{25}CH=CHCH_2SO_3)_2Ca$
Calcium bis(2-pentadecene-1-sulphonate)

$(C_{16}H_{33}SO_3)_2MV$
Methylviologen bishexadecanesulphonate

$(C_{18}H_{33}SO_3)_2MV$
Methylviologen bisoctadecanesulphonate

Sodium alkylarenesulphonates

1φC6
Sodium 4-(hexyl)benzenesulphonate

1φC7
Sodium 4-(heptylbenzenesulphonate

$(C_7H_{15})_2C_6H_3SO_3Na$
Sodium 2,5-diheptylbenzenesulphonate

1φC8
Sodium 4-(octyl)benzenesulphonate

2φC8
Sodium 4-(2-octyl)benzenesulphonate

$(C_8H_{17}OOCCH_2O)C_6H_4SO_3Na$
Sodium phenoxysulphonate ethanoic acid octyl ester

$(C_4H_9CH(C_2H_5)CH_2)C_6H_4SO_3Na$
Sodium 4-(2-ethyl-hexyl)benzenesulphonate

1φC9
Sodium 4-(nonyl)benzenesulphonate

3φC9
Sodium 4-(3-nonyl)benzenesulphonate

$(C_9H_{19}OOCCH_2)C_6H_4SO_3Na$
Sodium phenylsulphonate ethanoic acid nonyl ester

o-1φC10
Sodium 2-(decyl)benzenesulphonate

1φC10
Sodium 4-(decyl)benzenesulphonate

2φC10
Sodium 4-(2-decyl)benzenesulphonate

3φC10
    Sodium 4-(3-decyl)benzenesulphonate
5φC10
    Sodium 4-(5-decyl)benzenesulphonate
(C$_5$H$_{11}$CH(C$_3$H$_7$)CH$_2$)C$_6$H$_4$SO$_3$Na
    Sodium 4-(2-propyl-heptyl)benzenesulphonate
C$_{10}$LAS
    Sodium decylbenzenesulphonate (technical grade, purified)
NaC$_{10}$BS-in
    Sodium decylbenzenesulphonate (isomeric mixture of 3-phenyl, 4-phenyl, etc.)
NaC$_{10}$BS-mx
    Sodium decylbenzenesulphonate (isomeric mixture of 2-phenyl, 3-phenyl, etc.)
(C$_{10}$H$_{21}$OOC)C$_6$H$_4$SO$_3$Na
    Sodium decyl-4-sulphobenzoate
C$_{11}$H$_{23}$C$_6$H$_4$SO$_3$Na
    Sodium 4-(undecyl)benzenesulphonate
(C$_9$H$_{19}$CH(CH$_3$))C$_6$H$_4$SO$_3$Na
    Sodium 4-(2-undecyl)benzenesulphonate
C$_{11}$H$_{23}$(C$_3$H$_7$)C$_6$H$_3$SO$_3$Na
    Sodium 2-(undecyl)-5-propylbenzenesulphonate
C$_{11}$CH$_{23}$(C$_3$H$_7$)C$_6$H$_3$SO$_3$Na
    Sodium 2-propyl-5-undecylbenzenesulphonate
(C$_{11}$H$_{23}$S)C$_6$H$_4$SO$_3$Na
    Sodium 4-(thioundecyl)benzenesulphonate
C$_{13}$H$_{27}$(CH$_3$)C$_6$H$_3$SO$_3$Na
    Sodium 2-(tridecyl)-5-methylbenzenesulphonate
(C$_{10}$H$_{21}$COO)C$_6$H$_4$SO$_3$Na
    Sodium 4-(undecanoic acid)benzenesulphonate ester
(C$_{10}$H$_{21}$CONH)C$_6$H$_4$SO$_3$Na
    Sodium undecanoyl-4-aminobenzenesulphonate
(C$_{11}$H$_{23}$O)C$_6$H$_4$SO$_3$Na
    Sodium 4-(undecoxy)benzenesulphonate
(C$_{11}$H$_{23}$NH)C$_6$H$_4$SO$_3$Na
    Sodium 4-(undecylamino)benzenesulphonate
(C$_{11}$H$_{23}$SO)C$_6$H$_4$SO$_3$Na
    Sodium 4-(undecylsulphoxy)benzenesulphonate
(C$_{11}$H$_{23}$SO$_2$)C$_6$H$_4$SO$_3$Na
    Sodium 4-(undecylsulphonyl)benzenesulphonate
1φC12
    Sodium 4-(dodecyl)benzenesulphonate
2φC12
    Sodium 4-(2-dodecyl)benzenesulphonate
3φC12
    Sodium 4-(3-dodecyl)benzenesulphonate
4φC12
    Sodium 4-(4-dodecyl)benzenesulphonate
6φC12
    Sodium 4-(6-dodecyl)benzenesulphonate
(C$_6$H$_{13}$CH(C$_4$H$_9$)CH$_2$)C$_6$H$_4$SO$_3$Na
    Sodium 4-(2(butyl)octyl)benzenesulphonate
C$_{12}$H$_{25}$(C$_2$H$_5$)C$_6$H$_3$SO$_3$Na
    Sodium 2-(dodecyl)-5-etyhylbenzenesulphonate
C$_{12}$H$_{25}$(C$_2$H$_5$)C$_6$H$_3$SO$_3$Na
    Sodium 2-(ethyl)-5-dodecylbenzenesulphonate

C$_{12}$LAS
  Sodium dodecylbenzenesulphonate (technical grade, purified)
NaC$_{12}$BS-in
  Sodium dodecylbenzenesulphonate (isomeric mixture of 3-phenyl,
    4-phenyl, etc.)
NaC$_{12}$BS-mx
  Sodium dodecylbenzenesulphonate (isomeric mixture of 2-phenyl,
    3-phenyl, etc.)
C$_{12}$H$_{25}$(CH$_3$)$_2$C$_6$H$_2$SO$_3$Na
  Sodium 2,3-dimethyl-4-(dodecyl)benzenesulphonate
3C12-o-AXS
  Sodium 2,3-dimethyl-4-(3-dodecyl)benzenesulphonate
3φC12-m-ATS
  Sodium 3-methyl-4-(3-dodecyl)benzenesulphonate
1φC13
  Sodium 4-(1-tridecyl)benzenesulphonate
(C$_{11}$H$_{23}$CH(CH$_3$))C$_6$H$_4$SO$_3$Na
  Sodium 4-(2-tridecyl)benzenesulphonate
NaC$_{13}$BS-mx
  Sodium tridecylbenzenesulphonate (isomeric mixture of 2-phenyl,
    3-phenyl, etc.)
1φC14
  Sodium 4-(1-tetradecyl)benzenesulphonate
(C$_7$H$_{15}$CH(C$_5$H$_{11}$)CH$_2$)C$_6$H$_4$SO$_3$Na
  Sodium 4-(2-pentyl-nonyl)benzenesulphonate
C$_{14}$LAS
  Sodium tetradecylbenzenesulphonate (technical grade, purified)
NaC$_{14}$BS-mx
  Sodium tetradecylbenzenesulphonate (isomeric mixture of 2-phenyl,
    3-phenyl, etc.)
1φC15
  Sodium 4-(pentadecyl)benzenesulphonate
(C$_{13}$H$_{27}$CH(CH$_3$))C$_6$H$_4$SO$_3$Na
  Sodium 4-(2-pentadecyl)benzenesulphonate
NaC$_{15}$BS-mx
  Sodium pentadecylbenzenesulphonate (isomeric mixture of 2-phenyl,
    3-phenyl, etc.)
1φC16
  Sodium 4-(hexadecyl)benzenesulphonate
8φC16
  Sodium 4-(8-hexadecyl)benzenesulphonate
C$_{16}$LAS
  Sodium hexadecylbenzenesulphonate (technical grade, purified)
(C$_{15}$H$_{31}$CH(CH$_3$))C$_6$H$_4$SO$_3$Na
  Sodium 4-(2-heptadecyl)benzenesulphonate
1φC18
  Sodium 4-(octadecyl)benzenesulphonate
1φC20
  Sodium 4-(eicosyl)benzenesulphonate

Alkylarenesulphonates, other counterions

$C_{12}H_{25}C_6H_4SO_3Li$
    Lithium 4-(dodecyl)benzenesulphonate

$C_{12}H_{25}C_6H_4SO_3K$
    Potassium 4-(dodecyl)benzenesulphonate

$C_{12}H_{25}C_6H_4SO_3Rb$
    Rubidium 4-(dodecyl)benzenesulphonate

$C_6H_{13}CH(C_5H_{11})C_6H_4SO_3NH_2(CH_3)_2$
    Dimethylammonium 4-(6-dodecyl)benzenesulphonate

$C_6H_{13}CH(C_5H_{11})C_6H_4SO_3NH_2(C_2H_5)_2$
    Diethylammonium 4-(6-dodecyl)benzenesulphonate

$C_6H_{13}CH(C_5H_{11})C_6H_4SO_3NH_2(C_3H_7)_2$
    Dipropylammonium 4-(6-dodecyl)benzenesulphonate

$(C_8H_{17}C_6H_4SO_3)_2Ca$
    Calcium bis(octylbenzenesulphonate)

$(C_{11}H_{23}NHC_6H_4SO_3)_2Ca$
    Calcium bis-N-undecyl-4-aminobenzenesulphonate

$(C_{12}H_{25}C_6H_4SO_3)_2Ca$
    Calcium bis(dodecylbenzenesulphonate)

$C_{12}H_{25}C_6H_4SO_3Cs$
    Cesium dodecylbenzenesulphonate

Sodium alkylnaphtalenesulphonates

$C_8H_{17}C_{10}H_6SO_3Na$
    Sodium octylnaphtalenesulphonate

$C_{10}H_{21}C_{10}H_6SO_3Na$
    Sodium decylnaphtalenesulphonate

$C_{12}H_{25}C_{10}H_6SO_3Na$
    Sodium dodecylnaphtalenesulphonate

$C_{14}H_{29}C_{10}H_6SO_3Na$
    Sodium tetradecylnaphtalenesulphonate

$C_{16}H_{33}C_{10}H_6SO_3Na$
    Sodium hexadecylnaphtalenesulphonate

Cationic surfactants

Alkyltrimethylammonium salts

$C_8H_{17}N(CH_3)_3Br$
  Octyltrimethylammonium bromide
$C_{10}H_{21}N(CH_3)_3Br$
  Decyltrimethylammonium bromide
$(C_{10}H_{21})_2N(CH_3)_2Br$
  Didecyldimethylammonium bromide
$C_{12}H_{25}N(CH_3)_3Cl$
  Dodecyltrimethylammonium chloride
$C_{12}H_{25}N(CH_3)_3Br$
  Dodecyltrimethylammonium bromide
$C_{12}H_{25}N(CH_3)_3OH$
  Dodecyltrimethylammonium hydroxide
$C_{12}H_{25}C_2H_5N(CH_3)_2Br$
  Dodecyl-ethyldimethylammonium bromide
$C_{12}H_{25}C_3H_7N(CH_3)_2Br$
  Dodecylpropyldimethylammonium bromide
$C_{12}H_{25}C_4H_9N(CH_3)_2Br$
  Dodecylbutyldimethylammonium bromide
$C_{12}H_{25}C_8H_{17}N(CH_3)_2$
  Dodecyloctyldimethylammonium bromide
$C_{12}H_{25}C_{10}H_{21}N(CH_3)_2Br$
  Dodecyldecyldimethylammonium bromide
$(C_{12}H_{25})_2N(CH_3)_2Br$
  Didodecyldimethylammonium bromide
$C_{14}H_{29}N(C_2H_5)_3Br$
  Tetradecyltriethylammonium bromide
$C_{14}H_{29}N(CH_3)_3Br$
  Tetradecyltrimethylammonium bromide
$C_{14}H_{29}N(CH_3)_3Cl$
  Tetradecyltrimethylammonium chloride
$C_{14}H_{29}N(CH_3)_3NO_3$
  Tetradecyltrimethylammonium nitrate
$C_{14}H_{29}N(CH_3)_3OH$
  Tetradecyltrimethylammonium hydroxide
$C_{14}H_{29}N(C_3H_7)_3Br$
  Tetradecyltripropylammonium bromide
$C_{14}H_{29}N(C_4H_9)_3Br$
  Tetradecyltributylammonium bromide
$(C_{14}H_{29})_2N(CH_3)_2Br$
  Ditetradecyldimethylammonium bromide
$C_{16}H_{33}N(CH_3)_3Cl$
  Hexadecyltrimethylammonium chloride
$C_{16}H_{33}N(CH_3)_3Br$
  Hexadecyltrimethylammonium bromide
$C_{16}H_{33}N(CH_3)_3NO_3$
  Hexadecyltrimethylammonium nitrate
$C_{16}H_{33}N(CH_3)_3OH$
  Hexadecyltrimethylammonium hydroxide
$C_{16}H_{33}C_8H_{17}N(CH_3)_2Br$
  Hexadecyloctyldimethylammonium bromide

$(C_{16}H_{33}N(CH_3)_3)_2CO_3$
  Bis(hexadecyltrimethylammonium)carbonate
$(C_{16}H_{33}N(CH_3)_3)_2SO_4$
  Bis(hexadecyltrimethylammonium)sulphate
$(C_{16}H_{33})_2N(CH_3)_2Br$
  Bishexadecyldimethylammonium bromide
$C_{18}H_{37}N(CH_3)_3Br$
  Octadecyltrimethylammonium bromide
$C_{18}H_{37}N(CH_3)_3Cl$
  Octadecyltrimethylammonium chloride
$C_{20}H_{41}N(CH_3)_3Cl$
  Eicosanetrimethylammonium chloride

Alkylpyridinium salts

$C_6H_{13}NC_5H_5Br$
  N-Hexylpyridinium bromide
$C_8H_{17}NC_5H_5Br$
  N-Octylpyridinium bromide
$C_{10}H_{21}NC_5H_5Cl$
  N-Decylpyridinium chloride
$(CH_3CH_2)_2OH(CH_2)_7C_5H_4NCH_3I$
  N-Methyl-4-(8-ethyldecyl)pyridinium iodide
$(CH_3)_3C(CH_2)_8C_5H_4NCH_3I$
  N-Methyl-4-(9,9-dimethyldecyl)pyridinium iodide
$C_{11}H_{23}NC_5H_5Br$
  N-Undecylpyridinium bromide
$C_{10}H_{21}CHCH_3C_5H_4NCH_3I$
  N-Methyl-4-(1-methyl-undecyl)pyridinium iodide
$C_{12}H_{25}NC_5H_5Br$
  N-Dodecylpyridinium bromide
$C_{12}H_{25}NC_5H_5Cl$
  N-Dodecylpyridinium chloride
$C_{12}H_{25}NC_5H_5I$
  N-Dodecylpyridinium iodide
$C_{12}H_{25}NC_5H_4OMeBr$
  N-Dodecyl-4-methoxypyridinium bromide
$C_{12}H_{25}NC_5H_4OMeCl$
  N-Dodecyl-4-methoxypyridinium chloride
$C_{12}H_{25}C_5H_4NC_4H_9I$
  N-Butyl-4-dodecylpyridinium iodide
$C_{12}H_{25}C_5H_4NC_2H_5I$
  N-Ethyl-4-dodecylpyridinium iodide
$C_{12}H_{25}C_5H_4NCH_3I$
  N-Methyl-4-dodecylpyridinium iodide
$C_{12}H_{25}C_5H_4NCH(CH_3)_2I$
  N-2-Propyl-4-dodecylpyridinium iodide
$C_{12}H_{25}C_5H_4NC_3H_7I$
  N-Propyl-4-dodecylpyridinium iodide
$C_6H_{13}C{\equiv}CC_4H_9C_5H_4NCH_3I$
  N-Methyl-4-(dodec-5-ynyl)pyridinium iodide
$C_{12}H_{25}OC_5H_4NCH_3Br$
  N-Methyl-4-dodecyloxypyridinium bromide

$C_{13}H_{27}NC_5H_5Br$
  N-Tridecylpyridinium bromide
$C_{14}H_{29}NC_5H_4OCH_3Br$
  N-Tetradecyl-4-methoxypyridinium bromide
$C_{14}H_{29}NC_5H_5Br$
  N-Tetradecylpyridinium bromide
$C_{14}H_{29}NC_5H_5Cl$
  N-Tetradecylpyridinium chloride
$C_{14}H_{29}NC_5H_5I$
  N-Tetradecylpyridinium iodide
$C_{16}H_{33}NC_5H_5Br$
  N-Hexadecylpyridinium bromide
$C_{16}H_{33}NC_5H_5Cl$
  N-Hexadecylpyridinium chloride
$C_{16}H_{33}NC_5H_5H_2PO_4$
  N-Hexadecylpyridinium phosphate
$C_{16}H_{33}NC_5H_5HSO_4$
  N-Hexadecylpyridinium bisulphate
$C_{16}H_{33}NC_5H_5I$
  N-Hexadecylpyridinium iodide
$C_{16}H_{33}NC_5H_5NO_3$
  N-Hexadecylpyridinium nitrate
$C_{16}H_{33}NC_5H_5Tos$
  N-Hexadecylpyridinium tosylate
$C_{18}H_{37}NC_5H_5Br$
  N-Octadecylpyridinium bromide
$C_{18}H_{37}NC_5H_5I$
  N-Octadecylpyridinium iodide

Nonionic surfactants

Alkylpolyoxyethylene compounds

C_2E_1
  Ethyl monooxyethylene glycol ether
C_2E_2
  Ethyl dioxyethylene glycol ether
C_2E_3
  Ethyl trioxyethylene glycol ether
C_2E_4
  Ethyl tetraoxyethylene glycol ether
C_2E_5
  Ethyl pentaoxyethylene glycol ether
C_2E_6
  Ethyl hexaoxyethylene glycol ether
C_2E_7
  Ethyl heptaoxyethylene glycol ether
C_2E_8
  Ethyl octaoxyethylene glycol ether
C_2E_9
  Ethyl nonaoxyethylene glycol ether
C_2E_{10}
  Ethyl decaoxyethylene glycol ether
C_3E_1
  Propylmonooxyethylene glycol ether
$C_3H_8O_2$
  Methylmonooxyethylene glycol ether
C_4E_1
  Butylmonooxyethylene glycol ether
C_4E_6
  Butylhexaoxyethylene glycol ether
C_5E_2
  Pentyldioxyethylene glycol ether
C_6E_1
  Hexylmonooxyethylene glycol ether
C_6E_2
  Hexyldioxyethylene glycol ether
C_6E_3
  Hexyltrioxyethylene glycol ether
C_6E_4
  Hexyltetraoxyethylene glycol ether
C_6E_5
  Hexylpentaoxyethylene glycol ether
C_6E_6
  Hexylhexaoxyethylene glycol ether
C_8E_1
  Octylmonooxyethylene glycol ether
C_8E_2
  Octyldioxyethylene glycol ether
C_8E_3
  Octyltrioxyethylene glycol ether
C_8E_4
  Octyltetraoxyethylene glycol ether

C_8E_5
  Octylpentaoxyethylene glycol ether
C_8E_6
  Octylhexaoxyethylene glycol ether
C_8E_8
  Octyloctaoxyethylene glycol ether
C_8E_9
  Octylnonaoxyethylene glycol ether
C_8E_{12}
  Octyldodecaoxyethylene glycol ether
C_9E_4
  Nonyltetraoxyethylene glycol ether
C_9E_5
  Nonylpentaoxyethylene glycol ether
C_9E_6
  Nonylhexaoxyethylene glycol ether
C_9E_8
  Nonyloctaoxyethylene glycol ether
$C_{10}E_1$
  Decylmonooxyethylene glycol ether
$C_{10}E_2$
  Decyldioxyethylene glycol ether
$C_{10}E_3$
  Decyltrioxyethylene glycol ether
$C_{10}E_4$
  Decyltetraoxyethylene glycol ether
$C_{10}E_5$
  Decylpentaoxyethylene glycol ether
$C_{10}E_6$
  Decylhexaoxyethylene glycol ether
$C_{10}E_8$
  Decyloctaoxyethylene glycol ether
$C_{10}E_9$
  Decylnonaoxyethylene glycol ether
$C_{10}E_{10}$
  Decyldecaoxyethylene glycol ether
$C_{10}E_{11}$
  Decylundecaoxyethylene glycol ether
$C_{10}E_{16}$
  Decylhexadecaoxyethylene glycol ether
$C_{11}E_4$
  Undecyltetraoxyethylene glycol ether
$C_{11}E_5$
  Undecylpentaoxyethylene glycol ether
$C_{11}E_6$
  Undecylhexaoxyethylene glycol ether
$C_{11}E_8$
  Undecyloctaoxyethylene glycol ether
$C_{12}E_1$
  Dodecylmonooxyethylene glycol ether
$C_{12}E_2$
  Dodecyldioxyethylene glycol ether
$C_{12}E_3$
  Dodecyltrioxyethylene glycol ether

$C_{12}E_4$
    Dodecyltetraoxyethylene glycol ether

$C_{12}E_5$
    Dodecylpentaoxyethylene glycol ether

$C_{12}E_6$
    Dodecylhexaoxyethylene glycol ether

$C_{12}E_7$
    Dodecylheptaoxyethylene glycol ether

$C_{12}E_8$
    Dodecyloctaoxyethylene glycol ether

$C_{12}E_9$
    Dodecylnonaoxyethylene glycol ether

$C_{12}E_{10}$
    Dodecyldecaoxyethylene glycol ether

$C_{12}E_{11}$
    Dodecylundecaoxyethylene glycol ether

$C_{12}E_{12}$
    Dodecyldodecaoxyethylene glycol ether

$C_{12}E_{14}$
    Dodecyltetradecaoxyethylene glycol ether

$C_{12}E_{14.8}$
    Dodecylpoly(oxyethylene glycol)$_{14.8}$ ether

$C_{12}E_{15}$
    Dodecylpentadecaoxyethylene glycol ether

$C_{12}E_{18}$
    Dodecyloctadecaoxyethylene glycol ether

$C_{12}E_{19.3}$
    Dodecylpoly(oxyethylene glycol)$_{19.3}$ ether

$C_{12}E_{23}$
    Dodecyltricosaoxyethylene glycol ether

$C_{12}E_{23.9}$
    Dodecylpoly(oxyethylene glycol)$_{23.9}$ ether

$C_{12}E_{30}$
    Dodecyltriacontaoxyethylene glycol ether

$C_{12}E_{31}$
    Dodecylhentriacontaoxyethylene glycol ether

$C_{13}E_5$
    Tridecylpentaoxyethylene glycol ether

$C_{13}E_6$
    Tridecylhexaoxyethylene glycol ether

$C_{13}E_8$
    Tridecyloctaoxyethylene glycol ether

$C_{13}E_{10}$
    Tridecyldecaoxyethylene glycol ether

$C_{13}E_{12}$
    Tridecyldodecaoxyethylene glycol ether

$C_{13}E_{15}$
    Tridecylpentadecaoxyethylene glycol ether

$C_{13}E_{22}$
    Tridecyldocosaoxyethylene glycol ether

$C_{14}E_1$
    Tetradecylmonooxyethylene glycol ether

$C_{14}E_2$
    Tetradecyldioxyethylene glycol ether

$C_{14}E_3$
  Tetradecyltrioxyethylene glycol ether
$C_{14}E_4$
  Tetradecyltetraoxyethylene glycol ether
$C_{14}E_5$
  Tetradecylpentaoxyethylene glycol ether
$C_{14}E_6$
  Tetradecylhexaoxyethylene glycol ether
$C_{14}E_7$
  Tetradecylheptaoxyethylene glycol ether
$C_{14}E_8$
  Tetradecyloctaoxyethylene glycol ether
$C_{15}E_6$
  Pentadecylhexaoxyethylene glycol ether
$C_{15}E_8$
  Pentadecyloctaoxyethylene glycol ether
$C_{16}E_1$
  Hexadecylmonooxyethylene glycol ether
$C_{16}E_2$
  Hexadecyldioxyethylene glycol ether
$C_{16}E_3$
  Hexadecyltrioxyethylene glycol ether
$C_{16}E_4$
  Hexadecyltetraoxyethylene glycol ether
$C_{16}E_5$
  Hexadecylpentaoxyethylene glycol ether
$C_{16}E_6$
  Hexadecylhexaoxyethylene glycol ether
$C_{16}E_7$
  Hexadecylheptaoxyethylene glycol ether
$C_{16}E_8$
  Hexadecyloctaoxyethylene glycol ether
$C_{16}E_9$
  Hexadecylnonaoxyethylene glycol ether
$C_{16}E_{12}$
  Hexadecyldodecaoxyethylene glycol ether
$C_{16}E_{15}$
  Hexadecylpentadecaoxyethylene glycol ether
$C_{16}E_{21}$
  Hexadecylheneicosaoxyethylene glycol ether
$C_{16}E_{30}$
  Hexadecyltriacontaoxyethylene glycol ether
$C_{18}E_1$
  Octadecylmonooxyethylene glycol ether
$C_{18}E_2$
  Octadecyldioxyethylene glycol ether
$C_{18}E_3$
  Octadecyltrioxyethylene glycol ether
$C_{18}E_4$
  Octadecyltetraoxyethylene glycol ether
$C_{18}E_8$
  Octadecyloctaoxyethylene glycol ether

Alkylphenolethoxylate compounds

$C_6\phi EO_{7.5}$
Hexylphenol(ethylene oxide)$_{7.5}$ ether
$C_8\phi EO_1$
Octylphenol(ethylene oxide)$_1$ ether
$C_8\phi EO_2$
Octylphenol(ethylene oxide)$_2$ ether
$C_8\phi EO_{2.98}$
Octylphenol(ethylene oxide)$_{2.98}$ ether
$C_8\phi EO_3$
Octylphenol(ethylene oxide)$_3$ ether
$C_8\phi EO_4$
Octylphenol(ethylene oxide)$_4$ ether
$C_8\phi EO_{4.07}$
Octylphenol(ethylene oxide)$_{4.07}$ ether
$C_8\phi EO_5$
Octylphenol(ethylene oxide)$_5$ ether
$C_8\phi EO_{5.01}$
Octylphenol(ethylene oxide)$_{5.01}$ ether
$C_8\phi EO_6$
Octylphenol(ethylene oxide)$_6$ ether
$C_8\phi EO_{6.03}$
Octylphenol(ethylene oxide)$_{6.03}$ ether
$C_8\phi EO_7$
Octylphenol(ethylene oxide)$_7$ ether
$C_8\phi EO_{7.05}$
Octylphenol(ethylene oxide)$_{7.05}$ ether
$C_8\phi EO_{7.3}$
Octylphenol(ethylene oxide)$_{7.3}$ ether
$C_8\phi EO_{7.5}$
Octylphenol(ethylene oxide)$_{7.5}$ ether
$C_8\phi EO_{7.7}$
Octylphenol(ethylene oxide)$_{7.7}$ ether
$C_8\phi EO_8$
Octylphenol(ethylene oxide)$_8$ ether
$C_8\phi EO_{8.03}$
Octylphenol(ethylene oxide)$_{8.03}$ ether
$C_8\phi EO_{8.4}$
Octylphenol(ethylene oxide)$_{8.4}$ ether
$C_8\phi EO_{8.5}$
Octylphenol(ethylene oxide)$_{8.5}$ ether
$C_8\phi EO_{8.8}$
Octylphenol(ethylene oxide)$_{8.8}$ ether
$C_8\phi EO_{9.0}$
Octylphenol(ethylene oxide)$_{9.0}$ ether
$C_8\phi EO_{9.06}$
Octylphenol(ethylene oxide)$_{9.06}$ ether
$C_8\phi EO_{9.5}$
Octylphenol(ethylene oxide)$_{9.5}$ ether
$C_8\phi EO_{9.6}$
Octylphenol(ethylene oxide)$_{9.6}$ ether
$C_8\phi EO_{9.7}$
Octylphenol(ethylene oxide)$_{9.7}$ ether

C$_8$$\phi$EO$_{9.93}$
Octylphenol(ethylene oxide)$_{9.93}$ ether

C$_8$$\phi$EO$_{10}$
Octylphenol(ethylene oxide)$_{10}$ ether

C$_8$$\phi$EO$_{12.5}$
Octylphenol(ethylene oxide)$_{12.5}$ ether

C$_8$$\phi$EO$_{13}$
Octylphenol(ethylene oxide)$_{13}$ ether

C$_8$$\phi$EO$_{16}$
Octylphenol(ethylene oxide)$_{16}$ ether

C$_8$$\phi$EO$_{17}$
Octylphenol(ethylene oxide)$_{17}$ ether

C$_8$$\phi$EO$_{20}$
Octylphenol(ethylene oxide)$_{20}$ ether

C$_8$$\phi$EO$_{30}$
Octylphenol(ethylene oxide)$_{30}$ ether

C$_8$$\phi$EO$_{40}$
Octylphenol(ethylene oxide)$_{40}$ ether

C$_9$$\phi$EO$_5$
Nonylphenol(ethylene oxide)$_5$ ether

C$_9$$\phi$EO$_6$
Nonylphenol(ethylene oxide)$_6$ ether

C$_9$$\phi$EO$_{6.2}$
Nonylphenol(ethylene oxide)$_{6.2}$ ether

C$_9$$\phi$EO$_7$
Nonylphenol(ethylene oxide)$_7$ ether

C$_9$$\phi$EO$_{7.4}$
Nonylphenol(ethylene oxide)$_{7.4}$ ether

C$_9$$\phi$EO$_{7.5}$
Nonylphenol(ethylene oxide)$_{7.5}$ ether

C$_9$$\phi$EO$_8$
Nonylphenol(ethylene oxide)$_8$ ether

C$_9$$\phi$EO$_{8.1}$
Nonylphenol(ethylene oxide)$_{8.1}$ ether

C$_9$$\phi$EO$_{8.5}$
Nonylphenol(ethylene oxide)$_{8.5}$ ether

C$_9$$\phi$EO$_{8.6}$
Nonylphenol(ethylene oxide)$_{8.6}$ ether

C$_9$$\phi$EO$_9$
Nonylphenol(ethylene oxide)$_9$ ether

C$_9$$\phi$EO$_{9.5}$
Nonylphenol(ethylene oxide)$_{9.5}$ ether

C$_9$$\phi$EO$_{9.6}$
Nonylphenol(ethylene oxide)$_{9.6}$ ether

C$_9$$\phi$EO$_{10}$
Nonylphenol(ethylene oxide)$_{10}$ ether

C$_9$$\phi$EO$_{10.3}$
Nonylphenol(ethylene oxide)$_{10.3}$ ether

C$_9$$\phi$EO$_{10.5}$
Nonylphenol(ethylene oxide)$_{10.5}$ ether

C$_9$$\phi$EO$_{11}$
Nonylphenol(ethylene oxide)$_{11}$ ether

C$_9$$\phi$EO$_{12.5}$
Nonylphenol(ethylene oxide)$_{12.5}$ ether

C$_9$$\phi$EO$_{13}$
Nonylphenol(ethylene oxide)$_{13}$ ether
C$_9$$\phi$EO$_{14}$
Nonylphenol(ethylene oxide)$_{14}$ ether
C$_9$$\phi$EO$_{15}$
Nonylphenol(ethylene oxide)$_{15}$ ether
C$_9$$\phi$EO$_{15.8}$
Nonylphenol(ethylene oxide)$_{15.8}$ ether
C$_9$$\phi$EO$_{17.7}$
Nonylphenol(ethylene oxide)$_{17.7}$ ether
C$_9$$\phi$EO$_{18}$
Nonylphenol(ethylene oxide)$_{18}$ ether
C$_9$$\phi$EO$_{20}$
Nonylphenol(ethylene oxide)$_{20}$ ether
C$_9$$\phi$EO$_{30}$
Nonylphenol(ethylene oxide)$_{30}$ ether
C$_9$$\phi$EO$_{31}$
Nonylphenol(ethylene oxide)$_{31}$ ether
C$_9$$\phi$EO$_{40}$
Nonylphenol(ethylene oxide)$_{40}$ ether
C$_9$$\phi$EO$_{50}$
Nonylphenol(ethylene oxide)$_{50}$ ether
C$_9$$\phi$EO$_{100}$
Nonylphenol(ethylene oxide)$_{100}$ ether
C$_{10}$$\phi$EO$_9$
Decyl phenol(ethylene oxide)$_9$ ether
C$_{12}$$\phi$EO$_{5.3}$
Dodecylphenol(ethylene oxide)$_{5.3}$ ether
C$_{12}$$\phi$EO$_8$
Dodecylphenol(ethylene oxide)$_8$ ether
C$_{12}$$\phi$EO$_{8.2}$
Dodecylphenol(ethylene oxide)$_{8.2}$ ether
C$_{12}$$\phi$EO$_9$
Dodecylphenol(ethylene oxide)$_9$ ether
C$_{12}$$\phi$EO$_{9.4}$
Dodecylphenol(ethylene oxide)$_{9.4}$ ether
C$_{12}$$\phi$EO$_{9.7}$
Dodecylphenol(ethylene oxide)$_{9.7}$ ether
C$_{16}$$\phi$EO$_{12.4}$
Hexadecyl(ethylene oxide)$_{12.4}$ ether

Phosphoxides

C$_{12}$H$_{25}$PO(CH$_3$)$_2$
Dodecyldimethylphosphoxide

516

Alcohols

C$_2$H$_5$OH
  Ethanol
C$_3$H$_7$OH
  1-Propanol
CH$_3$CH(OH)CH$_2$OH
  1,2-Propanediol
HOCH$_2$CH$_2$CH$_2$OH
  1,3-Propanediol
C$_4$H$_9$OH
  1-Butanol
C$_5$H$_{11}$OH
  1-Pentanol
C$_6$H$_{13}$OH
  1-Hexanol
C$_7$H$_{15}$OH
  1-Heptanol
C$_8$H$_{17}$OH
  1-Octanol
C$_{10}$H$_{21}$OH
  1-Decanol
C$_{10}$H$_{25}$OH
  Dodecanol
C$_{14}$H$_{29}$OH
  Tetradecanol
C$_{16}$H$_{33}$OH
  Hexadecanol
C$_{18}$H$_{37}$OH
  Octadecanol
(CH$_3$)$_3$COH
  2-Methyl-2-propanol
HOCH$_2$CH$_2$OH
  1,2-Ethanediol
CH$_3$(CH$_2$)$_3$OH
  Butanol
CH$_3$CH(OH)CH$_2$CH$_2$OH
  1,3-Butanediol
HOCH$_2$(CH$_2$)$_2$CH$_2$OH
  1,4-Butanediol
C$_6$H$_5$CH$_2$OH
  Benzylalcohol

Alkanes - Cycloalkanes - Alkenes

C_5H_{12}
  Pentane
C_6H_{14}
  Hexane
C_7H_{16}
  Heptane
C_8H_{18}
  Octane
$(CH_3)_3CCH_2CH(CH_3)_2$
  Isooctane
C_9H_{20}
  Nonane
$C_{10}H_{22}$
  Decane
$C_{12}H_{26}$
  Dodecane
$C_{14}H_{30}$
  Tetradecane
$C_{15}H_{32}$
  Pentadecane
$C_{16}H_{34}$
  Hexadecane
$C_{30}H_{62}$
  Squalane

$c-C_6H_{12}$
  Cyclohexane

$CH_3(CH_2)_3CH=CH_2$
  Hexene
$CH_3(CH_2)_5CH=CH_2$
  1-Octene
$CH_3(CH_2)_3C(C_2H_5)=CH_2$
  2-Ethylhexene
$CH_3(CH_2)_9CH=CH_2$
  Dodecene
$CH_3(CH_2)_{11}CH=CH_2$
  Tetradecene-1

Deuterated Compounds

CD₃OD
  Deuteromethanol
CDCl₃
  Deuterochloroform
C₁₆N(CD₃)₃Br
  Hexadecyl-nona-deuteriotrimethylammonium bromide
D₂O
  Deuterium oxide

Fluorinated Compounds

Cl₂FC₂Cl₂F
  Difluorotetrachloroethane
C₇F₁₅COOH
  Sodium pentadecafluorooctanoate
C₁₆·NC₅H₅BF₄
  N-Hexadecylpyridinium tetrafluoroborate

Miscellaneous Inorganic Compounds
(in alphabetical order by formula)

$AgNO_3$
   Silver nitrate
$AlCl_3$
   Aluminium chloride
$BaCl_2$
   Barium chloride
$CaCl_2$
   Calcium chloride
H_2SO_4
   Sulphuric acid

K_2SO_4
   Potassium sulphate
KBr
   Potassium bromide
KCl
   Potassium chloride
KI
   Potassium iodide
KNO_3
   Potassium nitrate
KCNS
   Potassium thiocyanate
LiCl
   Lithium chloride
Li_2SO_4
   Lithium sulphate
$MgCl_2$
   Magnesium chloride
NaCNS
   Sodium thiocyanate
Na_2CO_3
   Sodium carbonate
Na_2SO_4
   Sodium sulphate
NaBr
   Sodium bromide
$NaBrO_3$
   Sodium bromate
NaCl
   Sodium chloride
NaI
   Sodium iodide
NaF
   Sodium fluoride
$NaNO_3$
   Sodium nitrate
NaOH
   Sodium hydroxide
NH_4Br
   Ammonium bromide

NH₄Cl
  Ammonium chloride
(NH₄)₂SO₄
  Ammonium sulphate
SrCl₂
  Strontium chloride
Zn(NO₃)₂
  Zinc nitrate

Miscellaneous Organic Compounds
(in alphabetical order by name)

CH_3CONH_2
Acetamide

CH_3COCH_3
Aceton

$HOCH_2CH_2NH_2$
Aminoethanol

$H_2NCH_2CHOHCH_3$
1-Amino-2-propanol

$C_6H_5NH_2$
Aniline

C_6H_6
Benzene

$C_6H_5CH_2OH$
Benzyl alcohol

C_6H_5Br
Bromobenzene

$CH_3COO(CH_2)_3CH_3$
Butylacetate

$C_6H_5(CH_2)_3CH_3$
Butylbenzene

$HOOCCH_2C(OH)(COOH)CH_2COOH$
Citric acid

$(CH_3)_2C_6H_2(OH)Cl$
4-Chloro-3,5-xylenol

$(C_2H_5)_2NCONH_2$
1,1-Diethylurea

$C_1E_1C_1$
Dimethylmonooxyethylene glycol ether

$(CH_3)_2NCONH_2$
1,1-Dimethylurea

$CH_3NHCONHCH_3$
1,3-Dimethylurea

$CH_3NHCSNHCH_3$
1,3-Dimethylthiourea

$c-C_4H_8O_2$
Dioxane

$CH_3CH_2NH_3NO_3$
Ethylammonium nitrate

$H_2NCH_2CH_2NH_2$
Ethylenediamine

$ClCH_2Cl$
Ethylene dichloride

$HOCH_2CH_2OH$
Ethylene glycol

$C_{15}H_{31}COOC_2H_5$
Ethyl oleate

$C_2H_5NHCONH_2$
Ethylurea

$HCONH_2$
Formamide

HCOOH
  Formic acid
HOCH$_2$CH(OH)CH$_2$OH
  Glycerol
C(NH$_2$)$_3$Cl
  Guanidinium chloride
HOCH$_2$CH$_2$SH
  2-Mercaptoethanol
CH$_3$NHCSNH$_2$
  Methylthiourea
CH$_3$NHCONH$_2$
  Methylurea
C$_6$H$_5$NO$_2$
  Nitrobenzene
CH$_3$(CH$_2$)$_7$CH=CH(CH$_2$)$_7$COOH
  Oleic acid
C$_6$H$_5$OH
  Phenol
C$_6$H$_5$NHCONH$_2$
  Phenylurea
Cl$_2$C=CCl$_2$
  Tetrachloroethene
C$_{14}$H$_{29}$SH
  Tetradecylmercaptan
(CH$_3$)$_4$NCl
  Tetramethylammonium chloride
(CH$_3$)$_2$NCON(CH$_3$)$_2$
  Tetramethylurea
CH$_3$C$_6$H$_5$
  Toluene
H$_2$NCSNH$_2$
  Thiourea
H$_2$NCONH$_2$
  Urea
CH$_3$C$_6$H$_4$CH$_3$
  Xylene
1,3-(CH$_3$)$_2$C$_6$H$_4$
  m-Xylene

CROSS INDEX

A
  Anthracene
ACA
  Acetamide
ANS
  Sodium 1-anilino-8-naphthalenesulphonate
2AP
  2-(9-Anthroyloxy)palmitic acid
6AS
  6-(9-Anthroyloxy)stearic acid
9AS
  9-(9-Anthroyloxy)stearic acid
12AS
  12-(9-Anthroyloxy)stearic acid
n-Bu$_4$NBr
  Tetrabutylammonium bromide
n-BuOH
  1-Butanol
t-BuOH
  2-Methyl-2-propanol
tert-Butanol
  2-Methyl-2-propanol
BZ-9-Ant
  Benzyl-9-anthroate
BZ-1-Np
  Benzyl-2-naphthoate
BZ-1-Pn
  Benzyl-1-pyrenoate
Carbamide
  Urea
Carbonyl diamide
  Urea
CD
  α-Cyclodextrin
Cetane
  Hexadecane
Cetylpyridinium bromide
  N-Hexadecylpyridinium bromide
Cetylpyridinium chloride
  N-Hexadecylpyridinium chloride
Cetylpyridinium iodide
  N-Hexadecylpyridinium iodide
Cetyltrimethylammonium bromide
  Hexadecyltrimethylammonium bromide
Cetyltrimethylammonium chloride
  Hexadecyltrimethylammonium chloride
Cetyltrimethylammonium hydroxide
  Hexadecyltrimethylammonium hydroxide

Cetyltrimethylammonium nitrate
  Hexadecyltrimethylammonium nitrate
CPB
  N-Hexadecylpyridinium bromide
CPC
  N-Hexadecylpyridinium chloride
CPI
  N-Hexadecylpyridinium iodide
$C_{14}PyBr$
  N-Tetradecylpyridinium bromide
$C_{16}PyBr$
  N-Hexadecylpyridinium bromide
$C_{14}PyCl$
  N-Tetradecylpyridinium chloride
$C_{16}PyCl$,
  N-Hexadecylpyridinium chloride
$C_{14}PyI$
  N-Tetradecylpyridinium iodide
$C_{16}PyI$
  N-Hexadecylpyridinium iodide
CTAB
  Hexadecyltrimethylammonium bromide
CTAC
  Hexadecyltrimethylammonium chloride
CTAN
  Hexadecyltrimethylammonium nitrate
CTAOH
  Hexadecyltrimethylammonium hydroxide
DCQEB
  2{6-(2,2-Dicyanovinyl)3,4-dihydro-2,2,4-trimethyl-1(2H)quinoyl}ethyl
  benzoate
DDMABr
  Didecyldimethylammonium bromide
DDPB
  N-Dodecylpyridinium bromide
DDPC
  N-Dodecylpyridinium chloride
DDTAB
  Dodecyltrimethylammonium bromide
DDTAC
  Dodecyltrimethylammonium chloride
Decaoxyethylene glycol monodecyl ether
  Decyldecaoxyethylene glycol ether
Decaoxyethylene glycol monododecyl ether
  Dodecyldecaoxyethylene glycol ether
Decaoxyethylene glycol monoethyl ether
  Ethyldecaoxyethylene glycol ether
Decaoxyethylene glycol monotridecyl ether
  Tridecyldecaoxyethylene glycol ether
Dimethylbenzene
  Xylene
Dioxyethylene glycol monodecyl ether
  Decyldioxyethylene glycol ether

Dioxyethylene glycol monododecyl ether
  Dodecyldioxyethylene glycol ether
Dioxyethylene glycol monoethyl ether
  Ethyldioxyethylene glycol ether
Dioxyethylene glycol monohexadecyl ether
  Hexadecyldioxyethylene glycol ether
Dioxyethylene glycol monohexyl ether
  Hexyldioxyethylene glycol ether
Dioxyethylene glycol monooctadecyl ether
  Octadecyldioxyethylene glycol ether
Dioxyethylene glycol monooctyl ether
  Octyldioxyethylene glycol ether
Dioxyethylene glycol monopentyl ether
  Pentyldioxyethylene glycol ether
Dioxyethylene glycol monotetradecyl ether
  Tetradecyldioxyethylene glycol ether
DMA
  N,N-Dimethylaniline
DMPC
  Dimyristoylphosphatidyl choline
Docosaoxyethylene glycol monotridecyl ether
  Tridecyldocosaoxyethylene glycol ether
DODAB
  Didodecyldimethylammonium bromide
Dodecaoxyethylene glycol monododecyl ether
  Dodecyldodecaoxoethylene glycol ether
Dodecaoxoethylene glycol monohexadecyl ether
  Hexadecyldodecaoxoethylene glycol ether
Dodecaoxoethylene glycol monooctyl ether
  Octyldodecaoxoethylene glycol ether
Dodecaoxoethylene glycol monotridecyl ether
  Tridecyldodecaoxoethylene glycol ether
DPC
  N-Dodecylpyridinium chloride
DPPC
  Dipalmitoylphosphatidyl choline
DTAB
  Decyltrimethylammonium bromide
DTAC
  Decyltrimethylammonium chloride
EAN
  Ethylammonium nitrate
EG
  Ethylene glycol
ET(30)
  2,6-Diphenyl-4-(2,4,6-triphenyl-1-pyridinio)phenoxide
1,2-Ethanediol
  Ethylene glycol
Et$_4$NBr
  Tetraethylammonium bromide
EtOH
  Ethanol
Heavy water
  Deuterium oxide

Heneicosaoxyethylene glycol monohexadecyl ether
  Hexadecylheneicosaoxyethylene glycol ether
Hentriacontaoxyethylene glycol monododecyl ether
  Dodecylhentriacontaoxyethylene glycol ether
Heptaoxyethylene glycol monododecyl ether
  Dodecylheptaoxyethylene glycol ether
Heptaoxyethylene glycol monoethyl ether
  Ethylheptaoxyethylene glycol ether
Heptaoxyethylene glycol monohexadecyl ether
  Hexadecylheptaoxyethylene glycol ether
Heptaoxyethylene glycol monotetradecyl ether
  Tetradecylheptaoxyethylene glycol ether
Hexadecaoxyethylene glycol monodecyl ether
  Decylhexadecaoxyethylene glycol ether
c-Hexane
  Cyclohexane
Hexaoxyethylene glycol monobutyl ether
  Butylhexaoxyethylene glycol ether
Hexaoxyethylene glycol monodecyl ether
  Decylhexaoxyethylene glycol ether
Hexaoxyethylene glycol monododecyl ether
  Dodecylhexaoxyethylene glycol ether
Hexaoxyethylene glycol monoethyl ether
  Ethylhexaoxyethylene glycol ether
Hexaoxyethylene glycol monohexadecyl ether
  Hexadecylhexaoxyethylene glycol ether
Hexaoxyethylene glycol monohexyl ether
  Hexylhexaoxyethylene glycol ether
Hexaoxyethylene glycol monononyl ether
  Nonylhexaoxyethylene glycol ether
Hexaoxyethylene glycol monooctyl ether
  Octylhexaoxyethylene glycol ether
Hexaoxyethylene glycol monopentadecyl ether
  Pentadecylhexaoxyethylene glycol ether
Hexaoxyethylene glycol monotetradecyl ether
  Tetradecylhexaoxyethylene glycol ether
Hexaoxyethylene glycol monotridecyl ether
  Tridecylhexaoxyethylene glycol ether
Hexaoxyethylene glycol monoundecyl ether
  Undecylhexaoxyethylene glycol ether
HHC
  4-Heptadecyl-7-hydroxycoumarin
Hydroxybenzene
  Phenol
α-Hydroxytoluene
  Benzyl alcohol
4-Hydroxytoluene
  p-Cresol
Igepal CO-610
  Nonylphenol(ethylene oxide)$_{8.5}$ ether
Igepal CO-630
  Nonylphenol(ethylene oxide)$_9$ ether
Igepal CO-660
  Nonylphenol(ethylene oxide)$_{10}$ ether

Igepal CO-710
  Nonylphenol(ethylene oxide)$_{10.5}$ ether
Igepal CO-730
  Nonylphenol(ethylene oxide)$_{15}$ ether
Igepal CO-850
  Nonylphenol(ethylene oxide)$_{20}$ ether
Igepal CO-880
  Nonylphenol(ethylene oxide)$_{30}$ ether
Igepal CO-890
  Nonylphenol(ethylene oxide)$_{40}$ ether
Igepal CO-970
  Nonylphenol(ethylene oxide)$_{50}$ ether
In
  Indole
Irium
  Sodium dodecylsulphate
LPC
  Lysophosphatidyl choline
2MA
  2-Methylanthracene
9MA
  9-Methylanthracene
Methylbenzene
  Toluene
4-Methylphenol
  p-Cresol
Monooxyethylene glycol monomethyl ether
  Methylmonooxyethylene glycol ether
Monooxyethylene glycol dimethyl ether
  Dimethylmonooxyethylene glycol ether
Monooxyethylene glycol monobutyl ether
  Butylmonooxyethylene glycol ether
Monooxyethylene glycol monodecyl ether
  Decylmonooxyethylene glycol ether
Monooxyethylene glycol monododecyl ether
  Dodecylmonooxyethylene glycol ether
Monooxyethylene glycol monoethyl ether
  Ethylmonooxyethylene glycol ether
Monooxyethylene glycol monohexadecyl ether
  Hexadecylmonooxyethylene glycol ether
Monooxyethylene glycol monohexyl ether
  Hexylmonooxyethylene glycol ether
Monooxyethylene glycol monooctadecyl ether
  Octadecylmonooxyethylene glycol ether
Monooxyethylene glycol monooctyl ether
  Octylmonooxyethylene glycol ether
Monooxyethylene glycol monopropyl ether
  Propylmonooxyethylene glycol ether
Monooxyethylene glycol monotetradecyl ether
  Tetradecylmonooxyethylene glycol ether
NaC
  Sodium cholate
NaDxS
  Sodium dextran sulphate

Na₃ citrate
  Sodium citrate
NaOAc
  Sodium acetate
NaTos
  Sodium 4-methylbenzenesulphonate
Nonaoxyethylene glycol monodecyl ether
  Decylnonaoxyethylene glycol ether
Nonaoxyethylene glycol monoethyl ether
  Ethylnonaoxyethylene glycol ether
Nonaoxyethylene glycol monohexadecyl ether
  Hexadecylnonaoxyethylene glycol ether
Nonaoxyethylene glycol monooctyl ether
  Octylnonaoxyethylene glycol ether
Octadecaoxyethylene glycol monododecyl ether
  Dodecyloctadecaoxyethylene glycol ether
Octaoxyethylene glycol monodecyl ether
  Decyloctaoxyethylene glycol ether
Octaoxyethylene glycol monododecyl ether
  Dodecyloctaoxyethylene glycol ether
Octaoxyethylene glycol monoethyl ether
  Ethyloctaoxyethylene glycol ether
Octaoxyethylene glycol monohexadecyl ether
  Hexadecyloctaoxyethylene glycol ether
Octaoxyethylene glycol monononyl ether
  Nonyloctaoxyethylene glycol ether
Octaoxyethylene glycol monooctadecyl ether
  Octadecyloctaoxyethylene glycol ether
Octaoxyethylene glycol monooctyl ether
  Octyloctaoxyethylene glycol ether
Octaoxyethylene glycol monopentadecyl ether
  Pentadecyloctaoxyethylene glycol ether
Octaoxyethylene glycol monotetradecyl ether
  Tetradecyloctaoxyethylene glycol ether
Octaoxyethylene glycol monotridecyl ether
  Tridecyloctaoxyethylene glycol ether
Octaoxyethylene glycol monoundecyl ether
  Undecyloctaoxyethylene glycol ether
PA
  Palmitic acid
PE
  Phosphatidyl ethanolamine
PEG
  Poly(ethylene glycol)
Pentadecaoxyethylene glycol monododecyl ether
  Dodecylpentadecaoxyethylene glycol ether
Pentadecaoxyethylene glycol monohexadecyl ether
  Hexadecylpentadecaoxyethylene glycol ether
Pentadecaoxyethylene glycol monotridecyl ether
  Tridecylpentadecaoxyethylene glycol ether
Pentaoxyethylene glycol monodecyl ether
  Decylpentaoxyethylene glycol ether
Pentaoxyethylene glycol monododecyl ether
  Dodecylpentaoxyethylene glycol ether

Pentaoxyethylene glycol monoethyl ether
  Ethylpentaoxyethylene glycol ether
Pentaoxyethylene glycol monohexadecyl ether
  Hexadecylpentaoxyethylene glycol ether
Pentaoxyethylene glycol monohexyl ether
  Hexylpentaoxyethylene glycol ether
Pentaoxyethylene glycol monononyl ether
  Nonylpentaoxyethylene glycol ether
Pentaoxyethylene glycol monooctyl ether
  Octylpentaoxyethylene glycol ether
Pentaoxyethylene glycol monotetradecyl ether
  Tetradecylpentaoxyethylene glycol ether
Pentaoxyethylene glycol monotridecyl ether
  Tridecylpentaoxyethylene glycol ether
Pentaoxyethylene glycol monoundecyl ether
  Undecylpentaoxyethylene glycol ether
PER
  Perylene
Polytergent B-200
  Nonylphenol(ethylene oxide)$_7$ ether
Polytergent G-200
  Octylphenol(ethylene oxide)$_7$ ether
n-PrOH
  Propanol
PS
  Phosphatidyl serine
PSS
  Sodium poly(styrenesulphonate)
PY
  Pyrene
PYCHO
  1-Pyrene-3-carboxaldehyde
PY(3)PY
  1,3-Di(1-pyrenyl)propane
PY(10)PY
  1,10-Di(1-pyrenyl)decane
SDBS
  Sodium dodecylbenzenesulphonate (isomeric mixture)
  Sodium dodecylbenzenesulphonate (inner isomeric mixture)
  Sodium dodecylbenzenesulphonate (80% para, 20% ortho isomers)
  Sodium dodecylbenzenesulphonate (technical grade, purified)
  Sodium 4-(dodecyl)benzenesulphonate
SDS
  Sodium dodecylsulphate
SDeS
  Sodium decylsulphate
Siponate DS-10
  Sodium dodecylbenzenesulphonate (80% para, 20% ortho isomers)
SLS
  Sodium dodecylsulphate
SM
  Sphingomyelin
SOBS
  Sodium 4-(octyl)benzenesulphonate

Sodium laurylsulphate
   Sodium dodecylsulphate
Sodium tosylate
   Sodium 4-methylbenzenesulphonate
SOS
   Sodium octylsulphate
Squalane
   2,6,10,15,19,23-Hexamethyltetracosane
STS
   Sodium tetradecylsulphate
Tetradecaoxyethylene glycol monododecyl ether
   Dodecyltetradecyloxyethylene glycol ether
Tetraoxyethylene glycol monodecyl ether
   Decyltetraoxyethylene glycol ether
Tetraoxyethylene glycol monododecyl ether
   Dodecyltetraoxyethylene glycol ether
Tetraoxyethylene glycol monoethyl ether
   Ethyltetraoxyethylene glycol ether
Tetraoxyethylene glycol monohexadecyl ether
   Hexadecyltetraoxyethylene glycol ether
Tetraoxyethylene glycol monohexyl ether
   Hexyltetraoxyethylene glycol ether
Tetraoxyethylene glycol monononyl ether
   Nonyltetraoxyethylene glycol ether
Tetraoxyethylene glycol monooctadecyl ether
   Octadecyltetraoxyethylene glycol ether
Tetraoxyethylene glycol monooctyl ether
   Octyltetraoxyethylene glycol ether
Tetraoxyethylene glycol monotetradecyl ether
   Tetradecyltetraoxyethylene glycol ether
Tetraoxyethylene glycol monoundecyl ether
   Undecyltetraoxyethylene glycol ether
TMACl
   Tetramethylammonium chloride
TPB
   N-Tetradecylpyridinium bromide
TPC
   N-Tetradecylpyridinium chloride
Tricontaoxyethylene glycol monododecyl ether
   Dodecyltriacontaoxyethylene glycol ether
Tricontaoxyethylene glycol monohexadecyl ether
   Hexadecyltriacontaoxyethylene glycol ether
Tricosaoxyethylene glycol monododecyl ether
   Dodecyltricosaoxyethylene glycol ether
Trioxyethylene glycol monodecyl ether
   Decyltrioxyethylene glycol ether
Trioxyethylene glycol monododecyl ether
   Dodecyltrioxyethylene glycol ether
Trioxyethylene glycol monoethyl ether
   Ethyltrioxyethylene glycol ether
Trioxyethylene glycol monohexadecyl ether
   Hexadecyltrioxyethylene glycol ether
Trioxyethylene glycol monohexyl ether
   Hexyltrioxyethylene glycol ether

Trioxyethylene glycol monooctadecyl ether
   Octadecyltrioxyethylene glycol ether
Trioxyethylene glycol monooctyl ether
   Octyltrioxyethylene glycol ether
Trioxyethylene glycol monotetradecyl ether
   Tetradecyltrioxyethylene glycol ether
Triton N-01
   Nonylphenol(ethylene oxide)$_{9.5}$ ether
Triton N-111
   Nonylphenol(ethylene oxide)$_{11}$ ether
Triton N-150
   Nonylphenol(ethylene oxide)$_{15}$ ether
Triton X-100
   Octylphenol(ethylene oxide)$_{9.5}$ ether
Triton X-102
   Octylphenol(ethylene oxide)$_{12.5}$ ether
Triton X-114
   Octylphenol(ethylene oxide)$_{7.5}$ ether
Triton X-165
   Octylphenol(ethylene oxide)$_{16}$ ether
Triton X-305
   Octylphenol(ethylene oxide)$_{30}$ ether
Triton X-405
   Octylphenol(ethylene oxide)$_{40}$ ether
TTAB
   Tetradecyltrimethylammonium bromide
TTAC
   Tetradecyltrimethylammonium chloride
Undecaoxyethylene glycol monodecyl ether
   Decylundecaoxyethylene glycol ether
Undecaoxyethylene glycol monododecyl ether
   Dodecylundecaoxyethylene glycol ether

PROPERTY INDEX

534

Aggregation number (continued)

in aqueous solutions of Na$_3$ citrate
 Dodecyltricosaoxyethylene glycol ether, $C_{12}H_{25}O(CH_2CH_2O)_{23}H$,
  $C_{12}E_{23}$ TIII.1.10

in aqueous solutions of NaCl
 Dodecyldodecaoxyethylene glycol ether, $C_{12}H_{25}O(CH_2CH_2O)_{12}H$,
  $C_{12}E_{12}$ TIII.1.10
 Dodecyloctadecaoxyethylene glycol ether, $C_{12}H_{25}O(CH_2CH_2O)_{18}H$,
  $C_{12}E_{18}$ TIII.1.10
 Dodecyloctaoxyethylene glycol ether, $C_{12}H_{25}O(CH_2CH_2O)_8H$, $C_{12}E_8$ TIII.1.10
 Dodecyltricosaoxyethylene glycol ether, $C_{12}H_{25}O(CH_2CH_2O)_{23}H$,
  $C_{12}E_{23}$ TIII.1.10
 Dodecyltrimethylammonium chloride, $C_{12}H_{25}N(CH_3)_3Cl$ TII.1.3
 N-Hexadecylpyridinium chloride, $C_{16}H_{33}NC_5H_5Cl$ TII.2.5
 Hexadecyltrimethylammonium chloride, $C_{16}H_{33}N(CH_3)_3Cl$ TII.1.3
 Nonylphenol(ethylene oxide)$_{10}$ ether, $C_9H_{19}C_6H_4O(CH_2CH_2O)_{10}H$,
  $C_9\phi EO_{10}$ TIII.2.11
 Nonylphenol(ethylene oxide)$_{15}$ ether, $C_9H_{19}C_6H_4O(CH_2CH_2O)_{15}H$,
  $C_9\phi EO_{15}$ TIII.2.11
 Nonylphenol(ethylene oxide)$_{30}$ ether, $C_9H_{19}C_6H_4O(CH_2CH_2O)_{10}H$,
  $C_9\phi EO_{30}$ TIII.2.11
 Nonylphenol(ethylene oxide)$_{50}$ ether, $C_9H_{19}C_6H_4O(CH_2CH_2O)_{10}H$,
  $C_9\phi EO_{50}$ TIII.2.11
 Octylphenol(ethylene oxide)$_{9.5}$ ether, $C_8H_{17}C_6H_4O(CH_2CH_2O)_{9.5}H$,
  $C_8\phi EO_{9.5}$ TIII.2.11
 Sodium 2-(decyl)benzenesulphonate, $C_{10}H_{21}C_6H_4SO_3Na$ TI.3.17
 Sodium 4-(2-decyl)benzenesulphonate, $C_8H_{17}CH(CH_3)C_6H_4SO_3Na$,
  $2\phi C10$ TI.3.20
 Sodium 4-(5-decyl)benzenesulphonate, $C_4H_9CH(C_5H_{12})C_6H_4SO_3Na$,
  $5\phi C10$ TI.3.20
 Sodium decylsulphate, $C_{10}H_{21}OSO_3Na$ TI.1.12
 Sodium 4-(3-dodecyl)benzenesulphonate, $C_9H_{19}CH(C_2H_5)C_6H_4SO_3Na$,
  $3\phi C12$ TI.3.20
 Sodium dodecylsulphate, $C_{12}H_{25}OSO_3Na$ TI.1.11
 Sodium nonylsulphate, $C_9H_{19}OSO_3Na$ TI.1.12
 Sodium octylsulphate, $C_8H_{17}OSO_3Na$ TI.1.12
 Sodium tetradecylsulphate, $C_{14}H_{29}OSO_3Na$ TI.1.12
 Sodium undecylsulphate, $C_{11}H_{23}OSO_3Na$ TI.1.12
 Tridecyldecaoxyethylene glycol ether, $C_{13}H_{27}O(CH_2CH_2O)_{10}H$,
  $C_{13}E_{10}$ TIII.1.10

in aqueous solutions of NaCNS
 Nonylphenol(ethylene oxide)$_{50}$ ether, $C_9H_{19}C_6H_4O(CH_2CH_2O)_{50}H$,
  $C_9\phi EO_{50}$ TIII.2.11

in aqueous solutions of Na$_2$SO$_4$
 Dodecyldodecaoxyethylene glycol ether, $C_{12}H_{25}O(CH_2CH_2O)_{12}H$,
  $C_{12}E_{12}$ TIII.1.10
 Dodecyloctadecaoxyethylene glycol ether, $C_{12}H_{25}O(CH_2CH_2O)_{18}H$,
  $C_{12}E_{18}$ TIII.1.10
 Dodecyloctaoxyethylene glycol ether, $C_{12}H_{25}O(CH_2CH_2O)_8H$, $C_{12}E_8$ TIII.1.10
 Dodecyltricosaoxyethylene glycol ether, $C_{12}H_{25}O(CH_2CH_2O)_{23}H$,
  $C_{12}E_{23}$ TIII.1.10
 Nonylphenol(ethylene oxide)$_{50}$ ether, $C_9H_{19}C_6H_4O(CH_2CH_2O)_{50}H$,
  $C_9\phi EO_{50}$ TIII.2.11

Aggregation number (to be continued)

Aggregation number (continued)

  in water

Nonylphenol(ethylene oxide)$_{15}$ ether, $C_9H_{19}C_6H_4O(CH_2CH_2O)_{15}H$,
$C_9\phi EO_{15}$ TIII.2.11

Nonylphenol(ethylene oxide)$_{18}$ ether, $C_9H_{19}C_6H_4O(CH_2CH_2O)_{18}H$,
$C_9\phi EO_{18}$ TIII.2.11

Nonylphenol(ethylene oxide)$_{20}$ ether, $C_9H_{19}C_6H_4O(CH_2CH_2O)_{20}H$,
$C_9\phi EO_{20}$ TIII.2.11

Nonylphenol(ethylene oxide)$_{30}$ ether, $C_9H_{19}C_6H_4O(CH_2CH_2O)_{30}H$,
$C_9\phi EO_{30}$ TIII.2.11

Nonylphenol(ethylene oxide)$_{40}$ ether, $C_9H_{19}C_6H_4O(CH_2CH_2O)_{40}H$,
$C_9\phi EO_{40}$ TIII.2.11

Nonylphenol(ethylene oxide)$_{50}$ ether, $C_9H_{19}C_6H_4O(CH_2CH_2O)_{50}H$,
$C_9\phi EO_{50}$ TIII.2.11

1,1'-(1,ω-Octanediyl)bispyridinium bis(tetradecane-
sulphonate), $C_5H_5N(CH_2)_8NC_5H_5(C_{14}H_{29}SO_3)_2$, $C_8BP(C_{14})_2$ TI.2.34

Octylhexaoxyethylene glycol ether, $C_8H_{17}O(CH_2CH_2O)_6H$, C_8E_6 TIII.1.10

Octylpentaoxyethylene glycol ether, $C_8H_{17}O(CH_2CH_2O)_5H$, C_8E_5 TIII.1.10

Octylphenol(ethylene oxide)$_9$ ether, $C_8H_{17}C_6H_4O(CH_2CH_2O)_9H$,
$C_8\phi EO_9$ TIII.2.11

Octylphenol(ethylene oxide)$_{9.5}$ ether, $C_8H_{17}C_6H_4O(CH_2CH_2O)_{9.5}H$,
$C_8\phi EO_{9.5}$ TIII.2.11

Octylphenol(ethylene oxide)$_{9.7}$ ether, $C_8H_{17}C_6H_4O(CH_2CH_2O)_{9.7}H$,
$C_8\phi EO_{9.7}$ TIII.2.11

Octylphenol(ethylene oxide)$_{10}$ ether, $C_8H_{17}C_6H_4O(CH_2CH_2O)_{10}H$,
$C_8\phi EO_{10}$ TIII.2.11

Octylphenol(ethylene oxide)$_{12.5}$ ether, $C_8H_{17}C_6H_4O(CH_2CH_2O)_{12.5}H$,
$C_8\phi EO_{12.5}$ TIII.2.11

Octylphenol(ethylene oxide)$_{16}$ ether, $C_8H_{17}C_6H_4O(CH_2CH_2O)_{16}H$,
$C_8\phi EO_{16}$ TIII.2.11

Octylphenol(ethylene oxide)$_{20}$ ether, $C_8H_{17}C_6H_4O(CH_2CH_2O)_{20}H$,
$C_8\phi EO_{20}$ TIII.2.11

Octylphenol(ethylene oxide)$_{30}$ ether, $C_8H_{17}C_6H_4O(CH_2CH_2O)_{30}H$,
$C_8\phi EO_{30}$ TIII.2.11

N-Octylpyridinium bromide, $CH_3(CH_2)_7NC_5Br$ TII.2.5

Sodium decanesulphonate, $C_{10}H_{21}SO_3Na$ TI.2.33

Sodium 2-(decyl)benzenesulphonate, $C_{10}H_{21}C_6H_4SO_3Na$ TI.3.17

Sodium 4-(decyl)benzenesulphonate, $C_{10}H_{21}C_6H_4SO_3Na$, $1\phi C10$ TI.3.18
     & TI.3.19

Sodium 4-(2-decyl)benzenesulphonate, $C_8H_{17}CH(CH_3)_6H_4SO_3Na$,
$2\phi C10$ TI.3.18

Sodium 4-(3-decyl)benzenesulphonate, $C_7H_{15}CH(C_2H_5)C_6H_4SO_3Na$,
$3\phi C10$ TI.3.18

Sodium 4-(5-decyl)benzenesulphonate, $C_4H_9CH(C_5H_{11})C_6H_4SO_3Na$
$5\phi C10$ TI.3.18

Sodium decylsulphate, $C_{10}H_{21}OSO_3Na$ TI.1.10

Sodium 2,5-diheptylbenzenesulphonate, $(C_7H_{15})_2C_6H_3SO_3Na$ TI.3.23

Sodium dodecanesulphonate, $C_{12}H_{25}SO_3Na$ TI.2.33

Sodium 4-(dodecyl)benzenesulphonate, $C_{12}H_{25}C_6H_4SO_3Na$, $1\phi C12$ TI.3.19

Sodium 4-(3-dodecyl)benzenesulphonate, $C_9H_{19}CH(C_2H_5)C_6H_4SO_3Na$,
$3\phi C12$ TI.3.18

Sodium 4-(4-dodecyl)benzenesulphonate, $C_8H_{17}CH(C_3H_7)C_6H_4SO_3Na$,
$4\phi C12$ TI.3.18

Aggregation number (to be continued)

Boiling point (continued)

Boiling point (to be continued)

Clouding temperature (continued)

in aqueous solutions of NH_4Br
  Octylphenol(ethylene oxide)$_{9.5}$ ether, $C_8H_{17}C_6H_4O(CH_2CH_2O)_{9.5}H$,
  $C_8\phi EO_{9.5}$ FIII.2.3

in aqueous solutions of NH_4Cl
  Octylphenol(ethylene oxide)$_{9.5}$ ether, $C_8H_{17}C_6H_4O(CH_2CH_2O)_{9.5}H$,
  $C_8\phi EO_{9.5}$ FIII.2.3

in aqueous solutions of $(NH_4)_2SO_4$
  Octylphenol(ethylene oxide)$_{9.5}$ ether, $C_8H_{17}C_6H_4O(CH_2CH_2O)_{9.5}H$,
  $C_8\phi EO_{9.5}$ FIII.2.3

in aqueous solutions of Ni^{2+}
  Octylphenol(ethylene oxide)$_{9.5}$ ether, $C_8H_{17}C_6H_4O(CH_2CH_2O)_{9.5}H$,
  $C_8\phi EO_{9.5}$ TIII.2.1

in aqueous solutions of Pb^{2+}
  Octylphenol(ethylene oxide)$_{9.5}$ ether, $C_8H_{17}C_6H_4O(CH_2CH_2O)_{9.5}H$,
  $C_8\phi EO_{9.5}$ TIII.2.1

in aqueous solutions of PO_4^{3-}
  Octylphenol(ethylene oxide)$_{9.5}$ ether, $C_8H_{17}C_6H_4O(CH_2CH_2O)_{9.5}H$,
  $C_8\phi EO_{9.5}$ TIII.2.1

in aqueous solutions of SCN^-
  Octylphenol(ethylene oxide)$_{9.5}$ ether, $C_8H_{17}C_6H_4O(CH_2CH_2O)_{9.5}H$,
  $C_8\phi EO_{9.5}$ TIII.2.1

in aqueous solutions of SO_4^{2-}
  Octylphenol(ethylene oxide)$_{9.5}$ ether, $C_8H_{17}C_6H_4O(CH_2CH_2O)_{9.5}H$,
  $C_8\phi EO_{9.5}$ TIII.2.1

in aqueous solutions of $SrCl_2$
  Nonylphenol(ethylene oxide)$_{15}$ ether, $C_9H_{19}C_6H_4O(CH_2CH_2O)_{15}H$,
  $C_9\phi EO_{15}$ FIII.2.9

in aqueous solutions of Zn^{2+}
  Octylphenol(ethylene oxide)$_{9.5}$ ether, $C_8H_{17}C_6H_4O(CH_2CH_2O)_{9.5}H$,
  $C_8\phi EO_{9.5}$ TIII.2.1

in aqueous solutions of $Zn(NO_3)_2$
  Octylphenol(ethylene oxide)$_{9.5}$ ether, $C_8H_{17}C_6H_4O(CH_2CH_2O)_{9.5}H$,
  $C_8\phi EO_{9.5}$ FIII.2.3

in D_2O
  Dodecyloctaoxyethylene glycol ether, $C_{12}H_{25}O(CH_2CH_2O)_8H$, $C_{12}E_8$ TIII.1.3
  Octylpentaoxyethylene glycol ether, $C_8H_{17}O(CH_2CH_2O)_5H$, C_8E_5 TIII.1.3

in D_2O/ sodium dodecylsulphate
  Butylmonooxyethylene glycol ether, $C_4H_9OCH_2CH_2OH$, C_4E_1 FIII.1.38

in water
  Butylmonooxyethylene glycol ether, $C_4H_9OCH_2CH_2OH$, C_4E_1 TIII.1.2
  Decyldecaoxyethylene glycol ether, $C_{10}H_{21}O(CH_2CH_2O)_{10}H$, $C_{10}E_{10}$ TIII.1.2
  Decylhexaoxyethylene glycol ether, $C_{10}H_{21}O(CH_2CH_2O)_6H$, $C_{10}E_6$ TIII.1.2
  Decyloctaoxyethylene glycol ether, $C_{10}H_{21}O(CH_2CH_2O)_8H$, $C_{10}E_8$ TIII.1.2
  Decylpentaoxyethylene glycol ether, $C_{10}H_{21}O(CH_2CH_2O)_5H$, $C_{10}E_5$ TIII.1.2
  Decylphenol(ethylene oxide)$_9$ ether, $C_{10}H_{21}C_6H_4O(CH_2CH_2O)_9H$,
  $C_{10}\phi EO_9$ TIII.2.1
  Decyltetraoxyethylene glycol ether, $C_{10}H_{21}O(CH_2CH_2O)_4H$, $C_{10}E_4$ TIII.1.2
  Decyltrioxyethylene glycol ether, $C_{10}H_{21}O(CH_2CH_2O)_3H$, $C_{10}E_3$ TIII.1.2
  Dodecyldecaoxyethylene glycol ether, $C_{12}H_{25}O(CH_2CH_2O)_{10}H$,
  $C_{12}E_{10}$ TIII.1.2
  Dodecyldodecaoxyethylene glycol ether, $C_{12}H_{25}O(CH_2CH_2O)_{12}H$,
  $C_{12}E_{12}$ TIII.1.2
  Dodecylhexaoxyethylene glycol ether, $C_{12}H_{25}O(CH_2CH_2O)_6H$, $C_{12}E_6$ TIII.1.2

Clouding temperature (to be continued)

Clouding temperature (continued)

in water

Dodecylheptaoxyethylene glycol ether, $C_{12}H_{25}O(CH_2CH_2O)_7H$,
$C_{12}E_7$ TIII.1.2

Dodecylnonaoxyethylene glycol ether, $C_{12}H_{25}O(CH_2CH_2O)_9H$,
$C_{12}E_9$ TIII.1.2

Dodecyloctaoxyethylene glycol ether, $C_{12}H_{25}O(CH_2CH_2O)_8H$,
$C_{12}E_8$ TIII.1.2

Dodecylpentaoxyethylene glycol ether, $C_{12}H_{25}O(CH_2CH_2O)_5H$,
$C_{12}E_5$ TIII.1.2

Dodecyltetraoxyethylene glycol ether, $C_{12}H_{25}O(CH_2CH_2O)_4H$,
$C_{12}E_4$ TIII.1.2

Dodecyltrioxyethylene glycol ether, $C_{12}H_{25}O(CH_2CH_2O)_3H$, $C_{12}E_3$ TIII.1.2

Dodecylundecaoxyethylene glycol ether, $C_{12}H_{25}O(CH_2CH_2O)_{11}H$,
$C_{12}E_{11}$ TIII.1.2

Hexadecyldodecaoxyethylene glycol ether, $C_{16}H_{33}O(CH_2CH_2O)_{12}H$,
$C_{16}E_{12}$ TIII.1.2

Hexadecylheptaoxyethylene glycol ether, $C_{16}H_{33}O(CH_2CH_2O)_7H$,
$C_{16}E_7$ TIII.1.2

Hexadecylhexaoxyethylene glycol ether, $C_{16}H_{33}O(CH_2CH_2O)_6H$,
$C_{16}E_6$ TIII.1.2

Hexadecylnonaoxyethylene glycol ether, $C_{16}H_{33}O(CH_2CH_2O)_9H$,
$C_{16}E_9$ TIII.1.2

Hexadecyloctaoxyethylene glycol ether, $C_{16}H_{33}O(CH_2CH_2O)_8H$,
$C_{16}E_8$ TIII.1.2

Hexadecyltetraoxyethylene glycol ether, $C_{16}H_{33}O(CH_2CH_2O)_4H$,
$C_{16}E_4$ TIII.1.2

Hexadecyltrioxyethylene glycol ether, $C_{16}H_{33}O(CH_2CH_2O)_3H$,
$C_{16}E_3$ TIII.1.2

Hexyldioxyethylene glycol ether, $C_6H_{13}O(CH_2CH_2O)_2H$, C_6E_2 TIII.1.2

Hexylhexaoxyethylene glycol ether, $C_6H_{13}O(CH_2CH_2O)_6H$, C_6E_6 TIII.1.2

Hexylpentaoxyethylene glycol ether, $C_6H_{13}O(CH_2CH_2O)_5H$, C_6E_5 TIII.1.2

Hexyltetraoxyethylene glycol ether, $C_6H_{13}O(CH_2CH_2O)_4H$, C_6E_4 TIII.1.2

Hexyltrioxyethylene glycol ether, $C_6H_{13}O(CH_2CH_2O)_3H$, C_6E_3 TIII.1.2

Nonylhexaoxyethylene glycol ether, $C_9H_{19}O(CH_2CH_2O)_6H$, C_9E_6 TIII.1.2

Nonylpentaoxyethylene glycol ether, $C_9H_{19}O(CH_2CH_2O)_5H$, C_9E_5 TIII.1.2

Nonylphenol(ethylene oxide)$_8$ ether, $C_9H_{19}C_6H_4O(CH_2CH_2O)_8H$,
$C_9\phi EO_8$ TIII.2.1

Nonylphenol(ethylene oxide)$_9$ ether, $C_9H_{19}C_6H_4O(CH_2CH_2O)_9H$,
$C_9\phi EO_9$ TIII.2.1

Nonylphenol(ethylene oxide)$_{30}$ ether, $C_9H_{19}C_6H_4O(CH_2CH_2O)_{30}H$,
$C_9\phi EO_{30}$ TIII.2.1

Nonyltetraoxyethylene glycol ether, $C_9H_{19}O(CH_2CH_2O)_4H$, C_9E_4 TIII.1.2

Octyldodecaoxyethylene glycol ether, $C_8H_{17}O(CH_2CH_2O)_{12}H$, C_8E_{12} TIII.1.2

Octylhexaoxyethylene glycol ether, $C_8H_{17}O(CH_2CH_2O)_6H$, C_8E_6 TIII.1.2

Octyloctaoxyethylene glycol ether, $C_8H_{17}O(CH_2CH_2O)_8H$, C_8E_8 TIII.1.2

Octylpentaoxyethylene glycol ether, $C_8H_{17}O(CH_2CH_2O)_5H$, C_8E_5 TIII.1.2

Octylphenol(ethylene oxide)$_{7.3}$ ether, $C_8H_{17}C_6H_4O(CH_2CH_2O)_{7.3}H$,
$C_8\phi EO_{7.3}$ TIII.2.1

Octylphenol(ethylene oxide)$_{7.7}$ ether, $C_8H_{17}C_6H_4O(CH_2CH_2O)_{7.7}H$,
$C_8\phi EO_{7.7}$ TIII.2.1

Octylphenol(ethylene oxide)$_8$ ether, $C_8H_{17}C_6H_4O(CH_2CH_2O)_8H$,
$C_8\phi EO_8$ TIII.2.1

Clouding temperature (to be continued)

Clouding temperature (continued)

in water

Octylphenol(ethylene oxide)$_{8.8}$ ether, $C_8H_{17}C_6H_4O(CH_2CH_2O)_{8.8}H$,
$C_8\phi EO_{8.8}$ TIII.2.1

Octylphenol(ethylene oxide)$_{9.5}$ ether, $C_8H_{17}C_6H_4O(CH_2CH_2O)_{9.5}H$,
$C_8\phi EO_{9.5}$ TIII.2.1

Octylphenol(ethylene oxide)$_{9.6}$ glycol ether,
$C_8H_{17}C_6H_4O(CH_2CH_2O)_{9.6}H$, $C_8\phi EO_{9.6}$ TIII.2.1

Octylphenol(ethylene oxide)$_{12.5}$ glycol ether,
$C_8H_{17}C_6H_4O(CH_2CH_2O)_{12.5}H$, $C_8\phi EO_{12.5}$ TIII.2.1

Octyltetraoxyethylene glycol ether, $C_8H_{17}O(CH_2CH_2O)_4H$, C_8E_4 TIII.1.2

Octyltrioxyethylene glycol ether, $C_8H_{17}O(CH_2CH_2O)_3H$, C_8E_3 TIII.1.2

Pentadecylhexaoxyethylene glycol ether, $C_{15}H_{31}O(CH_2CH_2O)_6H$,
$C_{15}E_6$ TIII.1.2

Pentadecyloctaoxyethylene glycol ether, $C_{15}H_{31}O(CH_2CH_2O)_8H$,
$C_{15}E_8$ TIII.1.2

Pentyldioxyethylene glycol ether, $C_5H_{11}O(CH_2CH_2O)_2H$, C_5E_2 TIII.1.2

Tetradecylheptaoxyethylene glycol ether, $C_{14}H_{27}O(CH_2CH_2O)_7H$,
$C_{14}E_7$ TIII.1.2

Tetradecylhexaoxyethylene glycol ether, $C_{14}H_{27}O(CH_2CH_2O)_6H$,
$C_{14}E_6$ TIII.1.2

Tetradecyloctaoxyethylene glycol ether, $C_{14}H_{29}O(CH_2CH_2O)_8H$,
$C_{14}E_8$ TIII.1.2

Tetradecylpentaoxyethylene glycol ether, $C_{14}H_{29}O(CH_2CH_2O)_5H$,
$C_{14}E_5$ TIII.1.2

Tetradecyltrioxyethylene glycol ether, $C_{14}H_{29}O(CH_2CH_2O)_3H$,
$C_{14}E_3$ TIII.1.2

Tridecylhexaoxyethylene glycol ether, $C_{13}H_{27}O(CH_2CH_2O)_6H$,
$C_{13}E_6$ TIII.1.2

Tridecyloctaoxyethylene glycol ether, $C_{13}H_{27}O(CH_2CH_2O)_8H$,
$C_{13}E_8$ TIII.1.2

Tridecylpentaoxyethylene glycol ether, $C_{13}H_{27}O(CH_2CH_2O)_5H$,
$C_{13}E_5$ TIII.1.2

Undecylhexaoxyethylene glycol ether, $C_{11}H_{23}O(CH_2CH_2O)_6H$, $C_{11}E_6$ TIII.1.2

Undecyloctaoxyethylene glycol ether, $C_{11}H_{23}O(CH_2CH_2O)_8H$, $C_{11}E_8$ TIII.1.2

Undecylpentaoxyethylene glycol ether, $C_{11}H_{23}O(CH_2CH_2O)_5H$,
$C_{11}E_5$ TIII.1.2

Undecyltetraoxyethylene glycol ether, $C_{11}H_{23}O(CH_2CH_2O)_4H$,
$C_{11}E_4$ TIII.1.2

in water/acetamide

Octylphenol(ethylene oxide)$_{9.5}$ ether, $C_8H_{17}C_6H_4O(CH_2CH_2O)_{9.5}H$,
$C_8\phi EO_{9.5}$ TIII.2.1

in water/acetone

Octylphenol(ethylene oxide)$_{9.5}$ ether, $C_8H_{17}C_6H_4O(CH_2CH_2O)_{9.5}H$,
$C_8\phi EO_{9.5}$ FIII.2.5

in water/aniline

Octylphenol(ethylene oxide)$_{9.5}$ ether, $C_8H_{17}C_6H_4O(CH_2CH_2O)_{9.5}H$,
$C_8\phi EO_{9.5}$ FIII.2.5

in water/benzene

Octylphenol(ethylene oxide)$_{9.5}$ ether, $C_8H_{17}C_6H_4O(CH_2CH_2O)_{9.5}H$,
$C_8\phi EO_{9.5}$ FIII.2.5

in water/butanol

Dodecylhexaoxyethylene glycol ether, $C_{12}H_{25}O(CH_2CH_2O)_6H$, $C_{12}E_6$ FIII.1.37

Clouding temperature (to be continued)

Clouding temperature (continued)

   in water/butylacetate

      Octylphenol(ethylene oxide)$_{9.5}$ ether, $C_8H_{17}C_6H_4O(CH_2CH_2O)_{9.5}H$,

      $C_8\phi EO_{9.5}$ FIII.2.5

   in water/CARBOWAX 4000, $H(OCH_2CH_2)_nOH$

      Dodecylhexaoxyethylene glycol ether, $C_{12}H_{25}O(CH_2CH_2O)_6H$, $C_{12}E_6$ FIII.1.37

   in water/cyclohexane

      Nonylphenol(ethylene oxide)$_{9.6}$ ether, $C_9H_{19}C_6H_4O(CH_2CH_2O)_{9.6}H$,

      $C_9\phi EO_{9.6}$ FIII.2.8

      Octylphenol(ethylene oxide)$_{9.5}$ ether, $C_8H_{17}C_6H_4O(CH_2CH_2O)_{9.5}H$,

      $C_8\phi EO_{9.5}$ FIII.2.5

   in water/decane

      Octylphenol(ethylene oxide)$_{9.5}$ ether, $C_8H_{17}C_6H_4O(CH_2CH_2O)_{9.5}H$,

      $C_8\phi EO_{9.5}$ FIII.2.5

   in water/decanol

      Dodecylhexaoxyethylene glycol ether, $C_{12}H_{25}O(CH_2CH_2O)_6H$, $C_{12}E_6$ FIII.1.37

   in water/1,1-diethylurea

      Octylphenol(ethylene oxide)$_{9.5}$ ether, $C_8H_{17}C_6H_4O(CH_2CH_2O)_{9.5}H$,

      $C_8\phi EO_{9.5}$ TIII.2.1

   in water/1,3-dimethylthiourea

      Octylphenol(ethylene oxide)$_{9.5}$ ether, $C_8H_{17}C_6H_4O(CH_2CH_2O)_{9.5}H$,

      $C_8\phi EO_{9.5}$ TIII.2.1

   in water/1,1-dimethylurea

      Octylphenol(ethylene oxide)$_{9.5}$ ether, $C_8H_{17}C_6H_4O(CH_2CH_2O)_{9.5}H$,

      $C_8\phi EO_{9.5}$ TIII.2.1

   in water/1,3-dimethylurea

      Octylphenol(ethylene oxide)$_{9.5}$ ether, $C_8H_{17}C_6H_4O(CH_2CH_2O)_{9.5}H$,

      $C_8\phi EO_{9.5}$ TIII.2.1

   in water/dodecane

      Dodecylhexaoxyethylene glycol ether, $C_{12}H_{25}O(CH_2CH_2O)_6H$, $C_{12}E_6$ FIII.1.36

      Dodecyltetraoxyethylene glycol ether, $C_{12}H_{25}O(CH_2CH_2O)_4H$,

      $C_{12}E_4$ FIII.1.34

      Octylphenol(ethylene oxide)$_{9.5}$ ether, $C_8H_{17}C_6H_4O(CH_2CH_2O)_{9.5}H$,

      $C_8\phi EO_{9.5}$ FIII.2.5

   in water/dodecanol

      Dodecylhexaoxyethylene glycol ether, $C_{12}H_{25}O(CH_2CH_2O)_6H$, $C_{12}E_6$ FIII.1.37

   in water-dodecanol-nitrobenzene

      Octylphenol(ethylene oxide)$_{9.5}$ ether, $C_8H_{17}C_6H_4O(CH_2CH_2O)_{9.5}H$,

      $C_8\phi EO_{9.5}$ FIII.2.5

   in water/ethylene dichloride

      Octylphenol(ethylene oxide)$_{9.5}$ ether, $C_8H_{17}C_6H_4O(CH_2CH_2O)_{9.5}H$,

      $C_8\phi EO_{9.5}$ FIII.2.5

   in water/2-ethylhexene

      Octylphenol(ethylene oxide)$_{9.5}$ ether, $C_8H_{17}C_6H_4O(CH_2CH_2O)_{9.5}H$,

      $C_8\phi EO_{9.5}$ FIII.2.5

   in water/ethylurea

      Octylphenol(ethylene oxide)$_{9.5}$ ether, $C_8H_{17}C_6H_4O(CH_2CH_2O)_{9.5}H$,

      $C_8\phi EO_{9.5}$ TIII.2.1

   in water/guanidinium chloride

      Octylphenol(ethylene oxide)$_{9.5}$ ether, $C_8H_{17}C_6H_4O(CH_2CH_2O)_{9.5}H$,

      $C_8\phi EO_{9.5}$ TIII.2.1

   in water/heptane

      Dodecylhexaoxyethylene glycol ether, $C_{12}H_{25}O(CH_2CH_2O)_6H$, $C_{12}E_6$ FIII.1.36

Clouding temperature (to be continued)

552

CMC (continued)

CMC (to be continued)

CMC (continued)

in water

Dodecylhentriacontaoxyethylene glycol ether, $C_{12}H_{25}O(CH_2CH_2O)_{31}H$,
$C_{12}E_{31}$ TIII.1.5
Dodecylheptaoxyethylene glycol ether, $C_{12}H_{25}O(CH_2CH_2O)_7H$,
$C_{12}E_7$ TIII.1.5
Dodecylhexaoxyethylene glycol ether, $C_{12}H_{25}O(CH_2CH_2O)_6H$, $C_{12}E_6$ TIII.1.5
Dodecylnonaoxyethylene glycol ether, $C_{12}H_{25}O(CH_2CH_2O)_9H$, $C_{12}E_9$ TIII.1.5
Dodecyloctaoxyethylene glycol ether, $C_{12}H_{25}O(CH_2CH_2O)_8H$, $C_{12}E_8$ TIII.1.5
Dodecylpentaoxyethylene glycol ether, $C_{12}H_{25}O(CH_2CH_2O)_5H$,
$C_{12}E_5$ TIII.1.5
N-Dodecylpyridinium bromide, $C_{12}H_{25}NC_5H_5Br$ TII.2.1
N-Dodecylpyridinium chloride, $C_{12}H_{25}NC_5H_5Cl$ TII.2.1
N-Dodecylpyridinium iodide, $C_{12}H_{25}NC_5H_5I$ TII.2.1
Dodecyltetradecaoxyethylene glycol ether, $C_{12}H_{25}O(CH_2CH_2O)_{14}H$,
$C_{12}E_{14}$ TIII.1.5
Dodecyltetraoxyethylene glycol ether, $C_{12}H_{25}O(CH_2CH_2O)_4H$,
$C_{12}E_4$ TIII.1.5
Dodecyltriacontaoxyethylene glycol ether, $C_{12}H_{25}O(CH_2CH_2O)_{30}H$,
$C_{12}E_{30}$ TIII.1.5
Dodecyltricosaoxyethylene glycol ether, $C_{12}H_{25}O(CH_2CH_2O)_{23}H$,
$C_{12}E_{23}$ TIII.1.5
Dodecyltrimethylammonium bromide, $C_{12}H_{25}N(CH_3)_3Br$ TII.1.1
1,1'(1,ω-Ethanediyl)bispyridinium bis(tetradecanesulphonate),
$C_5H_5N(CH_2)_2NC_5H_5(C_{14}H_{29}SO_3)_2$ TI.2.21
N-Ethyl-4-dodecylpyridinium iodide, $C_{12}H_{25}C_5H_4NC_2H_5I$ TII.2.1
Hexadecyldodecaoxyethylene glycol ether, $C_{16}H_{33}O(CH_2CH_2O)_{12}H$,
$C_{16}E_{12}$ TIII.1.5
Hexadecylheneicosaoxyethylene glycol ether, $C_{16}H_{33}O(CH_2CH_2O)_{21}H$,
$C_{16}E_{21}$ TIII.1.5
Hexadecylheptaoxyethylene glycol ether, $C_{16}H_{33}O(CH_2CH_2O)_7H$,
$C_{16}E_7$ TIII.1.5
Hexadecylhexaoxyethylene glycol ether, $C_{16}H_{33}O(CH_2CH_2O)_6H$,
$C_{16}E_6$ TIII.1.5
Hexadecyl-nona-deuteriotrimethylammonium bromide,
$C_{16}H_{33}N(CD_3)_3Br$ TII.1.1
Hexadecylnonaoxyethylene glycol ether, $C_{16}H_{33}O(CH_2CH_2O)_9H$,
$C_{16}E_9$ TIII.1.5
Hexadecylpentadecaoxyethylene glycol ether, $C_{16}H_{33}O(CH_2CH_2O)_{15}H$,
$C_{16}E_{15}$ TIII.1.5
N-Hexadecylpyridinium bisulphate, $C_{16}H_{33}NC_5H_5HSO_4$ TII.2.1
N-Hexadecylpyridinium bromide, $C_{16}H_{33}NC_5H_5Br$ TII.2.1
N-Hexadecylpyridinium chloride, $C_{16}H_{33}NC_5H_5Cl$ TII.2.1
N-Hexadecylpyridinium iodide, $C_{16}H_{33}NC_5H_5I$ TII.2.1
N-Hexadecylpyridinium nitrate, $C_{16}H_{33}NC_5H_5NO_3$ TII.2.1
N-Hexadecylpyridinium phosphate, $C_{16}H_{33}NC_5H_2PO_4$ TII.2.1
N-Hexadecylpyridinium tetrafluoroborate, $C_{16}H_{33}NC_5H_5BF_4$ TII.2.1
N-Hexadecylpyridinium tosylate, $C_{16}H_{33}NC_5H_5O_3SC_6H_4CH_3$ TII.2.1
Hexadecyltriacontaoxyethylene glycol ether, $C_{16}H_{33}O(CH_2CH_2O)_{30}H$,
$C_{16}E_{30}$ TIII.1.5
Hexadecyltrimethylammonium bromide, $C_{16}H_{33}N(CH_3)_3Br$ TII.1.1
Hexadecyltrimethylammonium chloride, $C_{16}H_{33}N(CH_3)_3Cl$ TII.1.1
Hexadecyltrimethylammonium hydroxide, $C_{16}H_{33}N(CH_3)_3OH$ TII.1.1
Hexadecyltrimethylammonium nitrate, $C_{16}H_{33}N(CH_3)_3NO_3$ TII.1.1

CMC (to be continued)

CMC (continued)

in water

1,1'(1,ω-Hexanediyl)bispyridinium bis(tetradecanesulphonate),
$C_5H_5N(CH_2)_6NC_5H_5(C_{14}H_{29}SO_3)_2$ TI.2.21

Hexylhexaoxyethylene glycol ether, $C_6H_{13}O(CH_2CH_2O)_6H$, C_6E_6 TIII.1.5

Hexylpentaoxyethylene glycol ether, $C_6H_{13}O(CH_2CH_2O)_5H$, C_6E_5 TIII.1.5

Hexyltetraoxyethylene glycol ether, $C_6H_{13}O(CH_2CH_2O)_4H$, C_6E_4 TIII.1.5

Hexyltrioxyethylene glycol ether, $C_6H_{13}O(CH_2CH_2O)_3H$, C_6E_3 TIII.1.5

Magnesium bis(decanesulphonate), $(C_{10}H_{21}SO_3)_2Mg$ TI.2.22

Magnesium bis(dodecanesulphonate), $(C_{12}H_{25}SO_3)_2Mg$ TI.2.22

Magnesium bis(octanesulphonate), $(C_8H_{17}SO_3)_2Mg$ TI.2.22

N-Methyl-4-(9,9-dimethyldecyl)pyridinium iodide,
$(CH_3)_3C(CH_2)_8C_5H_4NCH_3I$ TII.2.1

N-Methyl-4-(dodecyl)pyridinium iodide, $C_{12}H_{25}C_5H_4NCH_3I$ TII.2.1

N-Methyl-4-(2-dodecyl)pyridinium iodide,
$C_{10}H_{21}CH(CH_3)C_5H_4NCH_3I$ TII.2.1

N-Methyl-4-(dodec-5-ynyl)pyridinium iodide,
$C_6H_{13}C{\equiv}C(CH_2)_4C_5H_4NCH_3I$ TII.2.1

N-Methyl-4-(8-ethyldecyl)pyridinium iodide,
$(C_2H_5)_2CH(CH_2)_7C_5H_4NCH_3I$ TII.2.1

Methylviologen bis(decanesulphonate),
$CH_3NC_5H_4-C_5H_4NCH_3$ $(C_{10}H_{21}SO_3)_2$ TI.2.20

Methylviologen bis(dodecanesulphonate),
$CH_3NC_5H_4-C_5H_4NCH_3(C_{12}H_{25}SO_3)_2$ TI.2.20

Methylviologen bis(hexadecanesulphonate),
$CH_3NC_5H_4-C_5H_4NCH_3(C_{16}H_{33}SO_3)_2$ TI.2.20

Methylviologen bis(octadecanesulphonate),
$CH_3NC_5H_4-C_5H_4NCH_3(C_{18}H_{37}SO_3)_2$ TI.2.20

Methylviologen bis(tetradecanesulphonate),
$CH_3NC_5H_4-C_5H_4NCH_3(C_{14}H_{29}SO_3)_2$ TI.2.20

Nonyloctaoxyethylene glycol ether, $C_9H_{19}O(CH_2CH_2O)_8H$, C_9E_8 TIII.1.5

Nonylphenol(ethylene oxide)$_5$ ether, $C_9H_{19}C_6H_4O(CH_2CH_2O)_5H$,
$C_9\phi EO_5$ TIII.2.2

Nonylphenol(ethylene oxide)$_{8.5}$ ether, $C_9H_{19}C_6H_4O(CH_2CH_2O)_{8.5}H$,
$C_9\phi EO_{8.5}$ TIII.2.2

Nonylphenol(ethylene oxide)$_9$ ether, $C_9H_{19}C_6H_4O(CH_2CH_2O)_9H$,
$C_9\phi EO_9$ TIII.2.2
    & TIII.2.5

Nonylphenol(ethylene oxide)$_{9.5}$ ether, $C_9H_{19}C_6H_4O(CH_2CH_2O)_{9.5}H$,
$C_9\phi EO_{9.5}$ TIII.2.2

Nonylphenol(ethylene glycol)$_{10}$ ether,
$C_9H_{19}C_6H_4O(CH_2CH_2O)_{10}H$, $C_9\phi EO_{10}$ TIII.2.2

Nonylphenol(ethylene oxide)$_{10.5}$ ether, $C_9H_{19}C_6H_4O(CH_2CH_2O)_{10.5}H$,
$C_9\phi EO_{10.5}$ TIII.2.2

Nonylphenol(ethylene glycol)$_{15}$ ether, $C_9H_{19}C_6H_4O(CH_2CH_2O)_{15}H$,
$C_9\phi EO_{15}$ TIII.2.2

Nonylphenol(ethylene glycol)$_{20}$ ether, $C_9H_{19}C_6H_4O(CH_2CH_2O)_{20}H$,
$C_9\phi EO_{20}$ TIII.2.2

Nonylphenol(ethylene glycol)$_{30}$ ether, $C_9H_{19}C_6H_4O(CH_2CH_2O)_{30}H$,
$C_9\phi EO_{30}$ TIII.2.2

Nonylphenol(ethylene oxide)$_{31}$ ether, $C_9H_{19}C_6H_4O(CH_2CH_2O)_{31}H$,
$C_9\phi EO_{31}$ TIII.2.2

Nonylphenol(ethylene oxide)$_{40}$ ether, $C_9H_{19}C_6H_4O(CH_2CH_2O)_{40}H$,
$C_9\phi EO_{40}$ TIII.2.2

CMC (to be continued)

CMC (continued)

in water

CMC (to be continued)

CMC (continued)

   in water

CMC (to be continued)

CMC (continued)

in water

Sodium 4-(dodecyl)benzenesulphonate, $C_{12}H_{25}C_6H_4SO_3Na$, 1ϕC12	TI.3.10
Sodium 4-(2-dodecyl)benzenesulphonate, $C_{10}H_{21}CH(CH_3)C_6H_4SO_3Na$, 2$\phi$C12	TI.3.10
Sodium 4-(3-dodecyl)benzenesulphonate, $C_9H_{19}CH(C_2H_5)C_6H_4SO_3Na$, 3$\phi$C12	TI.3.10
Sodium 4-(4-dodecyl)benzenesulphonate, $C_8H_{17}CH(C_3H_7)C_6H_4SO_3Na$, 4$\phi$C12	TI.3.10
Sodium 4-(6-dodecyl)benzenesulphonate, $C_6H_{13}CH(C_5H_{11})C_6H_4SO_3Na$, 6$\phi$C12	TI.3.10
Sodium 2-(dodecyl)-5-ethylbenzenesulphonate, $C_{12}H_{25}(C_2H_5)C_6H_3SO_3Na$	TI.3.14
Sodium dodecylsulphate, $C_{12}H_{25}OSO_3Na$	TI.1.4
Sodium 2-ethyl-5-(dodecyl)benzenesulphonate, $C_{12}H_{25}(C_2H_5)C_6H_3SO_3Na$	TI.3.14
Sodium eicosylsulphate, $C_{20}H_{41}OSO_3Na$	TI.1.4
Sodium 3-ethoxydodecanesulphonate, $C_9H_{19}CH(OC_2H_5)C_2H_4SO_3Na$	TI.2.14
Sodium 3-(2-ethyl)hexoxydodecanesulphonate, $C_9H_{19}CH(OCH_2CH(C_2H_5)C_4H_9)C_2H_4SO_3Na$	TI.2.14
Sodium 4-(2-ethyl-hexyl)benzenesulphonate, $C_4H_9CH(C_2H_5)CH_2C_6H_4SO_3Na$	TI.3.10
Sodium heptadecanesulphonate, $C_{17}H_{35}SO_3Na$	TI.2.10
Sodium 2-heptadecylsulphate, $C_{15}H_{31}CH(CH_3)OSO_3Na$	TI.1.4
Sodium 4-(2-heptadecyl)benzenesulphonate, $C_{15}H_{31}CH(CH_3)C_6H_4SO_3Na$, 2$\phi$C17	TI.3.10
Sodium 9-heptadecylsulphate, $C_8H_{17}CH(C_8H_{17})OSO_3Na$	TI.1.4
Sodium 4-(heptyl)benzenesulphonate, $C_7H_{15}C_6H_4SO_3Na$, 1ϕC7	TI.3.10
Sodium hexadecanesulphonate, $C_{16}H_{33}SO_3Na$	TI.2.10
Sodium 7-hexadecanesulphonate, $C_9H_{19}CH(C_6H_{13})SO_3Na$	TI.2.13
Sodium 2-hexadecenesulphonate, $C_{13}H_{27}CH=CHCH_2SO_3Na$	TI.2.15
Sodium 4-(hexadecyl)benzenesulphonate, $C_{16}H_{33}C_6H_4SO_3Na$, 1ϕC16	TI.3.10
Sodium hexadecylsulphate, $C_{16}H_{33}OSO_3Na$	TI.1.4
Sodium 4-hexadecylsulphate, $C_{12}H_{25}CH(C_3H_7)OSO_3Na$	TI.1.4
Sodium 6-hexadecylsulphate, $C_{10}H_{21}CH(C_5H_{11})OSO_3Na$	TI.1.4
Sodium 8-hexadecylsulphate, $C_8H_{17}CH(C_7H_{15})OSO_3Na$	TI.1.4
Sodium hexanesulphonate, $C_6H_{13}SO_3Na$	TI.2.10
Sodium 3-(hexoxy)dodecanesulphonate, $C_9H_{19}CH(OC_6H_{13})C_2H_4SO_3Na$	TI.2.14
Sodium 4-(hexyl)benzenesulphonate, $C_6H_{13}C_6H_4SO_3Na$, 1ϕC6	TI.3.10
Sodium 3-hydroxydodecanesulphonate, $C_9H_{19}CH(OH)C_2H_4SO_3Na$	TI.2.14 & TI.2.15
Sodium 1-hydroxy-2-dodecanesulphonate, $C_{10}H_{21}CH(CH_2OH)SO_3Na$	TI.2.16
Sodium 3-hydroxyethoxyethoxytetradecanesulphonate, $C_{11}H_{23}CH(OC_2H_4OC_2H_4OH)C_2H_4SO_3Na$	TI.2.14
Sodium 3-hydroxyethoxytetradecanesulphonate, $C_{11}H_{23}CH(OC_2H_4OH)C_2H_4SO_3Na$	TI.2.14
Sodium 3-hydroxyhexadecanesulphonate, $C_{13}CH(OH)C_2H_4SO_3Na$	TI.2.14 & TI.2.15
Sodium 1-hydroxy-2-hexadecanesulphonate, $C_{13}H_{29}CH(CH_2OH)SO_3Na$	TI.2.16
Sodium 3-hydroxyoctadecanesulphonate, $C_{15}H_{31}CH(OH)C_2H_4SO_3Na$	TI.2.14 & TI.2.15
Sodium 1-hydroxy-2-octadecanesulphonate, $C_{16}H_{33}CH(CH_2OH)SO_3Na$	TI.2.16
Sodium 3-hydroxytetradecanesulphonate, $C_{11}H_{23}CH(OH)C_2H_4SO_3Na$	TI.2.14 & TI.2.15

CMC (to be continued)

CMC (continued)

in water

Sodium 1-hydroxy-2-tetradecanesulphonate,	
$C_{12}H_{25}CH(CH_2OH)SO_3Na$	TI.2.16
Sodium 3-methoxydodecanesulphonate, $C_9H_{19}CH(OCH_3)C_2H_4SO_3Na$	TI.2.14
Sodium 3-methoxyhexadecanesulphonate, $C_{13}H_{27}CH(OCH_3)C_2H_4SO_3Na$	TI.2.14
Sodium 3-morpholinotetradecanesulphonate,	
$C_{11}H_{23}CH(NC_4H_8O)C_2H_4SO_3Na$	TI.2.14
Sodium 15-nonacosylsulphate, $C_{14}H_{29}CH(C_{14}H_{29})OSO_3Na$	TI.1.4
Sodium 10-nonadecanesulphonate, $C_9H_{19}CH(C_9H_{19})SO_3Na$	TI.2.13
Sodium 5-nonadecylsulphate, $C_{14}H_{29}CH(C_4H_9)OSO_3Na$	TI.1.4
Sodium 10-nonadecylsulphate, $C_9H_{19}CH(C_9H_{19})OSO_3Na$	TI.1.4
Sodium 4-(nonyl)benzenesulphonate, $C_9H_{19}C_6H_4SO_3Na$, 1φC9	TI.3.10
Sodium 4-(3-nonyl)benzenesulphonate, $C_6H_{13}CH(C_2H_5)C_6H_3SO_3Na$,	
3φC9	TI.3.16
Sodium nonylsulphate, $C_9H_{19}OSO_3Na$	TI.1.4
Sodium octadecanesulphonate, $C_{18}H_{37}SO_3Na$	TI.2.10
Sodium 9-octadecanesulphonate, $C_9H_{19}CH(C_8H_{17})C_2H_4SO_3Na$	TI.2.13
Sodium 2-octadecenesulphonate, $C_{15}H_{31}CH=CHCH_2SO_3Na$	TI.2.15
Sodium 4-(octadecyl)benzenesulphonate, $C_{18}H_{37}C_6H_4SO_3Na$, 1φC18	TI.3.10
Sodium octadecylsulphate, $C_{18}H_{37}OSO_3Na$	TI.1.4
Sodium 2-octadecylsulphate, $C_{16}H_{33}CH(CH_3)OSO_3Na$	TI.1.4
Sodium 4-octadecylsulphate, $C_{13}H_{29}CH(C_3H_7)OSO_3Na$	TI.1.4
Sodium 6-octadecylsulphate, $C_{25}H_{25}CH(C_5H_{11})OSO_3Na$	TI.1.4
Sodium octanesulphonate, $C_8H_{17}SO_3Na$	TI.2.10
Sodium 3(octoxy)dodecanesulphonate, $C_9H_{19}CH(OC_8H_{17})C_2H_4SO_3Na$	TI.2.14
Sodium 4-(octyl)benzenesulphonate, $C_8H_{17}C_6H_4SO_3Na$, 1φC8	TI.3.10
Sodium (2-octyl)benzenesulphonate, $C_6H_{13}CH(CH_3)C_6H_4SO_3Na$,	
2φC8	TI.3.10
Sodium octylsulphate, $C_8H_{17}OSO_3Na$	TI.1.4
Sodium 3-oxododecanesulphonate, $C_9H_{19}COC_2H_4SO_3Na$	TI.2.15
Sodium 3-oxohexadecanesulphonate, $C_{13}H_{27}COC_2H_4SO_3Na$	TI.2.15
Sodium 3-oxooctadecanesulphonate, $C_{15}H_{31}COC_2H_4SO_3Na$	TI.2.15
Sodium 3-oxotetradecanesulphonate, $C_{11}H_{23}COC_2H_4SO_3Na$	TI.2.15
Sodium pentadecanesulphonate, $C_{15}H_{31}SO_3Na$	TI.2.10
Sodium 8-pentadecanesulphonate, $C_7H_{15}CH(C_7H_{15})SO_3Na$	TI.2.10
Sodium 2-pentadecylsulphate, $C_{13}H_{27}CH(CH_3)OSO_3Na$	TI.1.4
Sodium 3-pentadecylsulphate, $C_{12}H_{25}CH(C_2H_5)OSO_3Na$	TI.1.4
Sodium 5-pentadecylsulphate, $C_{10}H_{21}CH(C_4H_9)OSO_3Na$	TI.1.4
Sodium 8-pentadecylsulphate, $C_7H_{15}CH(C_7H_{15})OSO_3Na$	TI.1.4
Sodium 4-(2-pentadecyl)benzenesulphonate,	
$C_{13}H_{27}CH(CH_3)C_6H_4SO_3Na$, 2φC15	TI.3.10
Sodium 4-(2-pentylnonyl)benzenesulphonate,	
$C_7H_{15}CH(C_5H_{11})CH_2C_6H_4SO_3Na$	TI.3.10
Sodium 3-phenoxydodecanesulphonate, $C_9H_{19}CH(OC_6H_5)C_2H_4SO_3Na$	TI.2.14
Sodium phenoxysulphonate ethanoic acid octyl ester,	
$C_8H_{17}OOCCH_2OC_6H_4SO_3Na$	TI.3.11
Sodium phenylsulphonate ethanoic acid nonyl ester,	
$C_9H_{19}OOCCH_2C_6H_4SO_3Na$	TI.3.11
Sodium 3-phenoxytetradecanesulphonate, $C_{11}H_{23}CH(OC_6H_5)C_2H_4SO_3Na$	TI.2.14
Sodium 3-piperidinotetradecanesulphonate,	
$C_{11}H_{23}CH(NC_5H_{10})C_2H_4SO_3Na$	TI.2.14
Sodium 3-(propoxy)dodecanesulphonate,	
$C_9H_{19}CH(OC_3H_7)C_2H_4SO_3Na$	TI.2.14

CMC (to be continued)

CMC (continued)

   in water

Sodium 3-(2-propoxy)dodecanesulphonate,
$C_9H_{19}CH(OCH(CH_3)_2)C_2H_4SO_3Na$ TI.2.14

Sodium 3-(propoxy)hexadecanesulphonate,
$C_{13}H_{27}CH(OC_3H_7)C_2H_4SO_3Na$ TI.2.14

Sodium 3-propylaminotetradecanesulphonate,
$C_{11}H_{23}CH(NHC_3H_7)C_2H_4SO_3Na$ TI.2.14

Sodium 4-(2-propyl-heptyl)benzenesulphonate,
$C_5H_{11}CH(C_3H_7)CH_2C_6H_4SO_3Na$ TI.3.10

Sodium 2-propyl-5-(undecyl)benzenesulphonate,
$C_{11}H_{23}(C_3H_7)C_6H_3SO_3Na$ TI.3.14

Sodium tetradecanesulphonate, $C_{14}H_{29}SO_3Na$ TI.2.10
         & TI.2.24

Sodium 5-tetradecanesulphonate, $C_9H_{19}CH(C_4H_9)SO_3Na$ TI.2.13

Sodium 2-tetradecenesulphonate, $C_{11}H_{23}CH=CHCH_2SO_3Na$ TI.2.15

Sodium 4-(tetradecyl)benzenesulphonate, $C_{14}H_{29}C_6H_4SO_3Na$,
1φC14   TI.3.10

Sodium tetradecylsulphate, $C_{14}H_{29}OSO_3Na$ TI.1.4

Sodium 2-tetradecylsulphate, $C_{12}H_{25}CH(CH_3)OSO_3Na$ TI.1.4

Sodium 3-tetradecylsulphate, $C_{11}H_{23}CH(C_2H_5)OSO_3Na$ TI.1.4

Sodium 4-tetradecylsulphate, $C_{10}H_{21}CH(C_3H_7)OSO_3Na$ TI.1.4

Sodium 5-tetradecylsulphate, $C_9H_{19}CH(C_4H_9)OSO_3Na$ TI.1.4

Sodium 7-tetradecylsulphate, $C_7H_{15}CH(C_6H_{13})OSO_3Na$ TI.1.4

Sodium 4-(thioundecyl)benzenesulphonate, $C_{11}H_{23}SC_6H_4SO_3Na$ TI.3.11

Sodium 3-trichlorophenoxytetradecanesulphonate,
$C_{11}H_{23}CH(OC_6H_2Cl_3)C_2H_4SO_3Na$ TI.2.14

Sodium tridecanesulphonate, $C_{13}H_{27}SO_3Na$ TI.2.10

Sodium 2-tridecylbenzenesulphonate, $C_{11}H_{23}CH(CH3)C_6H_4SO_3Na$,
2φC13   TI.3.10

Sodium 2-(tridecyl)-5-methylbenzenesulphonate,
$C_{13}H_{27}(CH_3)C_6H_3SO_3Na$ TI.3.14

Sodium undecanesulphonate, $C_{11}H_{23}SO_3Na$ TI.2.10

Sodium 4-(undecanoic acid)benzenesulphonate ester,
$C_{10}H_{21}COOC_6H_4SO_3Na$ TI.3.11

Sodium undecanoyl-4-aminobenzenesulphonate,
$C_{10}H_{21}CONHC_6H_4SO_3Na$ TI.3.11

Sodium 4-(undecoxy)benzenesulphonate, $C_{11}H_{23}OC_6H_4SO_3Na$ TI.3.11

Sodium 4-(undecylamino)benzenesulphonate,
$C_{11}H_{23}NHC_6H_4SO_3Na$ TI.3.11

Sodium 4-(undecyl)benzenesulphonate, $C_{11}H_{23}C_6H_4SO_3Na$, 1φC11   TI.3.10

Sodium 2-undecylbenzenesulphonate, $C_9H_{19}CH(CH_3)C_6H_4SO_3Na$,
2φC11   TI.3.10

Sodium 2-(undecyl)-5-propylbenzenesulphonate,
$C_{11}H_{23}(C_3H_7)C_6H_3SO_3Na$ TI.3.14

Sodium undecylsulphate, $C_{11}H_{23}SO_4Na$ TI.1.4

Sodium 4-(undecylsulphonyl)benzenesulphonate,
$C_{11}H_{23}SO_2C_6H_4SO_3Na$ TI.3.11

Sodium 4-(undecylsulphoxy)benzenesulphonate,
$C_{11}H_{23}SOC_6H_4SO_3Na$ TI.3.11

1,1'(1,ω-Tetradecanediyl)bispyridinium bis(tetradecane-
sulphonate), $C_5H_5N(CH_2)_{14}NC_5H_5(C_{14}H_{29}SO_3)_2$, $C_{14}BP(C_{14})_2$ TI.2.21

Tetradecylhexaoxyethylene glycol ether, $C_{14}H_{29}O(CH_2CH_2O)_6H$,
$C_{14}E_6$ TIII.1.5

CMC (to be continued)

CMC (continued)

   in water

      Tetradecyloctaoxyethylene glycol ether, $C_{14}H_{29}O(CH_2CH_2O)_8H$,

        $C_{14}E_8$ TIII.1.5

      N-Tetradecylpyridinium bromide, $C_{14}H_{29}NC_5H_5Br$ TII.2.1

      Tetradecyltrimethylammonium bromide, $C_{14}H_{29}N(CH_3)_3Br$ TII.1.1

      Tetradecyltrimethylammonium hydroxide, $C_{14}H_{29}N(CH_3)_3OH$ TII.1.1

      Tetradecyltrimethylammonium nitrate, $C_{14}H_{29}N(CH_3)_3NO_3$ TII.1.1

      Tridecyldecaoxyethylene glycol ether, $C_{13}H_{27}O(CH_2CH_2O)_{10}H$,

        $C_{13}E_{10}$ TIII.1.5

      Tridecyldodecaoxyethylene glycol ether, $C_{13}H_{27}O(CH_2CH_2O)_{12}H$,

        $C_{13}E_{12}$ TIII.1.5

      Tridecyloctaoxyethylene glycol ether, $C_{13}H_{27}O(CH_2CH_2O)_8H$,

        $C_{13}E_8$ TIII.1.5

      N-Tridecylpyridinium bromide, $C_{13}H_{27}NC_5H_5Br$ TII.2.1

      Undecyloctaoxyethylene glycol ether, $C_{11}H_{23}O(CH_2CH_2O)_8H$, $C_{11}E_8$ TIII.1.5

      N-Undecylpyridinium bromide, $C_{11}H_{23}NC_5H_5Br$ TII.2.1

   in water/acetamide

      Octylphenol(ethylene oxide)$_{9.5}$ ether, $C_8H_{17}C_6H_4O(CH_2CH_2O)_{9.5}H$,

        $C_8\phi EO_{9.5}$ TIII.2.2

   in water/butanol

      Dodecyltrimethylammonium bromide, $C_{12}H_{25}N(CH_3)_3Br$ TII.2.1

   in water/butanol

      Hexadecyltrimethylammonium bromide, $C_{16}H_{33}N(CH_3)_3Br$ TII.2.1

   in water/α-cyclodextrin

      Octylphenol(ethylene oxide)$_{9.5}$ ether, $C_8H_{17}C_6H_4O(CH_2CH_2O)_{9.5}H$,

        $C_8\phi EO_{9.5}$ TIII.2.2

   in water/dioxane

      Dodecyltricosaoxyethylene glycol ether, $C_{12}H_{25}O(CH_2CH_2O)_{23}H$,

        $C_{12}E_{23}$ TIII.1.5

      Nonylphenol(ethylene oxide)$_{10}$ ether, $C_9H_{19}C_6H_4O(CH_2CH_2O)_{10}H$,

        $C_9\phi EO_{10}$ TIII.2.2

      Nonylphenol(ethylene oxide)$_{31}$ ether, $C_9H_{19}C_6H_4O(CH_2CH_2O)_{31}H$,

        $C_9\phi EO_{31}$ TIII.2.2

   in water - dodecyloctaoxyethylene glycol ether - NaCl

      Octadecyltrimethylammonium chloride, $C_{18}H_{37}N(CH_3)_3Cl$ TIII.1.5

   in water/eicosyltrimethylammonium cloride

      Dodecyloctaoxyethylene glycol ether, $C_{12}H_{25}O(CH_2CH_2O)_8H$,

        $C_{12}E_8$ TIII.1.5

   in water/ethanol

      Dodecylhexaoxyethylene glycol ether, $C_{12}H_{25}O(CH_2CH_2O)_6H$,

        $C_{12}E_6$ TIII.1.5

      Dodecyltricosaoxyethylene glycol ether, $C_{12}H_{25}O(CH_2CH_2O)_{23}H$,

        $C_{12}E_{23}$ TIII.1.5

      Dodecyltrimethylammonium bromide, $C_{12}H_{25}N(CH_3)_3Br$ TII.1.1

      Hexadecyltrimethylammonium bromide, $C_{16}H_{33}N(CH_3)_3Br$ TII.1.1

      Hexadecyltrimethylammonium chloride, $C_{16}H_{33}N(CH_3)_3Cl$ TII.1.1

      Hexadecyltrimethylammonium nitrate, $C_{16}H_{33}N(CH_3)_3NO_3$ TII.1.1

      Tetradecyltrimethylammonium bromide, $C_{14}H_{29}N(CH_3)_3Br$ TII.1.1

   in water/guanidinium chloride

      Nonylphenol(ethylene oxide)$_{10}$ ether, $C_9H_{19}C_6H_4O(CH_2CH_2O)_{10}H$,

        $C_9\phi EO_{10}$ TIII.2.2

      Nonylphenol(ethylene oxide)$_{31}$ ether, $C_9H_{19}C_6H_4O(CH_2CH_2O)_{31}H$,

        $C_9\phi EO_{31}$ TIII.2.2

CMC (to be continued)

CMC (continued)

   in water/hydrazine
     Sodium dodecylsulphate, $C_{12}H_{25}OSO_3Na$ TI.1.6
   in water - NaCl - KOH
     N-Hexadecylpyridinium chloride, $C_{16}H_{33}NC_5H_5Cl$ TII.2.1
   in water/octadecyltrimethylammonium chloride
     Dodecyloctaoxyethylene glycol ether, $C_{12}H_{25}O(CH_2CH_2O)_8H$,
       $C_{12}E_8$ TIII.1.5
   in water/polyethylene glycol
     Octylphenol(ethylene oxide)$_{9.5}$ ether, $C_8H_{17}C_6H_4O(CH_2CH_2O)_{9.5}H$,
       $C_8\phi EO_{9.5}$ TIII.2.2
   in water/sodium decylsulphate
     N-Hexylpyridinium bromide,$C_6H_{13}NC_5H_5Br$ TII.2.1
     N-Octylpyridinium bromide,$C_8H_{17}NC_5H_5Br$ TII.2.1
   in water - sodium decylsulphate - NaBr
     N-Hexylpyridinium bromide,$C_6H_{13}NC_5H_5Br$ TII.2.1
   in water/sodium dodecylsulphate
     N-Hexylpyridinium bromide,$C_6H_{13}NC_5H_5Br$ TII.2.1
     Nonylphenol(ethylene oxide)$_5$ ether, $C_9H_{19}C_6H_4O(CH_2CH_2O)_5H$,
       $C_9\phi EO_5$ TIII.2.2
     Nonylphenol(ethylene oxide)$_{10}$ ether, $C_9H_{19}C_6H_4O(CH_2CH_2O)_{10}H$,
       $C_9\phi EO_{10}$ TIII.2.2
     Nonylphenol(ethylene oxide)$_{15}$ ether, $C_9H_{19}C_6H_4O(CH_2CH_2O)_{15}H$,
       $C_9\phi EO_{15}$ TIII.2.2
   in water/sodium dodecylsulphate
     Nonylphenol(ethylene oxide)$_{20}$ ether, $C_9H_{19}C_6H_4O(CH_2CH_2O)_{20}H$,
       $C_9\phi EO_{20}$ TIII.2.2
     Nonylphenol(ethylene oxide)$_{30}$ ether, $C_9H_{19}C_6H_4O(CH_2CH_2O)_{30}H$,
       $C_9\phi EO_{30}$ TIII.2.2
     Octyldodecaoxyethylene glycol ether, $C_8H_{17}O(CH_2CH_2O)_{12}H$, C_8E_{12} TIII.1.5
     Octylhexaoxyethylene glycol ether, $C_8H_{17}O(CH_2CH_2O)_6H$, C_8E_6 TIII.1.5
     Octyltetraoxyethylene glycol ether, $C_8H_{17}O(CH_2CH_2O)_4H$, C_8E_4 TIII.1.5
   in water - sodium dodecylsulphate - NaCl
     Dodecyloctaoxyethylene glycol ether, $C_{12}H_{25}O(CH_2CH_2O)_8H$, $C_{12}E_8$ TIII.1.5
   in water/sodium 4-methylbenzenesulphonate (NaTos)
     N-Methyl-4-(dodecyl)pyridinium iodide, $C_{12}H_{25}C_5H_4NCH_3I$ TII.2.1
   in water/sodium octylsulphate
     N-Octylpyridinium bromide,$C_8H_{17}NC_5H_5Br$ TII.2.1
   in water - sodium octylsulphate - NaBr
     N-Octylpyridinium bromide,$C_8H_{17}NC_5H_5Br$ TII.2.1
   in water/sodium pentadecafluorooctanoate
     N-Hexylpyridinium bromide,$C_6H_{13}NC_5H_5Br$ TII.2.1
   in water - sodium pentadecafluorooctanoate - NaBr
     N-Hexylpyridinium bromide,$C_6H_{13}NC_5H_5Br$ TII.2.1
   in water/tetramethylammonium chloride
     Nonylphenol(ethylene oxide)$_{50}$ ether, $C_9H_{19}C_6H_4O(CH_2CH_2O)_{50}H$,
       $C_9\phi EO_{50}$ TIII.2.2
   in water/urea
     Dodecylheptaoxyethylene glycol ether, $C_{12}H_{25}O(CH_2CH_2O)_7H$,
       $C_{12}E_7$ TIII.1.5
     Dodecyltriacontaoxyethylene glycol ether, $C_{12}H_{25}O(CH_2CH_2O)_{30}H$,
       $C_{12}E_{30}$ TIII.1.5
     Hexadecyltriacontaoxyethylene glycol ether, $C_{16}H_{33}O(CH_2CH_2O)_{30}H$,
       $C_{16}E_{30}$ TIII.1.5

CMC (to be continued)

Enthalpy of micellization (continued)

   in aqueous solutions of NaCl

     Octylphenol(ethylene oxide)$_{16}$ ether, $C_8H_{17}C_6H_4O(CH_2CH_2O)_{16}H$,

      $C_8\phi EO_{16}$ TIII.2.8

     Octylphenol(ethylene oxide)$_{30}$ ether, $C_8H_{17}C_6H_4O(CH_2CH_2O)_{30}H$,

      $C_8\phi EO_{30}$ TIII.2.8

     Octylphenol(ethylene oxide)$_{40}$ ether, $C_8H_{17}C_6H_4O(CH_2CH_2O)_{40}H$,

      $C_8\phi EO_{40}$ TIII.2.8

     Sodium dodecylsulphate, $C_{12}H_{25}OSO_3Na$ TI.1.8

   in ethylammonium nitrate

     Octylphenol(ethylene oxide)$_{9.5}$ ether, $C_8H_{17}C_6H_4O(CH_2CH_2O)_{9.5}H$,

      $C_8\phi EO_{9.5}$ TIII.2.9

   in water

     1,1'(1,ω-Butanediyl)bispyridinium bis(tetradecanesulphonate),

      $C_5H_5N(CH_2)_4NC_5H_5(C_{14}H_{29}SO_3)_2$, $C_4BP(C_{14})_2$ TI.2.29

     Copper bis(tetradecanesulphonate), $(C_{14}H_{29}SO_3)_2Cu$ TI.2.29

     Decylhexaoxyethylene glycol ether, $C_{10}H_{21}O(CH_2CH_2O)_6H$, $C_{10}E_6$ TIII.1.8

     Decylnonaoxyethylene glycol ether, $C_{10}H_{21}O(CH_2CH_2O)_9H$, $C_{10}E_9$ TIII.1.8

     Decyloctaoxyethylene glycol ether, $C_{10}H_{21}O(CH_2CH_2O)_8H$, $C_{10}E_8$ TIII.1.8

     Decyltrimethylammonium bromide, $C_{10}H_{21}N(CH_3)_3Br$ TII.1.4

     Decyltrioxyethylene glycol ether, $C_{10}H_{21}O(CH_2CH_2O)_3H$, $C_{10}E_3$ TIII.1.8

     Dodecyldioxyethylene glycol ether, $C_{12}H_{25}O(CH_2CH_2O)_2H$, $C_{12}E_2$ TIII.1.8

     Dodecylheptaoxyethylene glycol ether, $C_{12}H_{25}O(CH_2CH_2O)_7H$,

      $C_{12}E_7$ TIII.1.8

     Dodecylhexaoxyethylene glycol ether, $C_{12}H_{25}O(CH_2CH_2O)_6H$, $C_{12}E_6$ TIII.1.8

     N-Dodecyl-4-methoxypyridinium bromide, $C_{12}H_{25}NC_5H_4OCH_3Br$ TII.2.13

     N-Dodecyl-4-methoxypyridinium chloride, $C_{12}H_{25}NC_5H_4OCH_3Cl$ TII.2.13

     Dodecyloctaoxyethylene glycol ether, $C_{12}H_{25}O(CH_2CH_2O)_8H$, $C_{12}E_8$ TIII.1.8

     Dodecylpentaoxyethylene glycol ether, $C_{12}H_{25}O(CH_2CH_2O)_5H$,

      $C_{12}E_5$ TIII.1.8

     N-Dodecylpyridinium bromide, $C_{12}H_{25}NC_5H_5Br$ TII.2.13

     N-Dodecylpyridinium chloride, $C_{12}H_{25}NC_5H_5Cl$ TII.2.13

     N-Dodecylpyridinium iodide, $C_{12}H_{25}NC_5H_5I$ TII.2.13

     Dodecyltetraoxyethylene glycol ether, $C_{12}H_{25}O(CH_2CH_2O)_4H$,

      $C_{12}E_4$ TIII.1.8

     Dodecyltriacontaoxyethylene glycol ether, $C_{12}H_{25}O(CH_2CH_2O)_{30}H$,

      $C_{12}E_{30}$ TIII.1.8

     Dodecyltrimethylammonium bromide, $C_{12}H_{25}N(CH_3)_3Br$ TII.1.4

     Dodecyltrioxyethylene glycol ether, $C_{12}H_{25}O(CH_2CH_2O)_3H$, $C_{12}E_3$ TIII.1.8

     1,1'(1,ω-Ethanediyl)bispyridinium bis(tetradecanesulphonate),

      $C_5H_5N(CH_2)_2NC_5H_5(C_{14}H_{29}SO_3)_2$, $C_2BP(C_{14})_2$ TI.2.29

     N-Hexadecylpyridinium bromide, $C_{16}H_{33}NC_5H_5Br$ FII.2.13

     N-Hexadecylpyridinium chloride, $C_{16}H_{33}NC_5H_5Cl$ FII.2.13

     Hexadecyltriacontaoxyethylene glycol ether,

      $C_{16}H_{33}O(CH_2CH_2O)_{30}H$, $C_{16}E_{30}$ TIII.1.8

     Hexadecyltrimethylammonium bromide, $C_{16}H_{33}N(CH_3)_3Br$ TII.1.4

     1,1'(1,ω-Hexanediyl)bispyridinium bis(tetradecanesulphonate),

      $C_5H_5N(CH_2)_6NC_5H_5(C_{14}H_{29}SO_3)_2$, $C_6BP(C_{14})_2$ TI.2.29

     Hexyltrioxyethylene glycol ether, $C_6H_{13}O(CH_2CH_2O)_3H$, C_6E_3 TIII.1.8

     N-Methyl-4-dodecyloxypyridinium bromide, $C_{12}H_{25}OC_5H_4NCH_3Br$ TII.2.13

     Methylviologen bis(tetradecanesulphonate),

      $CH_3NC_5H_4-C_5H_4NCH_3(C_{14}H_{29}SO_3)_2$ TI.2.29

     Nonylphenol(ethylene oxide)$_{9.5}$ ether, $C_9H_{19}C_6H_4O(CH_2CH_2O)_{10}H$,

      $C_9\phi EO_{9.5}$ TIII.2.8

Enthalpy of micellization (to be continued)

Enthalpy of micellization (continued)

   in water

      Nonylphenol(ethylene oxide)$_{11}$ ether, $C_9H_{19}C_6H_4O(CH_2CH_2O)_{11}H$,
      $C_9\phi EO_{11}$ TIII.2.8

      Nonylphenol(ethylene oxide)$_{15}$ ether, $C_9H_{19}C_6H_4O(CH_2CH_2O)_{15}H$,
      $C_9\phi EO_{15}$ TIII.2.8

      Nonylphenol(ethylene oxide)$_{30}$ ether, $C_9H_{19}C_6H_4O(CH_2CH_2O)_{30}H$,
      $C_9\phi EO_{30}$ TIII.2.8

      Octylhexaoxyethylene glycol ether, $C_8H_{17}O(CH_2CH_2O)_6H$, C_8E_6 TIII.1.8

      Octylnonaoxyethylene glycol ether, $C_8H_{17}O(CH_2CH_2O)_9H$, C_8E_9 TIII.1.8

      Octylphenol(ethylene oxide)$_{9.5}$ ether, $C_8H_{17}C_6H_4O(CH_2CH_2O)_{9.5}H$,
      $C_8\phi EO_{9.5}$ TIII.2.8

      Octylphenol(ethylene oxide)$_{12.5}$ ether, $C_8H_{17}C_6H_4O(CH_2CH_2O)_{12.5}H$,
      $C_8\phi EO_{12.5}$ TIII.2.8

      Octylphenol(ethylene oxide)$_{16}$ ether, $C_8H_{17}C_6H_4O(CH_2CH_2O)_{16}H$,
      $C_8\phi EO_{16}$ TIII.2.8

      Octylphenol(ethylene oxide)$_{30}$ ether, $C_8H_{17}C_6H_4O(CH_2CH_2O)_{30}H$,
      $C_8\phi EO_{30}$ TIII.2.8

      Octylphenol(ethylene oxide)$_{40}$ ether, $C_8H_{17}C_6H_4O(CH_2CH_2O)_{40}H$,
      $C_8\phi EO_{40}$ TIII.2.8

      Octyltrioxyethylene glycol ether, $C_8H_{17}O(CH_2CH_2O)_3H$, C_8E_3 TIII.1.8

      Pentadecyloctaoxyethylene glycol ether, $C_{15}H_{31}O(CH_2CH_2O)_8H$,
      $C_{15}E_8$ TIII.1.8

      Sodium decanesulphonate, $C_{10}H_{21}SO_3Na$ TI.2.26

      Sodium decylsulphate, $C_{10}H_{21}OSO_3Na$ TI.1.4

      Sodium 2-decylsulphate, $C_8H_{17}CH(CH_3)OSO_3Na$ TI.1.4

      Sodium dodecanesulphonate, $C_{12}H_{25}SO_3Na$ TI.2.26

      Sodium dodecylsulphate, $C_{12}H_{25}OSO_3Na$ TI.1.4

      Sodium hexadecanesulphonate, $C_{16}H_{33}SO_3Na$ TI.2.26

      Sodium hexanesulphonate, $C_6H_{13}SO_3Na$ TI.2.26

      Sodium octanesulphonate, $C_8H_{17}SO_3Na$ TI.2.26

      Sodium octylsulphate, $C_8H_{17}OSO_3Na$ TI.1.4

      Sodium tetradecanesulphonate, $C_{14}H_{29}SO_3Na$ TI.2.26

      Sodium tetradecylsulphate, $C_{14}H_{29}OSO_3Na$ TI.1.4

      Sodium 2-tetradecylsulphate, $C_{12}H_{25}CH(CH_3)OSO_3Na$ TI.1.4

      Sodium 4-tetradecylsulphate, $C_{10}H_{21}CH(C_3H_7)OSO_3Na$ TI.1.4

      1,1'(1,ω-Tetradecanediyl)bispyridinium bis(tetradecane-
      sulphonate), $C_5H_5N(CH_2)_{14}H_{29}SO_3)_2$, $C_{14}BP(C_{14})_2$ TI.2.2

      N-Tetradecyl-4-methoxypyridinium bromide, $C_{14}H_{29}NC_5H_4OCH_3Br$ TII.2.13

      Tetradecyloctaoxyethylene glycol ether, $C_{14}H_{29}O(CH_2CH_2O)_8H$,
      $C_{14}E_8$ TIII.1.8

      Tetradecyltrimethylammonium bromide, $C_{14}H_{29}N(CH_3)_3Br$ TII.1.4

      Tridecyloctaoxyethylene glycol ether, $C_{13}H_{27}O(CH_2CH_2O)_8H$,
      $C_{13}E_8$ TIII.1.8

      Undecyloctaoxyethylene glycol ether, $C_{11}H_{23}O(CH_2CH_2O)_8H$, $C_{11}E_8$ TIII.1.8

   in water/acetamide

      Octylphenol(ethylene oxide)$_{9.5}$ ether, $C_8H_{17}C_6H_4O(CH_2CH_2O)_{9.5}H$,
      $C_8\phi EO_{9.5}$ TIII.2.8

   in water/ethanol

      Dodecylhexaoxyethylene glycol ether, $C_{12}H_{25}O(CH_2CH_2O)_6H$, $C_{12}E_6$ TIII.1.8

   in water/ethylene glycol

      Nonylphenol(ethylene oxide)$_{30}$ ether, $C_9H_{19}C_6H_4O(CH_2CH_2O)_{30}H$,
      $C_9\phi EO_{30}$ TIII.2.8

Enthalpy of micellization (to be continued)

Gibbs free energy of micellization (continued)

   in water

Decylhexaoxyethylene glycol ether, $C_{10}H_{21}O(CH_2CH_2O)_6H$, $C_{10}E_6$	TIII.1.7
Decylnonaoxyethylene glycol ether, $C_{10}H_{21}O(CH_2CH_2O)_9H$, $C_{10}E_9$	TIII.1.7
Decyloctaoxyethylene glycol ether, $C_{10}H_{21}O(CH_2CH_2O)_8H$, $C_{10}E_8$	TIII.1.7
N-Decylpyridinium chloride, $C_{10}H_{21}NC_5H_5Cl$	TII.2.13
Decyltrioxyethylene glycol ether, $C_{10}H_{21}O(CH_2CH_2O)_3H$, $C_{10}E_3$	TIII.1.7
Dodecyldioxyethylene glycol ether, $C_{12}H_{25}O(CH_2CH_2O)_2H$, $C_{12}E_2$	TIII.1.7
Dodecylheptaoxyethylene glycol ether, $C_{12}H_{25}O(CH_2CH_2O)_7H$, $C_{12}E_7$	TIII.1.7
Dodecylhexaoxyethylene glycol ether, $C_{12}H_{25}O(CH_2CH_2O)_6H$, $C_{12}E_6$	TIII.1.7
N-Dodecyl-4-methoxypyridinium bromide, $C_{12}H_{25}NC_5H_4OCH_3Br$	TII.2.13
N-Dodecyl-4-methoxypyridinium chloride, $C_{12}H_{25}NC_5H_4OCH_3Cl$	TII.2.13
Dodecyloctaaoxyethylene glycol ether, $C_{12}H_{25}O(CH_2CH_2O)_8H$, $C_{12}E_8$	TIII.1.7
Dodecylpentaoxyethylene glycol ether, $C_{12}H_{25}O(CH_2CH_2O)_5H$, $C_{12}E_5$	TIII.1.7
N-Dodecylpyridinium bromide, $C_{12}H_{25}NC_5H_5Br$	TII.2.13
N-Dodecylpyridinium chloride, $C_{12}H_{25}NC_5H_5Cl$	TII.2.13
N-Dodecylpyridinium iodide, $C_{12}H_{25}NC_5H_5I$	TII.2.13
Dodecyltetraoxyethylene glycol ether, $C_{12}H_{25}O(CH_2CH_2O)_4H$, $C_{12}E_4$	TIII.1.7
Dodecyltrimethylammonium bromide, $C_{12}H_{25}N(CH_3)_3Br$	TII.1.4
Dodecyltrioxyethylene glycol ether, $C_{12}H_{25}O(CH_2CH_2O)_3H$, $C_{12}E_3$	TIII.1.7
N-Hexadecylpyridinium chloride, $C_{16}H_{33}NC_5H_5Cl$	TII.2.13
Hexadecyltriacontaoxyethylene glycol ether, $C_{16}H_{33}O(CH_2CH_2O)_{30}H$, $C_{16}E_{30}$	TIII.1.7
Hexadecyltrimethylammonium bromide, $C_{16}H_{33}N(CH_3)_3Br$	TII.1.4
Hexyltrioxyethylene glycol ether, $C_6H_{13}O(CH_2CH_2O)_3H$, C_6E_3	TIII.1.7
N-Methyl-4-dodecyloxypyridinium bromide, $C_{12}H_{25}OC_5H_4NCH_3Br$	TII.2.13
Nonylphenol(ethylene oxide)$_9$ ether, $C_9H_{19}C_6H_4O(CH_2CH_2O)_9H$, $C_9\phi EO_9$	TIII.2.7
Nonylphenol(ethylene oxide)$_{30}$ ether, $C_9H_{19}C_6H_4O(CH_2CH_2O)_{30}H$, $C_9\phi EO_{30}$	TIII.2.6
Octylhexaoxyethylene glycol ether, $C_8H_{17}O(CH_2CH_2O)_6H$, C_8E_6	TIII.1.7
Octylnonaoxyethylene glycol ether, $C_8H_{17}O(CH_2CH_2O)_9H$, C_8E_9	TIII.1.7
Octylphenol(ethylene oxide)$_{9.5}$ ether, $C_8H_{17}C_6H_4O(CH_2CH_2O)_{9.5}H$, $C_8\phi EO_{9.5}$	TIII.2.6
Octylphenol(ethylene oxide)$_{12.5}$ ether, $C_8H_{17}C_6H_4O(CH_2CH_2O)_{12.5}H$, $C_8\phi EO_{12.5}$	TIII.2.6
Octyltrioxyethylene glycol ether, $C_8H_{17}O(CH_2CH_2O)_3H$, C_8E_3	TIII.1.7
Pentadecyloctaoxyethylene glycol ether, $C_{15}H_{31}O(CH_2CH_2O)_8H$, $C_{15}E_8$	TIII.1.7
Sodium decanesulphonate, $C_{10}H_{21}SO_3Na$	TI.2.25 & TI.2.28
Sodium dodecanesulphonate, $C_{12}H_{25}SO_3Na$	TI.2.25
Sodium 3-dodecanesulphonate, $C_9H_{19}CH(C_2H_5)SO_3Na$	TI.2.28
Sodium 7-hexadecanesulphonate, $C_9H_{19}CH(C_6H_{13})SO_3Na$	TI.2.28
Sodium 10-nonadecanesulphonate, $C_9H_{19}CH(C_9H_{19})SO_3Na$	TI.2.28
Sodium octanesulphonate, $C_8H_{17}SO_3Na$	TI.2.25
Sodium 9-octadecanesulphonate, $C_9H_{19}CH(C_8H_{17})SO_3Na$	TI.2.28
Sodium 5-pentadecanesulphonate, $C_9H_{19}CH(C_4H_9)SO_3Na$	TI.2.28
N-Tetradecyl-4-methoxypyridinium bromide, $C_{14}H_{29}NC_5H_4OCH_3Br$	TII.2.13

Gibbs free energy of micellization (to be continued)

Krafft temperature (continued)

in water

Krafft temperature (to be continued)

Krafft temperature (continued)

in water

Sodium (dodecyl)naphthalenesulphonate, $C_{12}H_{25}C_{10}H_6SO_3Na$	TI.3.4
Sodium dodecylsulphate, $C_{12}H_{25}OSO_3Na$	TI.1.1
Sodium eicosanesulphonate, $C_{20}H_{41}SO_3Na$	TI.2.1
Sodium heptadecanesulphonate, $C_{17}H_{35}SO_3Na$	TI.2.1
Sodium heptadecylsulphate, $C_{17}H_{35}OSO_3Na$	TI.1.1
Sodium 4-(heptyl)benzenesulphonate, $C_7H_{15}C_6H_4SO_3Na$, 1φC7	TI.3.1
Sodium hexadecanesulphonate, $C_{16}H_{33}SO_3Na$	TI.2.1
Sodium 7-hexadecanesulphonate, $C_9H_{19}CH(C_6H_{13})SO_3Na$	TI.2.2
Sodium 2-hexadecenesulphonate, $C_{13}H_{27}CH=CHCH_2SO_3Na$	TI.2.4
Sodium 4-(hexadecyl)benzenesulphonate, $C_{16}H_{33}C_6H_4SO_3Na$, 1φC16	TI.3.1
Sodium (hexadecyl)naphthalenesulphonate, $C_{16}H_{33}C_{10}H_6SO_3Na$	TI.3.4
Sodium hexadecylsulphate, $C_{16}H_{33}OSO_3Na$	TI.1.1
Sodium 2-hydroxydecanesulphonate, $C_8H_{17}CHOHCH_2SO_3Na$	TI.2.4
Sodium 1-hydroxydodecanesulphonate, $C_{11}H_{23}CHOHSO_3Na$	TI.2.4
Sodium 2-hydroxydodecanesulphonate, $C_{10}H_{21}CHOH_2SO_3Na$	TI.2.4
Sodium 2-hydroxyheptadecanesulphonate, $C_{15}H_{31}CHOH_2SO_3Na$	TI.2.4
Sodium 1-hydroxyhexadecanesulphonate, $C_{15}H_{31}CHOHSO_3Na$	TI.2.4
Sodium 3-hydroxyhexadecanesulphonate, $C_{13}H_{27}CHOHCH_2CH_2SO_3Na$	TI.2.4
Sodium 1-hydroxyoctadecanesulphonate, $C_{17}H_{35}CHOHSO_3Na$	TI.2.4
Sodium 3-hydroxyoctadecanesulphonate, $C_{15}H_{31}CHOHCH_2CH_2SO_3Na$	TI.2.4
Sodium 2-hydroxypentadecanesulphonate, $C_{13}H_{27}CHOHCH_2SO_3Na$	TI.2.4
Sodium 3-hydroxypentadecanesulphonate, $C_{12}H_{25}CHOHCH_2CH_2SO_3Na$	TI.2.3
Sodium 1-hydroxytetradecanesulphonate, $C_{13}H_{27}CHOHSO_3Na$	TI.2.4
Sodium 3-hydroxytetradecanesulphonate, $C_{11}H_{23}CHOHCH_2CH_2SO_3Na$	TI.2.4
Sodium 2-hydroxytridecanesulphonate, $C_{11}H_{23}CHOHCH_2SO_3Na$	TI.2.4
Sodium 2-hydroxyundecanesulphonate, $C_9H_{19}CHOHCH_2SO_3Na$	TI.2.4
Sodium 2-methylheptadecylsulphate, $C_{15}H_{31}CH(CH_3)CH_2OSO_3Na$	TI.1.1
Sodium 2-methylpentadecylsulphate, $C_{13}H_{27}CH(CH_3)CH_2OSO_3Na$	TI.1.1
Sodium 2-methyltridecylsulphate, $C_{11}H_{23}CH(CH_3)CH_2OSO_3Na$	TI.1.1
Sodium 2-methylundecylsulphate, $C_9H_{19}CH(CH_3)CH_2OSO_3Na$	TI.1.1
Sodium 4-(3-nonyl)benzenesulphonate, $3-C_9H_{19}C_6H_4SO_3Na$, 3φC9	TI.3.1
Sodium 10-nonadecanesulphonate, $C_9H_{19}CH(C_9H_{19})SO_3Na$	TI.2.2
Sodium octadecanesulphonate, $C_{18}H_{37}SO_3Na$	TI.2.1
Sodium 9-octadecanesulphonate, $C_9H_{19}CH(C_8H_{17})SO_3Na$	TI.2.2
Sodium 2-octadecenesulphonate, $C_{15}H_{31}CH=CHCH_2SO_3Na$	TI.2.4
Sodium 4-(octadecyl)benzenesulphonate, $C_{18}H_{37}C_6H_4SO_3Na$, 1φC18	TI.3.1
Sodium octadecylsulphate, $C_{18}H_{37}OSO_3Na$	TI.1.1
Sodium 4-(octyl)benzenesulphonate, $C_8H_{17}C_6H_4SO_3Na$, 1φC8	TI.3.1
Sodium (octyl)naphthalenesulphonate, $C_8H_{17}C_{10}H_8SO_3Na$	TI.3.4
Sodium 2-oxododecanesulphonate, $C_{10}H_{21}COCH_2SO_3Na$	TI.2.4
Sodium 3-oxododecanesulphonate, $C_9H_{19}COCH_2CH_2SO_3Na$	TI.2.4
Sodium 2-oxoheptadecanesulphonate, $C_{15}H_{31}COCH_2SO_3Na$	TI.2.4
Sodium 3-oxohexadecanesulphonate, $C_{13}H_{27}COCH_2CH_2SO_3Na$	TI.2.4
Sodium 3-oxooctadecanesulphonate, $C_{15}H_{31}COCH_2CH_2SO_3Na$	TI.2.4
Sodium 2-oxopentadecanesulphonate, $C_{13}H_{27}COCH_2SO_3Na$	TI.2.4
Sodium 3-oxotetradecanesulphonate, $C_{11}H_{23}COCH_2CH_2SO_3Na$	TI.2.4
Sodium 2-oxotridecanesulphonate, $C_{11}H_{23}COCH_2SO_3Na$	TI.2.4
Sodium pentadecanesulphonate, $C_{15}H_{31}SO_3Na$	TI.2.1
Sodium 2-pentadecenesulphonate, $C_{12}H_{25}CH=CHCH_2SO_3Na$	TI.2.3
Sodium 4-(pentadecyl)benzenesulphonate, $C_{15}H_{31}C_6H_4SO_3Na$, 1φC15	TI.3.1
Sodium pentadecylsulphate, $C_{15}H_{31}OSO_3Na$	TI.1.1

Krafft temperature (to be continued)

Melting point (continued)

Dodecyldioxyethylene glycol ether, $C_{12}H_{25}O(CH_2CH_2O)_2H$, $C_{12}E_2$	TIII.1.1
1,1'-(1,ω-Dodecanediyl)bispyridinium bis(tetradecane- sulphonate), $C_5H_5N(CH_2)_{12}NC_5H_5(C_{14}H_{29}SO_3)_2$, $C_{12}BP(C_{14})_2$	TI.2.9
Dodecylhexaoxyethylene glycol ether, $C_{12}H_{25}O(CH_2CH_2O)_6H$, $C_{12}E_6$	TIII.1.1
Dodecylmonooxyethylene glycol ether, $C_{12}H_{25}OCH_2CH_2OH$, $C_{12}E_1$	TIII.1.1
Dodecylpentaoxyethylene glycol ether, $C_{12}H_{25}O(CH_2CH_2O)_5H$, $C_{12}E_5$	TIII.1.1
Dodecyltetraoxyethylene glycol ether, $C_{12}H_{25}O(CH_2CH_2O)_4H$, $C_{12}E_4$	TIII.1.1
Dodecyltrioxyethylene glycol ether, $C_{12}H_{25}O(CH_2CH_2O)_3H$, $C_{12}E_3$	TIII.1.1
1,1'-(1,ω-Ethanediyl)bispyridinium bis(tetradecanesulphonate), $C_5H_5N(CH_2)_2NC_5H_5(C_{14}H_{29}SO_3)_2$, $C_2BP(C_{14})_2$	TI.2.9
Hexadecanol, $C_{16}H_{33}OH$, $C_{16}E_0$	TIII.1.1
Hexadecyldioxyethylene glycol ether, $C_{16}H_{33}O(CH_2CH_2O)_2H$, $C_{16}E_2$	TIII.1.1
Hexadecyldodecaoxyethylene glycol ether, $C_{16}H_{33}O(CH_2CH_2O)_{12}H$, $C_{16}E_{12}$	TIII.1.1
Hexadecylheneicosaoxyethylene glycol ether, $C_{16}H_{33}O(CH_2CH_2O)_{21}H$, $C_{16}E_{21}$	TIII.1.1
Hexadecylheptaoxyethylene glycol ether, $C_{16}H_{33}O(CH_2CH_2O)_7H$, $C_{16}E_7$	TIII.1.1
Hexadecylhexaoxyethylene glycol ether, $C_{16}H_{33}O(CH_2CH_2O)_6H$, $C_{16}E_6$	TIII.1.1
Hexadecylmonooxyethylene glycol ether, $C_6H_{33}OCH_2CH_2OH$, C_6E_1	TIII.1.1
Hexadecylnonaoxyethylene glycol ether, $C_{16}H_{33}O(CH_2CH_2O)_9H$, $C_{16}E_9$	TIII.1.1
Hexadecyloctaoxyethylene glycol ether, $C_{16}H_{33}O(CH_2CH_2O)_8H$, $C_{16}E_8$	TIII.1.1
Hexadecylpentadecaoxyethylene glycol ether, $C_{16}H_{33}O(CH_2CH_2O)_{15}H$, $C_{16}E_{15}$	TIII.1.1
Hexadecylpentaoxyethylene glycol ether, $C_{16}H_{33}O(CH_2CH_2O)_5H$, $C_{16}E_5$	TIII.1.1
Hexadecyltetraoxyethylene glycol ether, $C_{16}H_{33}O(CH_2CH_2O)_4H$, $C_{16}E_4$	TIII.1.1
Hexadecyltrioxyethylene glycol ether, $C_{16}H_{33}O(CH_2CH_2O)_3H$, $C_{16}E_3$	TIII.1.1
Hexanol, $C_6H_{13}OH$, C_6E_0	TIII.1.1
Hexyldioxyethylene glycol ether, $C_6H_{13}O(CH_2CH_2O)_2H$, C_8E_2	TIII.1.1
1,1'-(1,ω-Hexanediyl)bispyridinium bis(tetradecanesulphonate), $C_5H_5N(CH_2)_6NC_5H_5(C_{14}H_{29}SO_3)_2$, $C_6BP(C_{14})_2$	TI.2.9
Hexylhexaoxyethylene glycol ether, $C_6H_{13}O(CH_2CH_2O)_6H$, C_8E_6	TIII.1.1
Hexylmonooxyethylene glycol ether, $C_6H_{13}OCH_2CH_2OH$, C_8E_1	TIII.1.1
Hexyltetraoxyethylene glycol ether, $C_6H_{13}O(CH_2CH_2O)_4H$, C_6E_4	TIII.1.1
Hexyltrioxyethylene glycol ether, $C_6H_{13}O(CH_2CH_2O)_3H$, C_6E_3	TIII.1.1
Methylviologen bistetradecanesulphonate, $CH_3NC_5H_4-C_5H_4NCH_3(C_{14}H_{29}SO_3)_2$	TI.2.9
Octadecanol, $C_{18}H_{37}OH$, $C_{18}E_0$	TIII.1.1
Octadecyldioxyethylene glycol ether, $C_{18}H_{37}O(CH_2CH_2O)_2H$, $C_{18}E_2$	TIII.1.1
Octadecylmonooxyethylene glycol ether, $C_{18}H_{37}OCH_2CH_2OH$, $C_{18}E_1$	TIII.1.1
Octadecyltetraoxyethylene glycol ether, $C_{18}H_{37}O(CH_2CH_2O)_4H$, $C_{18}E_4$	TIII.1.1
Octadecyltrioxyethylene glycol ether, $C_{18}H_{37}O(CH_2CH_2O)_3H$, $C_{18}E_3$	TIII.1.1
Octanol, $C_8H_{17}OH$, C_8E_0	TIII.1.1

Melting point (to be continued)

Micellar mass (continued)

   in aqueous solutions of NaCl

     Sodium pentadecanesulphonate, $C_{15}H_{31}SO_3Na$ TI.2.36

     Tridecyldecaoxyethylene glycol ether, $C_{13}H_{27}O(CH_2CH_2O)_{10}H$,
      $C_{13}E_{10}$ TIII.1.12

     Tridecyldocosaoxyethylene glycol ether, $C_{13}H_{27}O(CH_2CH_2O)_{22}H$,
      $C_{13}E_{22}$ TIII.1.12

     Tridecylpentadecaoxyethylene glycol ether, $C_{13}H_{27}O(CH_2CH_2O)_{15}H$,
      $C_{13}E_{15}$ TIII.1.12

   in aqueous solutions of Na_2SO_4

     Dodecyldodecaoxyethylene glycol ether, $C_{12}H_{25}O(CH_2CH_2O)_{12}H$,
      $C_{12}E_{12}$ TIII.1.12

     Dodecyloctadecaoxyethylene glycol ether, $C_{12}H_{25}O(CH_2CH_2O)_{18}H$,
      $C_{12}E_{18}$ TIII.1.12

     Dodecyloctaoxyethylene glycol ether, $C_{12}H_{25}O(CH_2CH_2O)_8H$, $C_{12}E_8$ TIII.1.12

     Dodecyltricosaoxyethylene glycol ether, $C_{12}H_{25}O(CH_2CH_2O)_{23}H$,
      $C_{12}E_{23}$ TIII.1.12

   in water

     Decylhexaoxyethylene glycol ether, $C_{10}H_{21}O(CH_2CH_2O)_6H$, $C_{10}E_6$ TIII.1.12

     Dodecyldodecaoxyethylene glycol ether, $C_{12}H_{25}O(CH_2CH_2O)_{12}H$,
      $C_{12}E_{12}$ TIII.1.12

     Dodecylhexaoxyethylene glycol ether, $C_{12}H_{25}O(CH_2CH_2O)_6H$, $C_{12}E_6$ TIII.1.12

     Dodecyloctadecaoxyethylene glycol ether, $C_{12}H_{25}O(CH_2CH_2O)_{18}H$,
      $C_{12}E_{18}$ TIII.1.12

     Dodecyloctaoxyethylene glycol ether, $C_{12}H_{25}O(CH_2CH_2O)_8H$, $C_{12}E_8$ TIII.1.12

     Dodecyltricosaoxyethylene glycol ether, $C_{12}H_{25}O(CH_2CH_2O)_{23}H$,
      $C_{12}E_{23}$ TIII.1.12

     Dodecyltrimethylammonium bromide, $C_{12}H_{25}N(CH_3)_3Br$ TII.1.11

     Hexadecyldodecaoxyethylene glycol ether, $C_{16}H_{33}O(CH_2CH_2O)_{12}H$,
      $C_{16}E_{12}$ TIII.1.12

     Hexadecylheinocosaoxyethylene glycol ether, $C_{16}H_{33}O(CH_2CH_2O)_{21}H$,
      $C_{16}E_{21}$ TIII.1.12

     Hexadecylheptaoxyethylene glycol ether, $C_{16}H_{33}O(CH_2CH_2O)_7H$,
      $C_{16}E_7$ TIII.1.12

     Hexadecylhexaoxyethylene glycol ether, $C_{16}H_{33}O(CH_2CH_2O)_6H$,
      $C_{16}E_6$ TIII.1.12

     Hexadecylnonaoxyethylene glycol ether, $C_{16}H_{33}O(CH_2CH_2O)_9H$,
      $C_{16}E_9$ TIII.1.12

     Hexadecyloctaoxyethylene glycol ether, $C_{16}H_{33}O(CH_2CH_2O)_8H$,
      $C_{16}E_8$ TIII.1.12

     Hexadecyltrimethylammonium bromide, $C_{16}H_{33}N(CH_3)_3Br$ TII.1.11

     Octylhexaoxyethylene glycol ether, $C_8H_{17}O(CH_2CH_2O)_6H$, C_8E_6 TIII.1.12

     Octyltetraoxyethylene glycol ether, $C_8H_{17}O(CH_2CH_2O)_4H$, C_8E_4 TIII.1.12

     Tetradecylhexaoxyethylene glycol ether, $C_{14}H29_{33}O(CH_2CH_2O)_6H$,
      $C_{14}E_6$ TIII.1.12

     Tetradecyltrimethylammonium bromide, $C_{14}H_{29}N(CH_3)_3Br$ TII.1.11

     Tridecyldecaoxyethylene glycol ether, $C_{13}H_{27}O(CH_2CH_2O)_{10}H$,
      $C_{13}E_{10}$ TIII.1.12

     Tridecyldocosaoxyethylene glycol ether, $C_{13}H_{27}O(CH_2CH_2O)_{22}H$,
      $C_{13}E_{22}$ TIII.1.12

     Tridecyldodecaoxyethylene glycol ether, $C_{13}H_{27}O(CH_2CH_2O)_{12}H$,
      $C_{13}E_{12}$ TIII.1.12

     Tridecyloctaoxyethylene glycol ether, $C_{13}H_{27}O(CH_2CH_2O)_8H$,
      $C_{13}E_8$ TIII.1.12

Micellar mass (to be continued)

594

Phase diagram (continued)

hexyltrioxyethylene glycol ether
    Hexyltrioxyethylene glycol ether ($C_6H_{13}O(CH_2CH_2O)_3H$, C_6E_3)/
    water   FIII.1.3
                                       & FIII.1.4

    Hexyltrioxyethylene glycol ether ($C_6H_{13}O(CH_2CH_2O)_3H$, C_6E_3)
    - water - decane   FIII.1.44
                                       & FIII.1.45

nonylphenol(ethylene oxide)$_{6.2}$ ether
    Nonylphenol(ethylene oxide)$_{6.2}$ ether ($C_9H_{19}C_6H_4O(CH_2CH_2O)_{6.2}H$,
    $C_9\phi EO_{6.2}$)/hexadecane   FIII.2.15
nonylphenol(ethylene oxide)$_{8.6}$ ether
    Nonylphenol(ethylene oxide)$_{8.6}$ ether ($C_9H_{19}C_6H_4O(CH_2CH_2O)_{8.6}H$,
    $C_9\phi EO_{8.6}$) - water - cyclohexane   FIII.2.21
nonylphenol(ethylene oxide)$_{9.6}$ ether
    Nonylphenol(ethylene oxide)$_{9.6}$ ether ($C_9H_{19}C_6H_4O(CH_2CH_2O)_{9.6}H$,
    $C_9\phi EO_{9.6}$)/hexadecane   FIII.2.15
nonylphenol(ethylene oxide)$_{10}$ ether
    Nonylphenol(ethylene oxide)$_{10}$ ether ($C_9H_{19}C_6H_4O(CH_2CH_2O)_{10}H$,
    $C_9\phi EO_{10}$)/water   FIII.2.6
                                       & FIII.2.7
octanol
    Hexyldioxyethylene glycol ether ($C_6H_{13}O(CH_2CH_2O)_2H$, C_6E_2)
    - water - octanol   FIII.1.43
    Hexylhexaoxyethylene glycol ether ($C_6H_{13}O(CH_2CH_2O)_6H$, C_6E_6)
    - water - octanol   FIII.1.47
    Octylhexaoxyethylene glycol ether ($C_8H_{17}O(CH_2CH_2O)_6H$, C_8E_6)
    - water - dodecane - octanol   FIII.1.50
octylhexaoxyethylene glycol ether
    Octylhexaoxyethylene glycol ether ($C_8H_{17}O(CH_2CH_2O)_6H$, C_8E_6)/
    water   FIII.1.8
                                       & FIII.1.9

    Octylhexaoxyethylene glycol ether ($C_8H_{17}O(CH_2CH_2O)_6H$, C_8E_6)
    - water - dodecane - octanol   FIII.1.50
octylpentaoxyethylene glycol ether
    Octylpentaoxyethylene glycol ether ($C_8H_{17}O(CH_2CH_2O)_5H$, C_8E_5)/
    water   FIII.1.5
                                       & FIII.1.7

octylphenol(ethylene oxide)$_{7.5}$ ether
    Octylphenol(ethylene oxide)$_{7.5}$ ether ($C_8H_{17}C_6H_4O(CH_2CH_2O)_{7.5}H$,
    $C_8\phi EO_{7.5}$)/water   FIII.2.1
octylphenol(ethylene oxide)$_{9.5}$ ether
    Octylphenol(ethylene oxide)$_{9.5}$ ether ($C_8H_{17}C_6H_4O(CH_2CH_2O)_{9.5}H$,
    $C_8\phi EO_{9.5}$)/deuterium oxide   FIII.2.2
    Octylphenol(ethylene oxide)$_{9.5}$ ether ($C_8H_{17}C_6H_4O(CH_2CH_2O)_{9.5}H$,
    $C_8\phi EO_{9.5}$)/water   FIII.2.2
    Octylphenol(ethylene oxide)$_{9.5}$ ether ($C_8H_{17}C_6H_4O(CH_2CH_2O)_{9.5}H$,
    $C_8\phi EO_{9.5}$) - water - benzene   FIII.2.18
    Octylphenol(ethylene oxide)$_{9.5}$ ether ($C_8H_{17}C_6H_4O(CH_2CH_2O)_{9.5}H$,
    $C_8\phi EO_{9.5}$) - water - decanol   FIII.2.20
    Octylphenol(ethylene oxide)$_{9.5}$ ether ($C_8H_{17}C_6H_4O(CH_2CH_2O)_{9.5}H$,
    $C_8\phi EO_{9.5}$) - water - toluene   FIII.2.19

Phase diagram (to be continued)

Phase inversion temperature (to be continued)

Phase inversion temperature (continued)

   in water/cyclohexane

      Hexadecylphenol(ethylene oxide)$_{12.4}$ ether,

        $C_{16}H_{33}C_6H_4O(CH_2CH_2O)_{12.4}H$, $C_{16}\phi EO_{12.4}$ TIII.2.25

      Hexylphenol(ethylene oxide)$_{7.5}$ ether, $C_6H_{13}C_6H_4O(CH_2CH_2O)_{7.5}H$,

        $C_6\phi EO_{7.5}$ TIII.2.25

      Nonylphenol(ethylene oxide)$_6$ ether, $C_9H_{19}C_6H_4O(CH_2CH_2O)_6H$,

        $C_9\phi EO_6$ in mixture with nonylphenol(ethylene oxide)$_8$ ether   FIII.2.31

      Nonylphenol(ethylene oxide)$_{6.2}$ ether, $C_9H_{19}C_6H_4O(CH_2CH_2O)_{6.2}H$,

        $C_9\phi EO_{6.2}$ in mixture with nonylphenol(ethylene oxide)$_{9.6}$

        ether   FIII.2.28

      Nonylphenol(ethylene oxide)$_8$ ether, $C_9H_{19}C_6H_4O(CH_2CH_2O)_8H$,

        $C_9\phi EO_8$ in mixture with nonylphenol(ethylene oxide)$_6$ ether   FIII.2.31

      Nonylphenol(ethylene oxide)$_{8.1}$ ether, $C_9H_{19}C_6H_4O(CH_2CH_2O)_{8.1}H$,

        $C_9\phi EO_{8.1}$ TIII.2.25

      Nonylphenol(ethylene oxide)$_{8.6}$ ether, $C_9H_{19}C_6H_4O(CH_2CH_2O)_{8.6}H$,

        $C_9\phi EO_{8.6}$ in mixture with calcium dodecylbenzenesulphonate   TIII.2.25

      Nonylphenol(ethylene oxide)$_{9.6}$ ether, $C_9H_{19}C_6H_4O(CH_2CH_2O)_{9.6}H$,

        $C_9\phi EO_{9.6}$ in mixture with nonylphenol(ethylene oxide)$_{6.2}$ ether   FIII.2.28

      Nonylphenol(ethylene oxide)$_{17.7}$ ether, $C_9H_{19}C_6H_4O(CH_2CH_2O)_{17.7}H$,

        $C_9\phi EO_{17.7}$ FIII.2.27

      Octylphenol(ethylene oxide)$_{8.4}$ ether, $C_8H_{17}C_6H_4O(CH_2CH_2O)_{8.4}H$,

        $C_8\phi EO_{8.4}$ TIII.2.25

      Octylphenol(ethylene oxide)$_{8.5}$ ether, $C_8H_{17}C_6H_4O(CH_2CH_2O)_{8.5}H$,

        $C_8\phi EO_{8.5}$ TIII.2.25

   in water/cyclohexane/heptane

      Nonylphenol(ethylene oxide)$_{6.2}$ ether, $C_9H_{19}C_6H_4O(CH_2CH_2O)_{6.2}H$,

        $C_9\phi EO_{6.2}$ FIII.2.29

      Nonylphenol(ethylene oxide)$_{7.4}$ ether, $C_9H_{19}C_6H_4O(CH_2CH_2O)_{7.4}H$,

        $C_9\phi EO_{7.4}$ FIII.2.29

      Nonylphenol(ethylene oxide)$_{8.1}$ ether, $C_9H_{19}C_6H_4O(CH_2CH_2O)_{8.1}H$,

        $C_9\phi EO_{8.1}$ FIII.2.29

      Nonylphenol(ethylene oxide)$_{9.6}$ ether, $C_9H_{19}C_6H_4O(CH_2CH_2O)_{9.6}H$,

        $C_9\phi EO_{9.6}$ FIII.2.29

      Nonylphenol(ethylene oxide)$_{15.8}$ ether, $C_9H_{19}C_6H_4O(CH_2CH_2O)_{15.8}H$,

        $C_9\phi EO_{15.8}$ FIII.2.29

   in water/decane

      Dodecylpentaoxyethylene glycol ether, $C_{12}H_{25}O(CH_2CH_2O)_5H$,

        $C_{12}E_5$ TIII.1.30

   in water/difluorotetrachloroethane

      Nonylphenol(ethylene oxide)$_{9.6}$ ether, $C_9H_{19}C_6H_4O(CH_2CH_2O)_{9.6}H$,

        $C_9\phi EO_{9.6}$ FIII.2.24

   in water/dodecene

      Nonylphenol(ethylene oxide)$_{9.6}$ ether, $C_9H_{19}C_6H_4O(CH_2CH_2O)_{9.6}H$,

        $C_9\phi EO_{9.6}$ FIII.2.23

   in water/ethyl oleate

      Nonylphenol(ethylene oxide)$_{9.6}$ ether, $C_9H_{19}C_6H_4O(CH_2CH_2O)_{9.6}H$,

        $C_9\phi EO_{9.6}$ FIII.2.24

   in water/heptane

      Dodecylphenol(ethylene oxide)$_9$ ether, $C_{12}H_{25}C_6H_4O(CH_2CH_2O)_9H$,

        $C_{12}\phi EO_9$ FIII.2.37

      Dodecyltetraoxyethylene glycol ether, $C_{12}H_{25}O(CH_2CH_2O)_4H$,

        $C_{12}\phi EO_4$ TIII.1.30

Phase inversion temperature (to be continued)

Phase inversion temperature (continued)

   in water/heptane

      Nonylphenol(ethylene oxide)$_{6.2}$ ether, $C_9H_{19}C_6H_4O(CH_2CH_2O)_{6.2}H$,

      $C_9\phi EO_{6.2}$ FIII.2.29

      Nonylphenol(ethylene oxide)$_{9.6}$ ether, $C_9H_{19}C_6H_4O(CH_2CH_2O)_{9.6}H$,

      $C_9\phi EO_{9.6}$ FIII.2.24

   in water/hexadecane

      Dodecylpentaoxyethylene glycol ether, $C_{12}H_{25}O(CH_2CH_2O)_5H$,

      $C_{12}E_5$ TIII.1.30

      Dodecylphenol(ethylene oxide)$_9$ ether, $C_{12}H_{25}C_6H_4O(CH_2CH_2O)_9H$,

      $C_{12}\phi EO_9$ FIII.2.37

      Dodecyltetraoxyethylene glycol ether, $C_{12}H_{25}O(CH_2CH_2O)_4H$,

      $C_{12}E_4$ TIII.1.30

      Nonylphenol(ethylene oxide)$_{6.2}$ ether, $C_9H_{19}C_6H_4O(CH_2CH_2O)_{6.2}H$,

      $C_9\phi EO_{6.2}$ FIII.2.30

      Nonylphenol(ethylene oxide)$_{7.4}$ ether, $C_9H_{19}C_6H_4O(CH_2CH_2O)_{7.4}H$,

      $C_9\phi EO_{7.4}$ FIII.2.30

      Nonylphenol(ethylene oxide)$_{9.6}$ ether, $C_9H_{19}C_6H_4O(CH_2CH_2O)_{9.6}H$,

      $C_9\phi EO_{9.6}$ FIII.2.26

                                                               & FIII.2.30

      Nonylphenol(ethylene oxide)$_{14}$ ether, $C_9H_{19}C_6H_4O(CH_2CH_2O)_{14}H$,

      $C_9\phi EO_{14}$ FIII.2.30

   in water/hexadecane/2,6,10,15,19,23-hexamethyltetracosane

      Dodecylpentaoxyethylene glycol ether, $C_{12}H_{25}O(CH_2CH_2O)_5H$,

      $C_{12}E_5$ TIII.1.30

      Dodecyltetraoxyethylene glycol ether, $C_{12}H_{25}O(CH_2CH_2O)_4H$,

      $C_{12}E_4$ TIII.1.30

   in water/hexene/heptane

      Nonylphenol(ethylene oxide)$_{9.6}$ ether, $C_9H_{19}C_6H_4O(CH_2CH_2O)_{9.6}H$,

      $C_9\phi EO_{9.6}$ FIII.2.25

   in water/paraffin/heptane

      Nonylphenol(ethylene oxide)$_{9.6}$ ether, $C_9H_{19}C_6H_4O(CH_2CH_2O)_{9.6}H$,

      $C_9\phi EO_{9.6}$ FIII.2.25

   in water/tetrachloroethene

      Nonylphenol(ethylene oxide)$_{9.6}$ ether, $C_9H_{19}C_6H_4O(CH_2CH_2O)_{9.6}H$,

      $C_9\phi EO_{9.6}$ FIII.2.24

   in water/tetradecane

      Dodecylpentaoxyethylene glycol ether, $C_{12}H_{25}O(CH_2CH_2O)_5H$,

      $C_{12}E_5$ TIII.1.30

   in water/toluene

      Dodecylphenol(ethylene oxide)$_9$ ether, $C_{12}H_{25}C_6H_4O(CH_2CH_2O)_9H$,

      $C_{12}\phi EO_9$ FIII.2.37

      Nonylphenol(ethylene oxide)$_{9.6}$ ether, $C_9H_{19}C_6H_4O(CH_2CH_2O)_{9.6}H$,

      $C_9\phi EO_{9.6}$ FIII.2.26

   in water/m-xylene

      Dodecylphenol(ethylene oxide)$_9$ ether, $C_{12}H_{25}C_6H_4O(CH_2CH_2O)_9H$,

      $C_{12}\phi EO_9$ FIII.2.37

      Nonylphenol(ethylene oxide)$_{9.6}$ ether, $C_9H_{19}C_6H_4O(CH_2CH_2O)_{9.6}H$,

      $C_9\phi EO_{9.6}$ FIII.2.26

   in water/xylene/heptane

      Nonylphenol(ethylene oxide)$_{9.6}$ ether, $C_9H_{19}C_6H_4O(CH_2CH_2O)_{9.6}H$,

      $C_9\phi EO_{9.6}$ FIII.2.26

Refractive index (continued)

Refractive index (to be continued)

Surface tension (continued)

   in water

Nonylphenol(ethylene oxide)$_5$ ether, $C_9H_{19}C_6H_4O(CH_2CH_2O)_5H$, $C_9\phi EO_5$	TIII.2.14
Nonylphenol(ethylene oxide)$_{10}$ ether, $C_9H_{19}C_6H_4O(CH_2CH_2O)_{10}H$, $C_9\phi EO_{10}$	TIII.2.14
Nonylphenol(ethylene oxide)$_{15}$ ether, $C_9H_{19}C_6H_4O(CH_2CH_2O)_{15}H$, $C_9\phi EO_{15}$	TIII.2.14
Nonylphenol(ethylene oxide)$_{20}$ ether, $C_9H_{19}C_6H_4O(CH_2CH_2O)_{20}H$, $C_9\phi EO_{20}$	TIII.2.14
Nonylphenol(ethylene oxide)$_{30}$ ether, $C_9H_{19}C_6H_4O(CH_2CH_2O)_{30}H$, $C_9\phi EO_{30}$	TIII.2.14
Octylphenol(ethylene oxide)$_1$ ether, $C_8H_{17}C_6H_4O(CH_2CH_2O)_1H$, $C_8\phi EO_1$	TIII.2.14
Octylphenol(ethylene oxide)$_2$ ether, $C_8H_{17}C_6H_4O(CH_2CH_2O)_2H$, $C_8\phi EO_2$	TIII.2.14
Octylphenol(ethylene oxide)$_3$ ether, $C_8H_{17}C_6H_4O(CH_2CH_2O)_3H$, $C_8\phi EO_3$	TIII.2.14
Octylphenol(ethylene oxide)$_4$ ether, $C_8H_{17}C_6H_4O(CH_2CH_2O)_4H$, $C_8\phi EO_4$	TIII.2.14
Octylphenol(ethylene oxide)$_5$ ether, $C_8H_{17}C_6H_4O(CH_2CH_2O)_5H$, $C_8\phi EO_5$	TIII.2.14
Octylphenol(ethylene oxide)$_6$ ether, $C_8H_{17}C_6H_4O(CH_2CH_2O)_6H$, $C_8\phi EO_6$	TIII.2.14
Octylphenol(ethylene oxide)$_7$ ether, $C_8H_{17}C_6H_4O(CH_2CH_2O)_7H$, $C_8\phi EO_7$	TIII.2.14
Octylphenol(ethylene oxide)$_8$ ether, $C_8H_{17}C_6H_4O(CH_2CH_2O)_8H$, $C_8\phi EO_8$	TIII.2.14
Octylphenol(ethylene oxide)$_9$ ether, $C_8H_{17}C_6H_4O(CH_2CH_2O)_9H$, $C_8\phi EO_9$	TIII.2.14
Octylphenol(ethylene oxide)$_{10}$ ether, $C_8H_{17}C_6H_4O(CH_2CH_2O)_{10}H$, $C_8\phi EO_{10}$	TIII.2.14
Octylphenol(ethylene oxide)$_{13}$ ether, $C_8H_{17}C_6H_4O(CH_2CH_2O)_{13}H$, $C_8\phi EO_{13}$	TIII.2.14
Octylphenol(ethylene oxide)$_{16}$ ether, $C_8H_{17}C_6H_4O(CH_2CH_2O)_{16}H$, $C_8\phi EO_{16}$	TIII.2.14
Octylphenol(ethylene oxide)$_{17}$ ether, $C_8H_{17}C_6H_4O(CH_2CH_2O)_{17}H$, $C_8\phi EO_{17}$	TIII.2.14
N-Octylpyridinium bromide, $C_8H_{17}NC_5H_5Br$	TII.2.8
Pentadecyloctaoxyethylene glycol ether, $C_{15}H_{31}O(CH_2CH_2O)_8H$, $C_{15}E_8$	TIII.1.14
Sodium decanesulphonate, $C_{10}H_{21}SO_3Na$	TI.2.24
Sodium hexadecanesulphonate, $C_{16}H_{33}SO_3Na$	TI.2.24
Sodium octadecanesulphonate, $C_{18}H_{37}SO_3Na$	TI.2.24
Sodium tetradecanesulphonate, $C_{14}H_{29}SO_3Na$	TI.2.24
Tetradecyloctaoxyethylene glycol ether, $C_{14}H_{29}O(CH_2CH_2O)_8H$, $C_{14}E_8$	TIII.1.14
Tridecyloctaoxyethylene glycol ether, $C_{13}H_{27}O(CH_2CH_2O)_8H$, $C_{13}E_8$	TIII.1.14
N-Tridecylpyridinium bromide, $C_{13}H_{27}NC_5H_5Br$	FII.2.1
Undecyloctaoxyethylene glycol ether, $C_{11}H_{23}O(CH_2CH_2O)_8H$, $C_{11}E_8$	TIII.1.14
N-Undecylpyridinium bromide, $C_{11}H_{23}NC_5H_5Br$	FII.2.1

Surface tension (to be continued)

ABBREVIATIONS

Abs	UV/Vis absorption
Ac	Activity
AK	Adsorption Kinetics
C	Capillary
Ca	Calorimetry
calc	calculated
Co	Electrical conductivity
Compr	Compressibility
Cryst	Crystallisation temperature of a 1 % solution
CV	Cyclic Voltammetry
De	Density
DE	Dielectric measurement
DS	Dye Solubilisation
DSC	Differential Scanning Calorimetry
DV	Drop Volume
EM	Electron Microscopy
Fl	Fluorescence
FP	Fluorescence Polarisation
FPD	Freezing Point Depression
GF	Gel Filtration
^{2}H-NMR	Deuterium NMR
I	Isotachophoresis
IFT	Interfacial Tension
IR	Infrared
IS	Iodine Solubilisation
K	Partition coefficient
KM	Kinetic Measurement
ln(c)-ln(c')	ln(CMC) vs. ln(counterion) plot
LS	Light Scattering
MA	Mass Action model
MC	Microcalorimetry
MO	Membrane Osmometry
MP	Melting Point Apparatus
NMR	Nuclear Magnetic Resonance
NS	Small Angle Neutron Scattering
OS	Osmometry
PG NMR	Fourier Transform Pulsed Gradient Spin-Echo NMR
pNa	Sodium selective electrode
Pot	Potentiometry
PS	Phase Separation model
Pyr	Fluorescence measurement using pyrene
QELS	Quasi Elastic Light Scattering
QENS	Quasi Elastic Neutron Scattering
R	Refractive Index
RT	Radio Tracer
SANS	Small Angle Neutron Scattering
SD	Spinning Drop

S&D	Sedimentation and Diffusion Coefficient
SE	Sedimentation Equilibrium
SEC	Size Exclusion Chromatography
SM	Stalagmometer
SO	Schlieren Optics
Sol	Solubility
Sol/CM	Intersection point of solubility and CMC curve
SPM	Semipermeable Membrane
ST	Surface tension
Tu	Turbidimetry
U	Unknown
UC	Ultracentrifuge
UR	Ultrasonic relaxation
UV	Ultraviolet Spectroscopy
Vis	Viscosity
Vol	Volume measurement
VP	Vapour Pressure
VPO	Vapour Pressure Osmometry
WP	Wilhelmy Plate

| ΔH | Extrapolation of enthalpy plot |
| ϕ_L | CMC is break in relative apparent molar enthalpy vs. molality curve |

PART V: ABBREVIATIONS